The R Primer

Second Edition

Chapman & Hall/CRC
The R Series

Series Editors

Aims and Scope

This book series reflects the recent rapid growth in the development and application of R, the programming language and software environment for statistical computing and graphics. R is now widely used in academic research, education, and industry. It is constantly growing, with new versions of the core software released regularly and more than 7,000 packages available. It is difficult for the documentation to keep pace with the expansion of the software, and this vital book series provides a forum for the publication of books covering many aspects of the development and application of R.

The scope of the series is wide, covering three main threads:

- Applications of R to specific disciplines such as biology, epidemiology, genetics, engineering, finance, and the social sciences.
- Using R for the study of topics of statistical methodology, such as linear and mixed modeling, time series, Bayesian methods, and missing data.
- The development of R, including programming, building packages, and graphics.

The books will appeal to programmers and developers of R software, as well as applied statisticians and data analysts in many fields. The books will feature detailed worked examples and R code fully integrated into the text, ensuring their usefulness to researchers, practitioners and students.

Published Titles

Stated Preference Methods Using R, *Hideo Aizaki, Tomoaki Nakatani, and Kazuo Sato*

Using R for Numerical Analysis in Science and Engineering, *Victor A. Bloomfield*

Event History Analysis with R, *Göran Broström*

Extending R, *John M. Chambers*

Computational Actuarial Science with R, *Arthur Charpentier*

Testing R Code, *Richard Cotton*

The R Primer, Second Edition, *Claus Thorn Ekstrøm*

Statistical Computing in C++ and R, *Randall L. Eubank and Ana Kupresanin*

Basics of Matrix Algebra for Statistics with R, *Nick Fieller*

Reproducible Research with R and RStudio, Second Edition, *Christopher Gandrud*

R and MATLAB® *David E. Hiebeler*

Statistics in Toxicology Using R *Ludwig A. Hothorn*

Nonparametric Statistical Methods Using R, *John Kloke and Joseph McKean*

Displaying Time Series, Spatial, and Space-Time Data with R, *Oscar Perpiñán Lamigueiro*

Programming Graphical User Interfaces with R, *Michael F. Lawrence and John Verzani*

Analyzing Sensory Data with R, *Sébastien Lê and Theirry Worch*

Parallel Computing for Data Science: With Examples in R, C++ and CUDA, *Norman Matloff*

Analyzing Baseball Data with R, *Max Marchi and Jim Albert*

Growth Curve Analysis and Visualization Using R, *Daniel Mirman*

R Graphics, Second Edition, *Paul Murrell*

Introductory Fisheries Analyses with R, *Derek H. Ogle*

Data Science in R: A Case Studies Approach to Computational Reasoning and Problem Solving, *Deborah Nolan and Duncan Temple Lang*

Multiple Factor Analysis by Example Using R, *Jérôme Pagès*

Customer and Business Analytics: Applied Data Mining for Business Decision Making Using R, *Daniel S. Putler and Robert E. Krider*

Implementing Reproducible Research, *Victoria Stodden, Friedrich Leisch, and Roger D. Peng*

Graphical Data Analysis with R, *Antony Unwin*

Using R for Introductory Statistics, Second Edition, *John Verzani*

Advanced R, *Hadley Wickham*

bookdown: Authoring Books and Technical Documents with R Markdown, *Yihui Xie*

Dynamic Documents with R and knitr, Second Edition, *Yihui Xie*

The R Primer

Second Edition

Claus Thorn Ekstrøm

University of Copenhagen
Denmark

CRC Press is an imprint of the
Taylor & Francis Group, an **informa** business
A CHAPMAN & HALL BOOK

CRC Press
Taylor & Francis Group
6000 Broken Sound Parkway NW, Suite 300
Boca Raton, FL 33487-2742

CRC Press is an imprint of Taylor & Francis Group, an Informa business

Printed in Canada on acid-free paper
Version Date: 20170110

International Standard Book Number-13: 978-1-138-63197-7 (Paperback)
International Standard Book Number-13: 978-1-4987-7255-6 (Hardback)

Visit the Taylor & Francis Web site at
http://www.taylorandfrancis.com

and the CRC Press Web site at
http://www.crcpress.com

Contents

Preface

A lot has happened to R and in the R community in the five years since the first edition of *The R Primer* was published. Changes in R itself, hundreds of new and improved packages, but most importantly that R has seen even more widespread acceptance and has been adopted by more and more research fields as the primary software tool for data analysis.

In addition, the recent focus in the scientific fields on research documentation and reproducible research has made R an even more valuable and ideal tool for data analysis and hence it made sense to update the current text to reflect some of the changes to R. The majority of the problems and cases covered in the book have been updated and the text has grown substantially to cover additional situations and problems.

The second edition of *The R Primer* follows the same format as the first edition in that it assumes that the reader knows what he or she wants and/or what analysis method is appropriate in a given situation, but that the reader needs help with both coding and interpreting the output from that model in R. The problems in this second edition have been — just as for the first edition — heavily inspired by the questions encountered at the statistical consultancy service at the University of Copenhagen. During these consultancy sessions we discuss statistical problems with clients and often provide them with example code that covers their situation. The examples and problems in the book have proven valuable as a reference point for the clients in this regard as it is easy to show them a working example and explain the changes they need to adapt to get the code working in their situation. Hopefully, others will also find the text useful as a collection of solutions to common situations for newcomers and intermediate users of R.

Claus Thorn Ekstrøm
Copenhagen 2016

Preface to the first edition

This book is not about statistical theory, neither is it meant to teach R programming. This book is intended for readers who know the basics of R, but find themselves with problems or situations that are commonly encountered by newcomers to R or for readers who want to see compact examples of different types of typical statistical analyses. In other words, if you understand basic statistics and already know a bit about R then this book is for you.

R has rapidly become the *lingua franca* of statistical computing; it is a free statistical programming software and it can be downloaded from `http://cran.r-project.org`. Many newcomers to R are often intimidated by the command-line interface, or the sheer number of functions and packages, or just trying to figure out how to import data and perform a simple statistical analysis.

The book consists of a number of examples that illustrate a specific situation, topic, or problem from data import over data management and classical statistical analyses to graphics. Each example is self-contained and provides R code that can be run exactly as shown and the results from the book can be reproduced. The only change — barring simulated data, machine set-up and small tweaks to make figures suitable for printing — is that some of the output lines have been removed for brevity.

This is not a "missing manual" or a thorough exploration of the functions used. Instead of trying to cover every possible option or special case that might be of interest, we focus on the common situations that most beginning users are likely to encounter. Thus we concentrate on the basics of getting things done and giving examples that can be used as a starting point for the reader rather than exploring the multitude of options available with every command and the ever-increasing number of packages. For most problems — and this is particularly true for a programming language like R— there is more than one way to solve a problem. Here, I have provided a single solution to most problems and have tried to use base R if at all possible. If there are other functions and/or packages available that cover or extend the same functionality, then some of them are listed at the end of each example.

The R list of frequently asked questions is highly recommended and covers a few of the same topics mentioned here. However, it does not cover examples of statistical analyses and it rarely covers some of the most basic problems new users encounter.

Base graphics are used throughout the book. More advanced graphics can be produced with the recent `lattice` and `ggplot2` packages (see Sarkar (2008), Wickham (2009), or Murrell (2011) for further information on advanced R graphics). A more complete coverage of R and/or statistics can be found in the books by Venables and Ripley (2002), Verzani (2005), Crawley (2007), Dalgaard (2008), and Everitt and Hothorn (2010). These books have a slightly different target audience than the present text and are all highly recommended.

The R Primer has a supporting web site at

`http://www.rprimer.dk`

where additional topics are covered and where the R code used in the book can be found.

I would like to thank all R developers and package writers for the enormous work they have done and continue to put into the R program and extensions. I appreciate all the helpful responses to my enquiries and suggestions. I am grateful to my colleagues at the Faculty of Life Sciences, University of Copenhagen, as well as Klaus K. Holst, Duncan Temple Lang, and Bendix Carstensen for their ideas, comments, suggestions, and encouragement on various stages of the manuscript. Many thanks to Tina Ekstrøm for once again creating a wonderful cover, and last, but not least, thanks to Marlene, Ellen, and Anna for bearing with me through yet another book.

Claus Thorn Ekstrøm
Frederiksberg 2011

CHAPTER 1

Importing and exporting data

The first step in most statistical analyses is to get the data into R. While it is possible to manually type the data directly into R using, say, the `c` and `data.frame` functions, for most serious work the data is typically stored in a file somewhere locally on the computer, a network, or somewhere on the Web.

Table 1.1 gives a quick overview of R functions and related packages for importing common data formats. Specific solutions to importing different types of data are presented in the rest of this chapter.

1.1 IMPORT AN R DATASET FROM A PACKAGE

Problem: You want to import an R dataset from a package

Solution: Many R packages (including those that are part of the default installation) come with several datasets that are ready to use. R datasets found in packages are imported with the `data` function and require that the package is loaded first.

```
> library(MESS)   # Load the package that contains the data
> data(bees)      # Load the data
> head(bees)      # Show the first couple of lines of data
    Locality Replicate  Color  Time       Type Number id
1  Havreholm         A  White july1 Bumblebees      1  1
4  Havreholm         A Yellow july1 Bumblebees      2  1
7  Havreholm         A   Blue july1 Bumblebees      0  1
10 Havreholm         A  White july1   Solitary      1  1
13 Havreholm         A Yellow july1   Solitary      4  1
16 Havreholm         A   Blue july1   Solitary      3  1
```

Table 1.1: Common file formats and R functions to read them

File type	Function	Package
R data file	`load` or `data`	
Text file	`read.table`	
CSV	`read.csv` or `read.csv2`	
JSON	`fromJSON`	`jsonlite`
Excel spreadsheet	`read_excel`	`readxl`
SAS	`read_sas`	`haven`
SPSS .SAV or .POR	`read_sav` or `read_por`	`haven`
Stata	`read_stata`	`haven`
html	`read_html`	`rvest`
xml	`xmlTreeParse`	`XML`

1.2 LOAD AND SAVE R DATA FILES

Problem: You want to load or save an external R dataset.

Solution: R has its own data file format where files typically have extension `RData` or `rda` and they are imported with the `load` function. A single R data files may contain multiple R objects, and the objects can be *any* type and not just data frames. The objects imported have the same name and structure as they have in the file, and we only see a list of the imported objects if the `verbose=TRUE` argument is included.

```
> load("penny.rda")                 # Load data
> load("penny.rda", verbose=TRUE)  # Load data and show objects
Loading objects:
  big
  bang
  theory
```

Beware that `load` overwrites existing objects with the same names without giving any warnings.

The `save` function is used to save R objects to a file. `save` takes any number of R objects as arguments and saves them to an external file, and the `file` argument specifies the name of the file.

```
> x <- c(1, 7, 3, 2)
> df <- data.frame(v1=c("A", "B", "C", "D"), v2=1:4)
> save(x, df, file="mydata.rda")      # Save two objects
```

See also: Problem 5.16 explains how to change the working directory.

1.3 READ AND WRITE TEXT FILES

Problem: You want to work with a dataset stored in a text file.

Solution: Data stored in simple text files can be read into R using the `read.table` function. By default, the observations should be listed in columns where the individual fields are separated by one or more white space characters (e.g., space, or tabulators), and where each line in the file corresponds to one row of the data frame. The columns do not need to be straight or formatted, but multi-word observations like `high income` need to be put in quotes or combined into a single word so they are not interpreted as two columns. If the text file `mydata.txt` has the following content

```
acid        digest   name
30.3        70.6     NA
29.8        67.5     Eeny
  NA        87.0     Meeny
 4.1        89.9     Miny
 4.4          .      Moe
 2.8        93.1      .
 3.8        96.7     ""
```

we can read the file into R with the following command:

```
> indata <- read.table("mydata.txt", header=TRUE)
> indata
  acid digest  name
1 30.3   70.6  <NA>
2 29.8   67.5  Eeny
3   NA   87.0 Meeny
4  4.1   89.9  Miny
5  4.4      .   Moe
6  2.8   93.1     .
7  3.8   96.7
```

The first argument is the name of the data file, and the second argument (`header=TRUE`) is optional and should be used only if the first line of the text file provides the variable/column names. If the first line does not contain the column names, the variables will be labeled consecutively `V1`, `V2`, `V3`, etc. Each line in the input file must contain the same number of columns for `read.table` to work. The `sep` option can be set to indicate which character that separates the columns. For example, use `sep="\t"` if the columns are separated by tabs. Data read with `read.table` are stored as a data frame within R.

The default code for missing observations is the character string NA which we can see works both for the first and third observation above (acid is read as a numeric vector and name as a factor). Empty character fields are read as empty character vectors, unless the argument na.strings contains the value "" in which case they become missing values. Empty numeric fields (for example if the columns are separated by tabs) are automatically considered missing.

```
> indata <- read.table("mydata.txt", header=TRUE,
+                      na.strings=c("NA", ""))
> indata
  acid digest  name
1 30.3   70.6  <NA>
2 29.8   67.5  Eeny
3   NA   87.0 Meeny
4  4.1   89.9  Miny
5  4.4      .   Moe
6  2.8   93.1     .
7  3.8   96.7  <NA>
```

Note that R considers the variable digest as a factor and not numeric since the period '.' for observation 5 is read as a character string. If periods should also be considered missing variables we need to include that in na.strings vector.

```
> indata <- read.table("mydata.txt", header=TRUE,
+                      na.strings=c("NA", "", "."))
> indata
  acid digest  name
1 30.3   70.6  <NA>
2 29.8   67.5  Eeny
3   NA   87.0 Meeny
4  4.1   89.9  Miny
5  4.4     NA   Moe
6  2.8   93.1  <NA>
7  3.8   96.7  <NA>
```

Data stored as simple text files can be read by most programs so it is often desirable to export a data frame as a simple text file. write.table writes a data frame or a matrix as a text file. The first argument should be the R object to write and the file argument sets the name of the output file as shown below.

```
> data(trees)
> trees$Girth[2] <- NA     # Set 2nd obs of Girth to missing
> write.table(trees, file="savedata.txt")
```

which produces the following file

```
"Girth" "Height" "Volume"
"1" 8.3 70 10.3
"2" NA 65 10.3
"3" 8.8 63 10.2
"4" 10.5 72 16.4
  .   .   .   .
  .   .   .   .
```

There are several important arguments to `write.table` that change the output. The `quote` option defaults to `TRUE` and determines if character or factor columns are surrounded by double quotes. `quote=FALSE` means that none of the columns are quoted and if set to a numeric vector then the values determine which columns are to be quoted. `sep` sets the field separator string and it defaults to a single white space character. `na` has a default value of `NA` and the option sets the string used for missing values in the file. The `dec` and `eol` arguments set the character and character(s) for decimal points and end-of-lines, respectively. The default arguments to `write.table` prints out the row names (the line numbers) as shown in the output above, but that can be circumvented with the `row.names=FALSE` argument.

The example below produces a file with comma as decimal point, where strings are not quoted, where a period represents `NA`'s, no row names, and where the end-of-line characters match the format used on machines running Windows.

```
> write.table(trees, file="mydata2.txt", row.names=FALSE,
+             dec=",", eol="\r\n", na=".", quote=FALSE)
```

which results in the following file

```
Girth Height Volume
8,3 70 10,3
. 65 10,3
8,8 63 10,2
10,5 72 16,4
10,7 81 18,8
   .   .   .
   .   .   .
```

See also: The `read_delim` function from the `readr` package reads text files much faster and often has more natural default arguments than `read.table`. Problem 5.16 explains how to change the working directory.

1.4 READ AND WRITE CSV FILES

Problem: You want to import or export a dataset stored as a comma-separated values (CSV) file.

Solution: A standard CSV file is a plain text file where each line in the file corresponds to an observation and where the variables are separated from each other by commas. Quotation marks are used to embed field values that contain the separator character.

`read.csv` is a wrapper function that sets the correct options for `read.table` in order to read in the delimited file. `read.csv2` is used to read in semicolon separated files which is the default CSV format in some locales where a comma is the decimal point character and where a semicolon is used as separator instead.

An example of the CSV file `mydata.csv` is shown below

```
"Id","Sex","Age","Score"
21,"Male",14,"Little"
26,"Male",13,"None"
27,"Male",13,"Moderate, severe"
29,"Female",13,"Little"
30,"Female",15,"Little"
31,"Male",14,"Moderate, severe"
```

which we import with

```
> indata <- read.csv("mydata.csv")
> head(indata)
  Id    Sex Age            Score
1 21   Male  14           Little
2 26   Male  13             None
3 27   Male  13 Moderate, severe
4 29 Female  13           Little
5 30 Female  15           Little
6 31   Male  14 Moderate, severe
```

Both `read.csv` and `read.csv2` assume by default that a header line is present in the CSV file (i.e., `header=TRUE` is the default). If no header line is present you need to specify the `header=FALSE` argument.

The `write.csv` function creates a delimited file, and the function works in the same way as `write.table` set with the correct arguments to obtain the CSV file format. `write.csv2` is used to create semicolon separated files.

The following code exports the `trees` data frame to the `mydata.csv` CSV file.

```
> data(trees)
> write.csv(trees, file = "savedata.csv")
```

The start of `savedata.csv` looks like this

```
"","Girth","Height","Volume"
"1",8.3,70,10.3
"2",8.6,65,10.3
"3",8.8,63,10.2
"4",10.5,72,16.4
  .     .    .    .
  .     .    .    .
```

The default arguments for `write.csv` writes the row names to the output file which is seldom, what you want in a CSV file. The `row.names=FALSE` argument is set to prevent row names from being written to the file, and `quote=FALSE` prevents the quoting of strings.

```
> data(trees)
> write.csv(trees, file = "savedata2.csv", row.names=FALSE, quote=FALSE)
```

This produces the following file which is in the standard CSV format.

```
Girth,Height,Volume
8.3,70,10.3
8.6,65,10.3
8.8,63,10.2
  .    .    .
  .    .    .
```

See also: The `readr` package provides the `read_csv` and `read_csv2` functions which read files much faster and often have more natural default arguments. The `fread` function from the `data.table` package is extremely fast for reading in very large CSV files.

1.5 READ DATA FROM THE CLIPBOARD

Problem: You have selected some data and copied them to the clipboard and want to import the selection into R.

Solution: Sometimes it is desirable to select data from a document, a web page, or from a spreadsheet and import the selection directly into R. This can be done by copying the selection to the clipboard and then subsequently importing the contents of the clipboard into R. This approach can be used on platforms that have the equivalent of a clipboard, which is the case for Windows, Mac OS X, and machines running the X Window System used on many Linux systems.

The contents of the clipboard can be read using the value `"clipboard"` for the `file` option to `read.table` under Windows and X, and by using `pipe("pbpaste")` under Mac OS X.

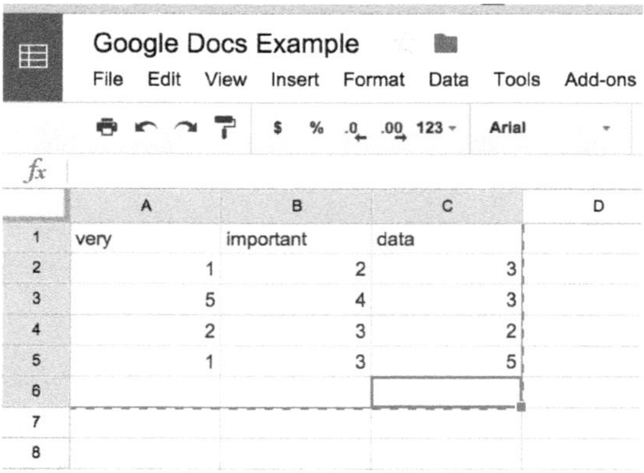

Figure 1.1: Selection of cells to be copied to the clipboard.

If we select some cells as shown in Figure 1.1 and copy the selection to the clipboard then we import the selection with the following code:

```
> mydata <- read.table(file="clipboard", header=TRUE) # Windows/X
> mydata <- read.table(pipe("pbpaste"), header=TRUE)  # Mac OsX
```

Note that the actual syntax depends on the operating system. The `read.table` reads and parses the input from the clipboard as if it had read the data from an ASCII file (see Problem 1.3). This may cause problems with empty cells or if there is text with spaces in any of the cells, which is not a problem in this case. Also, the

```
> mydata
  very important data
1    1         2    3
2    5         4    3
3    2         3    2
4    1         3    5
```

If the data are separated by tabs on the clipboard — which is generally the case for spreadsheet data — we can specify the field separator to be tabs in the call to `read.table`, which will handle these two situations.

```
> mydata <- read.table(file="clipboard", sep="\t", header=TRUE)
```

On machines running the X11 Windows system the `"X11_clipboard"` value can be used for the `file` option to copy from the clipboard.

See also: The help file for the `file` function lists information about clipboards. Spreadsheet data on the clipboard are often stored in the Data Interchange Format (DIF) and the `read.DIF` function can be used to read DIF formats directly. `read.DIF` is sometimes more robust than using `read.table` when there are empty cells.

SPREADSHEETS

1.6 READ AND WRITE DATA FROM A SPREADSHEET

Problem: You want to read data from a spreadsheet or write a data frame as a spreadsheet.

Solution: A simple way to transfer data from a spreadsheet into R is to export the spreadsheet to a delimited file like a comma-separated file and then import the CSV file as described in Problem 1.4.

A similar approach can be used to transfer an R data frame in the other direction from R to a spreadsheet. First, it should be saved as a comma-separated file as described in Problem 1.4. Then the CSV file can be imported into the spreadsheet. This will work for virtually any spreadsheet including Microsoft Excel, Google sheets, OpenOffice Calc, and LibreOffice Calc.

1.7 READ AND WRITE EXCEL FILES

Problem: You want to read or write a Microsoft Excel data file.

Solution: The `read_excel` function from the `readxl` package can be used to read Excel worksheets (both older and more recent Excel formats) directly into R.

The first argument to `read_excel` should be the path to the Excel spreadsheet, while the `sheet` argument expects a sheet name or a number representing which sheet to import from the Excel file. By default, the first line in the Excel sheet is assumed to be a header line, but that can be changed by setting the `col_names=FALSE` argument.

The following Excel worksheet saved as the file `cunningplan.xlsx` can be imported with the following code:

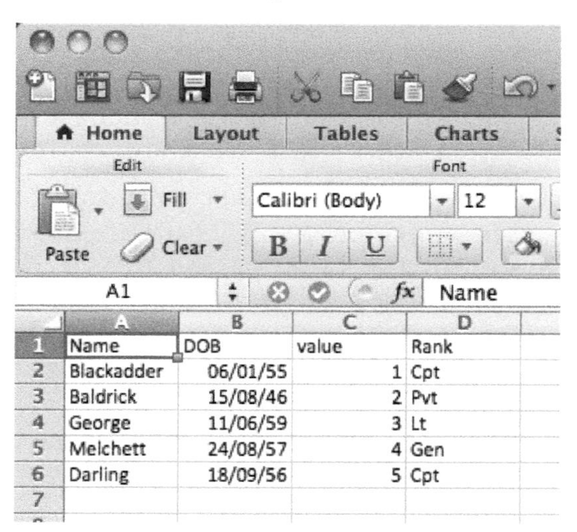


```
> library(readxl)
> goesforth <- read_excel("data/cunningplan.xlsx", sheet=1)
> goesforth
# A tibble: 5 × 4
        Name        DOB value  Rank
       <chr>     <dttm> <dbl> <chr>
1 Blackadder 1955-01-06     1   Cpt
2   Baldrick 1946-08-15     2   Pvt
3     George 1959-06-11     3    Lt
4   Melchett 1957-08-24     4   Gen
5    Darling 1956-09-18     5   Cpt
```

The `skip` argument can be set to a numeric value to indicate how many non-empty lines in the Excel file should be skipped before starting to read the file. Completely empty rows are silently removed when reading in the Excel file.

It is not possible to use the `readxl` package to save an R data frame to Excel yet, but the `xlsx` package provides the `write.xlsx` function which produces Excel 2007 spreadsheets. The following code saves the `trees` data frame to the `mydata.xls` Excel file.

```
> library(xlsx)
> data(trees)
> write.xlsx(trees, "mydata.xlsx")
```

By default the row and columns names are written to the Excel file. They are suppressed by setting the arguments `row.names` and `col.names` to `FALSE`, respectively.

```
> write.xlsx(trees, "mydata.xlsx", row.names=FALSE) # No row numbers
```

See also: The `xlsx` package also contains a function, `read.xlsx` for reading Excel files but `read_excel` is much faster, more robust at handling dates, and does not have any external requirements. Problems 1.4 and 1.6 explain how to use the CSV file format for transfering data between R and Excel.

STATISTICAL SOFTWARE PROGRAMS

1.8 IMPORT AND EXPORT SAS DATASETS

Problem: You want to import or export a SAS dataset.

Solution: SAS binary datasets typically have the extension `sas7bdat` but the file format is not unique and depends on the operating system and the version of SAS. The SAS system previously encouraged users to use the SAS transport file format to transfer data in order to prevent problems with the non-standard binary formats across different platforms.

`read_sas` from the `haven` package circumvents the need for the SAS transport file format since it imports binary SAS files directly, and it just requires the file name as input.

```
> library(haven)
> indata <- read_sas("hip.sas7bdat")
> head(indata)
# A tibble: 6 × 1
      Y
  <dbl>
```

```
1 14.96
2 17.34
3 16.40
4 19.33
5 17.69
6 17.50
```

To read SAS files stored in the SAS transport (XPORT) file format we use the `read.xport` function from the **foreign** package, which also just requires the file name as input.

```
> library(foreign)
> indata <- read.xport("somefile.xpt")
> head(indata)
  GIRTH HEIGHT VOLUME
1   8.3     70   10.3
2   8.6     65   10.3
3   8.8     63   10.2
4  10.5     72   16.4
5  10.7     81   18.8
6  10.8     83   19.7
```

The following code can be used from within SAS to store the SAS dataset `sasdata` in the XPORT format.

```
libname mydata xport "somefile.xpt";

/* Create a dataset in XPORT format and save it in
 * the somefile.xpt file. The file is referenced
 * internally in SAS by the name mydata.
 * Here we take an existing SAS dataset called
 * sasdata and put it into the mydata file.
 */

DATA mydata.thisdata;
  SET sasdata;

RUN;
```

R data frames can be easily exported to SAS through the `write.foreign` function from the **foreign** package. Besides the `package="SAS"` argument which tells the function to export the data frame to be read by SAS, it requires two additional arguments, `datafile` and `codefile`, which give the names of the data file and instructions to SAS for importing the data, respectively.

```
> library(foreign)
> data(trees)
> write.foreign(trees, datafile = "toSAS.dat",
+               codefile="toSAS.sas", package="SAS")
```

It is easy to get the data into SAS since the codefile contains a SAS script that can be read and interpreted directly by SAS.

1.9 IMPORT AND EXPORT SPSS DATASETS

Problem: You want to import or export an SPSS dataset.

Solution: Datasets stored by the SPSS "save" and "export" commands can be read by the `read_sav` function from the `haven` package.

SPSS files typically have the `sav` file extension and to import an SPSS dataset saved in the `cancer.sav` file we use the following command in R:

```
> library(haven)
> indata <- read_sav("cancer.sav")
> head(indata)
# A tibble: 6 × 9
     ID   TRT   AGE WEIGHIN STAGE TOTALCIN TOTALCW2 TOTALCW4
  <dbl> <dbl> <dbl>   <dbl> <dbl>    <dbl>    <dbl>    <dbl>
1     1     0    52   124.0     2        6        6        6
2     5     0    77   160.0     1        9        6       10
3     6     0    60   136.5     4        7        9       17
4     9     0    61   179.6     1        6        7        9
5    11     0    59   175.8     2        6        7       16
6    15     0    69   167.6     1        6        6        6
# ... with 1 more variables: TOTALCW6 <dbl>
```

In addition to the normal `.sav` file format, SPSS has also a portable format that works across different versions of SPSS and operating systems. These portable files typically have the `.por` extension and are imported with the `read_por` function which works similarly to code listed above.

SPSS data files are created in R with the `write_sav` function.

```
> data(trees)
> write_sav(trees, "myspssfile.sav")
```

See also: The `write.foreign` from the `foreign` package can also export to the SPSS file format.

1.10 IMPORT OR EXPORT A STATA DATASET

Problem: You want to import or export a Stata dataset.

Solution: Datasets stored by the "SAVE" command in Stata can be read in R by the `read_dta` function from the `haven` package.

To read a Stata dataset in the file `statafile.dta` we use the following commands in R:

```
> library(haven)
> indata <- read_dta("statafile.dta")
> head(indata)
# A tibble: 6 × 3
  Girth Height Volume
  <dbl>  <dbl>  <dbl>
1   8.3     70   10.3
2   8.6     65   10.3
3   8.8     63   10.2
4  10.5     72   16.4
5  10.7     81   18.8
6  10.8     83   19.7
```

The `haven` package also provides the `write_dta` function which exports a data frame to the Stata binary format.

```
> data(trees)
> write.dta(trees, file="mydata.dta")
```

In Stata you can read the exported file using the `use mydata.dta` command.

1.11 IMPORT A SYSTAT DATASET

Problem: You want to import a Systat dataset.

Solution: Rectangular datasets stored by the "SAVE" command in Systat are saved as `*.sys` or `*.syd` files and they can be read in R by the `read.systat` function from the `foreign` package. To read a Systat dataset saved as the file `Play4.syd` we use the following commands in R:

```
> library(foreign)
> indata <- read.systat("Play4.syd")
> head(indata)
     OBJECT.   X   Y
```

```
1 A              3.1 17.8
2 B              3.6  6.8
3 C              4.2  6.0
4 D              4.0  4.1
5 E              6.7  6.2
6 F             16.0  4.6
```

DATA EXCHANGE FORMATS AND DATABASES

1.12 IMPORT A JSON DATASET

Problem: You want to import a JSON dataset into R.

Solution: JSON stands for JavaScript Object Notation and it is a simple general purpose format for representing data. In some sense it serves the same purpose as XML but the format is vastly simpler and faster to access. JSON is widely popular and is often used for transfering data from the Web.

The JSON file format is a plain text file that accepts five data types ('null', 'true', 'false', number, and string) that can be combined to form either an ordered unnamed array (similar to an R vector) or a named array. The JSON is slightly simpler than the R data values, since the JSON format cannot faithfully represent all of the values in R. In particular, there is no way to distinguish between `NA`, `Inf`, or `NULL`, which are all coded as 'null' in the JSON format. Square brackets [ ] are used for ordered unnamed arrays while curly brackets `{` and `}` are used for named arrays with a colon separating the name from the data. Examples of the data types and containers are shown below.

```
[ "X", "Y", "Z", ...]
[ 300, 200, 500, ... ]
[ true, true, false, ... ]
{ "dollars" : 5, "euros" : 20, ... }
```

The data containers can be nested to form more complicated structures. Consider the following data table with header and a missing value for observation 3 of `qty`.

```
name grams qty new
X    300   4   true
Y    200   5   false
Z    500       true
```

It can be represented in two ways in JSON. Either column-wise (a named array of vectors)

```
{
"name": ["X", "Y", "Z"],
"grams": [300, 200, 500],
"qty": [4, 5, null],
"new": [true, false, true]
}
```

or row-wise (a vector of named arrays)

```
[
  { "name": "X",
    "grams": 300,
    "qty": 4,
    "new": true },
  { "name": "Y",
    "grams": 200,
    "qty": 5,
    "new": false },
  { "name": "Z",
    "grams": 500,
    "qty": null,
    "new": true }
]
```

R has several packages for working with JSON data. Here we will focus on `jsonlite` which provides two main functions, `toJSON` and `fromJSON` for conversion to and from data in JSON format, respectively.

The `fromJSON` requires either a string that contains the JSON data or a file name as its first argument. The `fromJSON` function imports the data as a list which allows for complex nested structures in the imported data (in which case the returned object is a nested list). The `flatten=TRUE` argument can be used to flatten nested lists into a non-nested list.

```
> library(jsonlite)
> indata <- fromJSON("exampledata.json")
> class(indata)        # Note a list is returned
[1] "list"
> indata
$name
[1] "X" "Y" "Z"

$grams
[1] 300 200 500
```

```
$qty
[1]  4  5 NA

$new
[1]  TRUE FALSE  TRUE
```

If the list elements are all vectors with the same length then the data can be converted directly to a data frame.

```
> df <- as.data.frame(indata)
> class(df)
[1] "data.frame"
> df
  name grams qty   new
1    X   300   4  TRUE
2    Y   200   5 FALSE
3    Z   500  NA  TRUE
```

The `toJSON` function converts standard R data types into JSON strings. `toJSON` takes an R object as its first argument and it accepts a number of optional arguments that determine how it encodes specific objects. The `pretty=TRUE` argument makes the resulting JSON string less compact to ease human reading.

```
> toJSON(1:5)   # JSON string for a numeric vector
[1,2,3,4,5]
> outdata <- toJSON(df, pretty=TRUE)
> class(outdata)
[1] "json"
> outdata # Default is row-wise
[
  {
    "name": "X",
    "grams": 300,
    "qty": 4,
    "new": true
  },
  {
    "name": "Y",
    "grams": 200,
    "qty": 5,
    "new": false
  },
  {
    "name": "Z",
    "grams": 500,
    "new": true
```

```
  }
]
```

Column-wise representation can be set with the `dataframe="column"` argument when the R object is a data frame. Similar arguments exist for other types of R objects.

```
> outdata <- toJSON(df, dataframe="column", pretty=TRUE)
> outdata
{
  "name": ["X", "Y", "Z"],
  "grams": [300, 200, 500],
  "qty": [4, 5, "NA"],
  "new": [true, false, true]
}
```

Note how the missing observation for `qty` in this representation is coded as the string "NA" and not as the special JSON-type 'null'. This can be set with the argument `na="null"`.

```
> toJSON(df, dataframe="column", pretty=TRUE, na="null")
{
  "name": ["X", "Y", "Z"],
  "grams": [300, 200, 500],
  "qty": [4, 5, null],
  "new": [true, false, true]
}
```

Finally we can use the `cat` function to write the JSON string to a file

```
> cat(outdata, file="output.json")
```

See also: The `jsonlite` package also provides two other functions, `stream_in` and `stream_out`, that allow for processing very large JSON files on computers with limited memory. The `rjson` and `RJSONIO` packages also read JSON formats.

1.13 READ DATA FROM A SIMPLE XML FILE

Problem: You want to import a dataset stored as a simple structure in the XML file format.

Solution: The XML (eXtensible Markup Language) was designed to transport and store data and XML has seen widespread use in interchanging data over the Internet.

An XML file consists of a series of elements which form a document tree. The tree starts at the root and branches to the lowest level of the tree. XML documents must contain a root node (or element) which is "the parent" of all other nodes, and all nodes can have their own subnodes ("child elements").

An example XML file is shown below where the tree data from the `trees` dataset are stored in XML format. The root node `<document>` has several child nodes (the `<rows>`) and each row has its own child elements corresponding to the variables in the data frame and their values.

```
<?xml version="1.0"?>
 <document>
   <row>
     <Girth>8.3</Girth>
     <Height>70</Height>
     <Volume>10.3</Volume>
   </row>
   <row>
     <Girth>8.6</Girth>
     <Height>65</Height>
     <Volume>10.3</Volume>
    .
    .
    .
     <Volume>77</Volume>
   </row>
 </document>
```

The `XML` package provides numerous tools for parsing and generating XML in R. Since XML is such a flexible format, the `XML` package primarily consists of functions that must be combined to parse and extract information from a specific type of XML structure.

XML document files with a simple structure can be imported and converted to a data frame directly using the `xmlToDataFrame` function. By simple, we mean a collection of nodes that have the same sub-nodes such that each node corresponds to an observation or row in the data frame and each of its sub-nodes contains primitive values corresponding to the variables. The data file shown above has such a simple structure.

```
> library(XML)
> url <- "http://www.rprimer.dk/download/mydata.xml"
> indata <- xmlToDataFrame(url)
> head(indata)
```

```
  Girth Height Volume
1   8.3     70   10.3
2   8.6     65   10.3
3   8.8     63   10.2
4  10.5     72   16.4
5  10.7     81   18.8
6  10.8     83   19.7
```

See also: Use Problem 1.14 to import XML files that do not have a simple structure.

1.14 READ DATA FROM AN XML FILE

Problem: You want to import a dataset stored in the XML file format by manually coding how to extract the relevant information.

Solution: Problem 1.13 showed how to import data from an XML (eXtensible Markup Language) file with a simple structure. Here we will try to import data from a more non-trivial situation, which `xmlToDataFrame` cannot handle.

As a more complex example we will try to import the following XML file that contains artificial data on currency exchange rates. The first couple of nodes are document creation data, while the actual exchange rates begin with the `<rates>` node. The exchange rates (measured against the euro) and dates are coded as tags to the `<exch>` and `<Date>` nodes while the bank source is a leaf node with no children. Also note, that information from both banks is not available for both dates and that not all exchange rates nodes may be present.

```
<?xml version="1.0"?>
<bankdata>
<author>Claus</author>
<valid>Not at all</valid>
<rates>
<Date time="2011-03-10">
  <bank>
    <source>Some bank</source>
    <exch currency="USD" rate="1.3817"/>
    <exch currency="DKK" rate="7.4581"/>
  </bank>
  <bank>
    <source>Some other bank</source>
    <exch currency="USD" rate="1.2382"/>
    <exch currency="DKK" rate="7.3312"/>
  </bank>
```

```
</Date>
<Date time="2011-03-09">
  <bank>
    <source>Some bank</source>
    <exch currency="USD" rate="1.3884"/>
  </bank>
</Date>
</rates>
</bankdata>
```

Recall that an XML tree structure consists of a series of nodes branching out from the root node, and that each of the nodes may itself have children. Data are stored either as values or as attributes/tags of a node.

The `xmlTreeParse` is the work-horse for importing general XML documents. `xmlTreeParse` parses an XML file and stores the tree in an R structure. We subsequently traverse the tree and extract data from the relevant nodes. `xmlTreeParse` requires a file name or location as input for where to find the XML file, and it returns an R XML object with the parsed XML file. The `useInternalNodes` option can be set to `TRUE` to increase parsing speed.

First, `xmlRoot` should be called to get a pointer to the top-level node or parent of the XML tree. The `skip` option can be set to `FALSE` to prevent R from skipping over document type definitions in the XML file if those are present.

The XML tree structure works like a recursive list-like object and the individual nodes in the tree are accessed using named or numbered indices, `[[]]`. The XML tree can be traversed with the proper indices and for each node we can get the parent and list of children sub-nodes using the `xmlParent` and `xmlChildren` functions, respectively.

Information can be extracted from a node using one of the `xmlName`, `xmlValue`, `xmlGetAttr`, and `xmlAttrs` functions, which return the node name, node contents, a named attribute, and all attributes, respectively.

```
> library(XML)
> # Location of the example XML file
> url <- "http://www.rprimer.dk/download/bank.xml"
> # Parse the tree
> doc <- xmlTreeParse(url, useInternalNodes=TRUE)
> top <- xmlRoot(doc)          # Identify the root node
> xmlName(top)                 # Show node name of root node
[1] "bankdata"
> names(top)                   # Name of root node children
  author    valid    rates
```

```
"author"  "valid"  "rates"
> xmlValue(top[[1]])           # Access first element
[1] "Claus"
> xmlValue(top[["author"]])    # First element with named index
[1] "Claus"
> names(top[[3]])              # Children of node 3
  Date   Date
"Date" "Date"
> xmlAttrs(top[[3]][[1]])      # Extract tags from a Date node
        time
"2011-03-10"
> top[["rates"]][[1]][[1]]     # Tree from first bank and date
<bank>
  <source>Some bank</source>
  <exch currency="USD" rate="1.3817"/>
  <exch currency="DKK" rate="7.4581"/>
</bank>
> xmlValue(top[[3]][[1]][[1]][[1]]) # Bank name is node value
[1] "Some bank"
> xmlAttrs(top[[3]][[1]][[1]][[1]]) # but has no tags
NULL
> xmlAttrs(top[[3]][[1]][[1]][[2]]) # The <exch> node has tags
currency     rate
   "USD" "1.3817"
> xmlValue(top[[3]][[1]][[1]][[2]]) # but no value
[1] ""
```

A function can be applied recursively to children of a node using the `xmlApply` and `xmlSApply` functions, which works similarly as `apply` and `sapply` except for XML tree structures. Extracting the individual exchange rates and combining them with the proper bank name and date can be quite cumbersome using indices, loops, and `xmlApply`. Instead, we can use the XML Path Language, XPath, to query and extract information from specific nodes in the XML tree structure. Table 1.2 shows examples of useful XPath query strings. These can be used with the `xpathApply` or `xpathSApply` functions, which accept a node from where to start the search as first argument, an XPath query string as second argument, and the function to apply as third argument.

```
> # Search tree for all source nodes and return their value
> xpathSApply(doc, "//source", xmlValue)
[1] "Some bank"       "Some other bank" "Some bank"
> # Search full tree for all exch nodes where currency is "DKK"
> xpathApply(doc, "//exch[@currency='DKK']", xmlAttrs)
[[1]]
currency     rate
```

Table 1.2: Examples of XPath search expression

Expression	Description
`/node`	top-level node only
`//node`	node at any level
`//node[@name]`	node with an attribute named "name"
`//node[@name="a"]`	node with named attr. with value "a"
`//node/@x`	value of attribute x in node with such attr.

```
   "DKK" "7.4581"

[[2]]
currency     rate
   "DKK" "7.3312"
```

Below we search through the complete XML tree for each node, `<exch>`, which may be found at any level. From the `<exch>` node we extract its attributes and get the bank name through its parent and the exchange rate date from the time attribute from its grandparent. The `do.call` function is used `rbind` to combine the resulting lists.

```
> res <- xpathApply(doc, "//exch",
+           function(ex) {
+              c(xmlAttrs(ex),
+                bank=xmlValue(xmlParent(ex)[["source"]]),
+                date=xmlGetAttr(xmlParent(xmlParent(ex)), "time"))
+             })
> result <- do.call(rbind, res)
> result
     currency rate     bank               date
[1,] "USD"    "1.3817" "Some bank"        "2011-03-10"
[2,] "DKK"    "7.4581" "Some bank"        "2011-03-10"
[3,] "USD"    "1.2382" "Some other bank" "2011-03-10"
[4,] "DKK"    "7.3312" "Some other bank" "2011-03-10"
[5,] "USD"    "1.3884" "Some bank"        "2011-03-09"
```

The variables in the resulting object can then be converted to their proper formats and combined in a data frame.

1.15 EXPORT A DATA FRAME TO XML

Problem: You want to export a data frame in the XML (eXtensible Markup Language) file format.

Solution: The XML (eXtensible Markup Language) was designed to transport and store data and XML has seen widespread use in interchanging data over the Internet.

The `write.xml` function from the `MESS` package takes an R data frame and writes it a file. Besides the R data frame it requires that the filename is given as the `file` argument.

```
> library(MESS)
> data(bees)
> write.xml(bees, file="bees.xml")
```

The resulting `bees.xml` looks as follows (shortened for brevity and with a few additional line breaks inserted to make the output more legible).

```
<?xml version=''1.0''?>
<document>
  <row>
    <Locality>Havreholm</Locality>
    <Replicate>A</Replicate>
    <Color>White</Color>
    <Time>july1</Time>
    <Type>Bumblebees</Type>
    <Number>1</Number>
    <id>1</id>
  </row>
  <row>
    <Locality>Havreholm</Locality>
    <Replicate>A</Replicate>
    <Color>Yellow</Color>
    .
    .
    .
   </row>
 </document>
```

See also: See Problem 1.13 for more information on the XML format and on installing the `XML` package. The `xml2` package also provides functions for parsing XML files.

List [edit]

Rank	Country/Territory	Capital	Population	Year
1	China	Beijing	20,693,000[1]	2012
2	Japan	Tokyo	13,189,000[2]	2011
3	Russia	Moscow	11,541,000[3]	2011
4	South Korea	Seoul	10,369,593[4]	2014
5	Indonesia	Jakarta	10,187,595[5]	2011
6	Mexico	Mexico City	8,851,080[6]	2010
7	United Kingdom	London	8,630,100[7]	2015
8	Peru	Lima	8,481,415[8]	2012
9	Thailand	Bangkok	8,249,117[9]	2010

Figure 1.2: Top of HTML table of capitals sorted by population size. Courtesy of Wikipedia.

1.16 IMPORT DATA FROM AN HTML TABLE

Problem: You want to import data from an HTML table directly into R.

Solution: The Web is a rich source of data and a lot of that information is stored in HTML tables. HTML tables are not forced to be perfectly well-formed so parsing them might result in some problems.

The `htmltab` packages tries to make it easy to extract data from HTML tables by providing robust methods for parsing the information. The `htmltab` function requires an URL as input and by default it reads and parses the first table found on the web page. If there are multiple tables on the page then the desired table is chosen with the `which` argument, which accepts either a number or an XPath character string.

The following code imports the Wikipedia HTML table shown in Figure 1.2 that lists capitals by city population size. The relevant table is the second table on the web page.

```
> library(htmltab)
> url <- paste0("http://en.wikipedia.org/wiki/",
+               "List_of_national_capitals_by_population")
> result <- htmltab(doc = url, which=2,
+                   rm_nodata_cols=FALSE)  # Select table 2
> head(result)
  Rank Country/Territory   Capital Population Year
2    1             China   Beijing 20,693,000 2012
3    2             India New Delhi 16,787,949 2014
```

```
4   3            Japan      Tokyo 13,189,000 2011
5   4      Philippines     Manila 12,877,253 2015
6   5           Russia     Moscow 11,541,000 2011
7   6            Egypt      Cairo 10,230,350 2012
  Percent of Population
2                 1.52%
3                 0.90%
4                10.32%
5                12.44%
6                 8.07%
7                11.10%
```

By default, htmltab removes extra information found as superscript or footnotes. The argument rm_superscript=FALSE keeps any superscript information found in the cells when parsing the content of the table while the rm_footnotes=FALSE argument ensures that any footnotes after the table are kept as well. The table contains empty cells for the year column which is why we set rm_nodata_cols=FALSE to prevent the column from being disregarded.

See also: If the HTML table is more complicated (e.g., there are rows that span multiple columns) and htmltab cannot correctly parse it then the body and header arguments can be provided to give information about which rows should be used to construct the body and the header of the table, respectively. These arguments either expect numbers representing the relevant rows or XPath expression to identify the appropriate rows. Problem 1.14 explains XPath expressions in more detail.

1.17 SCRAPE DATA FROM AN HTML WEB PAGE

Problem: You want to scrape data from an HTML web page and import data with the same structure into R.

Solution: Problem 1.16 showed how to import data from a web page HTML table, but sometimes the data you want is not in a form that is easily downloaded. In those situations we can use web-scraping to extract information from a web page based on the HTML tags contained in the web page.

The rvest package provides a relatively easy way to scrape data from a web page. For this example we will extract information from the "Top Rated Movies" list from the The Internet Movie Database at imdb.com, which is currently found at the web address listed in the code

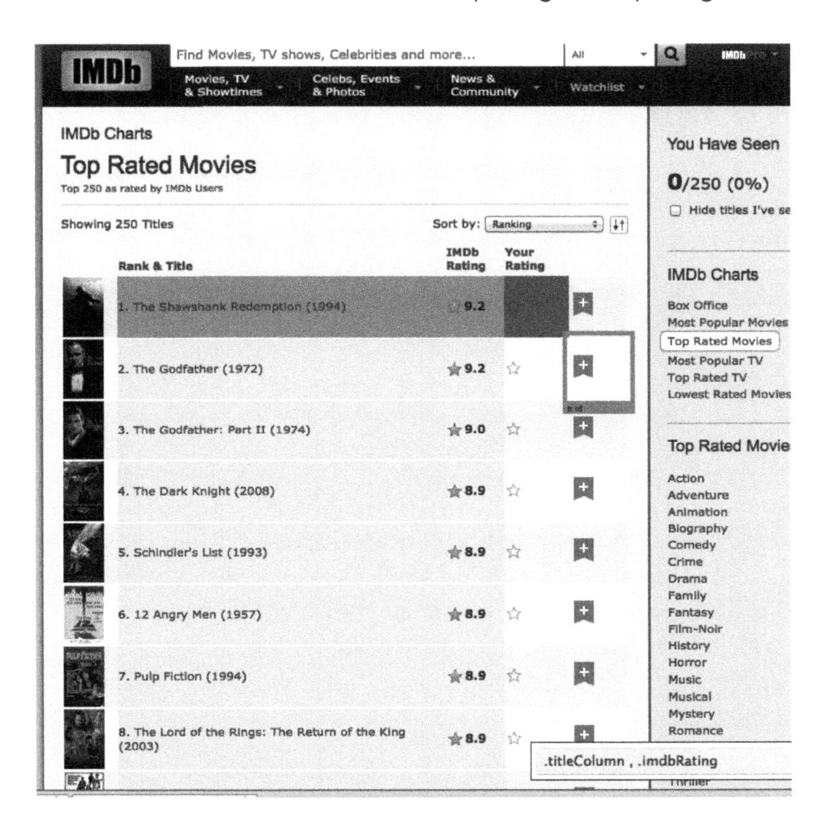

Figure 1.3: Selection of the relevant items from IMDb's top-rated movie list. The elements found by SelectorGadget are shown in the bottom right corner: "`.titleColumn`" and "`.imdbRating`".

below. First we read in the HTML page using the `read_html` function which returns an XML class.

```
> library(rvest)
> webaddress <- "http://www.imdb.com/chart/top?ref_=nv_mv_250_6"
> htmlpage <- read_html(webaddress)
```

The next step is crucial since this is where we specify which elements from the HTML page we wish to extract. You can identify the elements by looking at the HTML source code or you can use a tool like the SelectorGadget (see `SelectorGadget.com`) to click at the type of elements you wish to extract from the web page and see the corresponding tags and a visual representation of what will be extracted (see Figure 1.3).

Once we know which tags to select we use the `html_nodes` function to extract the relevant data. In the code below we specify the tags directly but it is also possible to specify them using XPath search expressions (see Problem 1.14).

From IMDb we extract the two pieces of information we are interested in — the titles and the movie ratings — separately to make the subsequent text conversion simpler.

```
> titlenodes  <- html_nodes(htmlpage, ".titleColumn")
> ratingnodes <- html_nodes(htmlpage, ".imdbRating")
> head(ratingnodes,3)
{xml_nodeset (3)}
[1] <td class="ratingColumn imdbRating">\n            <strong ...
[2] <td class="ratingColumn imdbRating">\n            <strong ...
[3] <td class="ratingColumn imdbRating">\n            <strong ...
```

The extracted nodes contain character strings that include all the HTML information from each node. To extract just the text without all the HTML codes we can use `html_text` which returns a vector of character strings — one for each of the extracted nodes. Depending on the string contents we may have to extract and convert the string to other data types.

```
> head(html_text(titlenodes))  # Show output
[1] "\n      1.\n      The Shawshank Redemption\n        (1994)\n    "
[2] "\n      2.\n      The Godfather\n        (1972)\n    "
[3] "\n      3.\n      The Godfather: Part II\n        (1974)\n    "
[4] "\n      4.\n      The Dark Knight\n        (2008)\n    "
[5] "\n      5.\n      Schindler's List\n        (1993)\n    "
[6] "\n      6.\n      12 Angry Men\n        (1957)\n    "
```

The IMDb data contains a number of newlines, `\n`, so we can use the `strsplit` function to split each character string into individual elements. The text from the HTML page contains a lot of starting and trailing white space which we remove using the `str_trim` function from the `stringr` package.

The titles and years become elements 3 and 4 after splitting the strings from the title nodes on newlines, and the ratings become the second element after splitting on newlines. All the results are combined into a data frame.

```
                    titles  years ratings
1 The Shawshank Redemption (1994)     9.2
```

```
2             The Godfather (1972)     9.2
3   The Godfather: Part II (1974)     9.0
4          The Dark Knight (2008)     8.9
5         Schindler's List (1993)     8.9
6             12 Angry Men (1957)     8.9
```

1.18 IMPORT FROM A MYSQL/POSTGRESQL DATABASE

Problem: You want to import data from a MySQL, PostgreSQL, or MariaDB database.

Solution: MySQL, PostgreSQL, and MariaDB are popular choices of free database systems and they are widely used for storing data. R has several packages that provide direct access to these relational databases. The RMySQL package can be used to access MySQL or MariaDB database systems, while the package RPostgreSQL is used for PostgreSQL. Here we show an example for a MySQL/MariaDB database but the commands are identical for a PostgreSQL database apart from the driver name.

Before we can send SQL query strings to the SQL server we need to open a connection. This is done with the dbConnect function, which requires a driver that matches the database type as well as information about the user, password, database name, and host name. For a MySQL database the driver is MySQL() and it is PostgreSQL() for a PostgreSQL database. The argument port can also be supplied if the default port for the server is not used.

```
> library(RMySQL)
> mydb <- dbConnect(MySQL(), user="rprimer", password="PASSword",
+                   dbname="rprimer", host="192.168.1.151")
```

Queries to the SQL server can then be sent to the connected server with the dbGetQuery function. dbGetQuery needs the connection object as first argument and the SQL query string as second argument, and it returns a data frame. Below we extract all the information from the students table.

```
> indata <- dbGetQuery(mydb, "select * from students")
> indata
          name  id    sex       exam score
1     John Doe 102   Male 2016-03-30    86
2     Jane Doe 103 Female 2016-02-03    87
3 Benny Ukendt 104   Male 2013-08-17    43
4  A. Einstein 105   Male 1921-01-01    99
```

Data frames are easily inserted into the SQL database with the `dbWriteTable` function. It requires at least two arguments besides the database connection: the name of the table in the database (argument `name`) and the data frame to insert (argument `value`). The arguments `append`, `row.names`, and `overwrite` all accept logicals and can be used with `dbWriteTable` to append data to an existing table, add a variable with the row names to the table, and overwrite an existing table, respectively.

Below we save the cherry trees data frame to the SQL database and use the `dbListTables` function to list all available tables in the database.

```
> data(trees)
> dbWriteTable(mydb, name="newTableName", value=trees, overwrite=TRUE)
[1] TRUE
> dbListTables(mydb)
[1] "newTableName" "students"
```

Finally, the `dbDisconnect` function closes the connection to the database.

```
> dbDisconnect(mydb)
[1] TRUE
```

See also: Problem 1.19 for examples of the more general `RODBC` package.

1.19 READ DATA FROM AN SQL DATABASE USING ODBC

Problem: Import data from an application that supports Open DataBase Connectivity.

Solution: Open DataBase Connectivity (ODBC) makes it possible for any application to access data from an SQL database regardless of which database management system is used to handle the data.

The `RODBC` package provides an interface to databases that support an ODBC interface, which includes most popular commercial and free databases such as MySQL, PostgreSQL, Microsoft SQL Server, Microsoft Access, and Oracle. Having one package with a common interface allows the same R code to access different database systems.

The `odbcConnect` function opens a connection to a database and returns an object which works as a handle for the connection. A character string containing the data source name (DSN) should be supplied as the first argument to `odbcConnect` to set the database server to connect to.

The function has two optional arguments, `uid` and `pwd`, which set the user id and password for authentication, respectively, if that is required by the database server, and is not provided by the DSN.

The data source name is located in a separate text file or in the registry and it contains the information that the ODBC driver needs in order to connect to a specific database. This includes the name, directory, and driver of the database, and possibly the user id and password. Each database requires a separate entry in the DSN, and DSN-less connections require that all the necessary information to be supplied within R (for example by using `odbcDriverConnect` instead of `odbcConnect`). The information for setting up the DSN should accompany your database software.

Once a connection to a database server is established, then the available database tables can be seen with the `sqlTables` function, with the proper handle as first argument. The `sqlFetch` function fetches the entire table from the SQL database and returns it as an R data frame. `sqlFetch` requires two arguments, where the first is the connection handle and the second is a character string containing the desired table to extract from the database.

The workhorse is the `sqlQuery` function which is used to make SQL queries directly to the database and return the results as R data frames. The first argument to `sqlQuery` sets the connection channel to use and the second argument is the selection string which should be specified as a regular SQL query string. Finally, the `odbcClose` function closes the connection to the channel specified by the first argument.

In the example below, we use an existing DSN to access a database called "myproject" which contains a "paper" table. The entire table is extracted as well as a selection of salespeople with sales larger than a given number.

```
> library(RODBC)
> # Connect to SQL database with username and password
> channel <- odbcConnect("mydata", uid="tv", pwd="office")
> sqlTables(channel)                    # List tables in the database
> mydata <- sqlFetch(channel, "paper")       # Fetch entire table
> mydata
> sqlQuery(channel,
+          "SELECT * FROM paper WHERE Sales>12 ORDER BY Person")
> odbcClose(channel)
```

CHAPTER 2

Manipulating data

R uses several basic data types that are frequently used in most R functions or calculations. Here we list the most fundamental data type objects:

- A *vector* can be either numeric, complex, a character vector, or a logical vector. These vectors correspond to a sequence of numbers, complex numbers, characters, or logical values, respectively.
- An *array* is a vector with a dimension attribute, where the dimension attribute is a vector of non-negative integers. If the length of the dimension vector is k, then the array is k-dimensional. The most commonly used array is a *matrix*, which is a two-dimensional array of numbers.
- A *factor* object is used to define categorical (nominal or ordered) variables. They can be viewed as integer vectors where each integer value has a corresponding label.
- A *list* is an ordered collection of objects. A vector can only contain elements of one type but a list can be used to create collections of vectors or objects of mixed type.
- A *data frame* is a list of vectors or factors all of the same length such that each "row" corresponds to an observation.

The following code uses the functions `c`, `factor`, `data.frame`, and `list` to show examples of the different basic data types.

```
> vec1 <- 4                          # Numeric vector of length 1
> vec1
[1] 4
> vec2 <- c(1, 2, 3.4, 5.6, 7)  # Numeric vector of length 5
> vec2
[1] 1.0 2.0 3.4 5.6 7.0
> # Create a character vector
> vec3 <- c("Spare", "a talent", "for an", "old", "ex-leper")
> vec3
[1] "Spare"    "a talent" "for an"   "old"      "ex-leper"
> # And a vector of logical values
> vec4 <- c(TRUE, TRUE, FALSE, TRUE, FALSE)
> vec4
[1]  TRUE  TRUE FALSE  TRUE FALSE
> f <- factor(vec3)              # Make factor based on vec3
> f
[1] Spare    a talent for an   old      ex-leper
Levels: Spare a talent ex-leper for an old
```

Complex numbers are specified by adding an imaginary part to a numeric number.

```
> x <- 1 + 2i                        # Enter a complex number
> x
[1] 1+2i
> sqrt(-1)                           # R does not see a complex number

Warning in sqrt(-1): NaNs produced

[1] NaN
> sqrt(-1 + 0i)                      # but it does here
[1] 0+1i
```

Matrices can be created from a vector and then specifying either the number of columns or rows using the `ncol` or `nrow` arguments, respectively. By default the matrices are filled column-wise but this can be changed by setting the `byrow=TRUE` argument.

```
> m <- matrix(1:6, ncol=2)                    # Create a matrix
> m
     [,1] [,2]
[1,]    1    4
[2,]    2    5
[3,]    3    6
> m2 <- matrix(1:6, ncol=2, byrow=TRUE)  # Fill the matrix row-wise
> m2
     [,1] [,2]
```

```
[1,]    1    2
[2,]    3    4
[3,]    5    6
```

Lists and data frames are created by providing a number of R objects that may or may not be named.

```
> list(vec2, comp=x, m)      # Combine 3 different objects, one named
[[1]]
[1] 1.0 2.0 3.4 5.6 7.0

$comp
[1] 1+2i

[[3]]
     [,1] [,2]
[1,]    1    4
[2,]    2    5
[3,]    3    6
> data.frame(vec2, f, vec4)  # Make a data frame
  vec2        f  vec4
1  1.0    Spare  TRUE
2  2.0 a talent  TRUE
3  3.4   for an FALSE
4  5.6      old  TRUE
5  7.0 ex-leper FALSE
```

2.1 USE MATHEMATICAL FUNCTIONS

Problem: You want to apply basic numeric mathematical function or use arithmetic operators in your calculations.

Solution: R contains a large number of arithmetic operators and mathematical functions and some of the most common are listed in Table 2.1. The code below shows examples where these functions are used.

```
> 5/2 + 2*(5.1 - 2.3)  # Add, subtract, multiply and divide
[1] 8.1
> 2**8                 # 2 to the power 8
[1] 256
> 1.61^5               # 1.61 to the power 5
[1] 10.81756
> 10 %% 3              # 10 modulus 3 has remainder 1
[1] 1
> 10 %/% 3             # integer division
```

Table 2.1: Mathematical operators and functions

Symbol/function	Description
+	addition
-	subtraction
*	multiplication
/	division
^ or **	exponentiation
%%	modulus
%/%	integer division
abs(x)	absolute value
sqrt(x)	square root
ceiling(x)	smallest integer not less than x
floor(x)	largest integer not greater than x
trunc(x)	truncate x by discarding decimals
round(x, digits=0)	round x to digits decimals
signif(x, digits=6)	round to digits significant digits
cos(x), sin(x) and tan(x)	cosine, sine, and tangent
log(x)	natural logarithm
log(x, base=2)	logarithm with base 2
log10(x)	common logarithm
exp(x)	exponential function
%*%	matrix multiplication

```
[1] 3
> abs(-3.1)           # absolute value of -3.1
[1] 3.1
> sqrt(5)             # square root of 5
[1] 2.236068
> ceiling(4.3)        # smallest integer larger than 4.3
[1] 5
> floor(4.3)          # largest integer smaller than 4.3
[1] 4
> trunc(4.3)          # remove decimals
[1] 4
> round(4.5)
[1] 4
> round(4.51)
[1] 5
> round(4.51, digits=1)
[1] 4.5
> # Angles for trigonometric functions use radians - not degrees
```

```
> cos(pi/2)              # cosine of pi/2.
[1] 6.123234e-17
> sin(pi/4)              # sine of pi/4
[1] 0.7071068
> tan(pi/6)              # tangent of (pi/6)
[1] 0.5773503
> log(5)                 # natural logarithm of 5
[1] 1.609438
> log(5, base=2)         # binary logarithm of 5
[1] 2.321928
> log10(5)               # common logarithm of 5
[1] 0.69897
> exp(log(5) + log(3)) # exponential function
[1] 15
```

Matrices are created from a numerical vector by specifying the number of rows or columns. By default the elements of the matrix are filled column-wise.

```
> # Create two matrices
> x <- matrix(1:6,ncol=3)
> y <- matrix(c(1, 1, 0, 0, 0, 1), ncol=2)
> x
     [,1] [,2] [,3]
[1,]    1    3    5
[2,]    2    4    6
> y
     [,1] [,2]
[1,]    1    0
[2,]    1    0
[3,]    0    1
> x %*% y          # Matrix multiplication
     [,1] [,2]
[1,]    4    5
[2,]    6    6
> y %*% x          # Matrix multiplication
     [,1] [,2] [,3]
[1,]    1    3    5
[2,]    1    3    5
[3,]    2    4    6
```

2.2 USE COMMON VECTOR OPERATIONS

Problem: You need to identify and/or use some of the common functions.

Table 2.2: Frequently used functions for vectors operations

Symbol/function	Description
`c(x, ...)`	concatenate elements into a vector
`head(x)`	print the head of the vector `x`
`tail(x)`	print the tail of the vector `x`
`length(x)`	length of vector `x`
`unique(x)`	unique elements of `x`
`duplicated(x)`	duplicated elements of `x`
`rev(x)`	reverse the vector `x`
`seq(x)`	create a sequence of numbers
`sort(x)`	sort a vector `x`
`order(x)`	order elements of vector `x`
`class(x)`	list the class of `x`
`any(x)`	are any elements of `x` `TRUE`?
`all(x)`	are all elements of `x` `TRUE`?
`which.min(x)`	index of smallest value of `x`
`which.max(x)`	index of largest value of `x`
`which(x)`	indices of `TRUE` elements of `x`
`is.na(x)`	element-wise test for missing observations of `x`
`is.`*Something*`(x)`	returns `TRUE` if the input `x` belongs to class *Something*

Solution: R is vector based and there are many functions available that simplify working with vectors. Here we will present a handful of ubiquitous functions for working with vectors. Table 2.2 gives an overview and examples are found below and in Problem 2.5.

The `c` function is ubiquitous in R and can be used to glue elements together to a vector. `head` and `tail` print the start and end of a vector, respectively, and they also work for most other object types. The `seq`, `length`, and `class` functions return a sequence of numbers, the length of a vector, and the class of an R object, respectively. `seq` either accepts a single number (in which case it produces a vector of integer values from 1 up to the input), two numbers (in which case it produces the sequence from the first number to the second number with a step increase of 1), or three numbers representing the from, to, and step size, respectively.

```
> seq(5)          # vector with sequence from 1 to 5
[1] 1 2 3 4 5
> seq(4, 6)       # from 4 to 6 with step size 1
[1] 4 5 6
```

```
> 4:6                # Alternative way to make sequence with step 1
[1] 4 5 6
> seq(6, 4)          # from 6 to 4 with step size 1
[1] 6 5 4
> seq(0, 3, .5)      # from 0 to 3 with steps 0.5
[1] 0.0 0.5 1.0 1.5 2.0 2.5 3.0
> x <- c(6, 8, 1:4, 9, 7, 5, 3)
> x
 [1] 6 8 1 2 3 4 9 7 5 3
> length(x)          # length of x
[1] 10
> head(x)            # Show the start of x
[1] 6 8 1 2 3 4
> tail(x)            # and the bottom
[1] 3 4 9 7 5 3
> rev(x)             # reverse the order of elements in x
 [1] 3 5 7 9 4 3 2 1 8 6
> class(1:3)         # class of a sequence (only has integers)
[1] "integer"
> class(x)           # class of x (numeric because of c())
[1] "numeric"
> class(c(1, 2.5))   # and now numeric variables
[1] "numeric"
> class(c(TRUE, TRUE, FALSE))
[1] "logical"
```

The `which.min` and `which.max` functions both return the index corresponding to the smallest and largest element, respectively. `unique` and `duplicated` return the unique element of a vector and the index corresponding to the duplicated values, respectively.

```
> x
 [1] 6 8 1 2 3 4 9 7 5 3
> which.min(x)  # The 3rd element is smallest (value 1)
[1] 3
> which.max(x)  # The 2nd element is largest (value 8)
[1] 7
> y <- c(2, 4, 6, 2, 4, 9, 2)
> which.min(y)  # Return the first element if there are ties
[1] 1
> which.min(c(FALSE, TRUE, FALSE)) # FALSE = 0, TRUE = 1
[1] 1
> which.max(c(FALSE, TRUE, FALSE))
[1] 2
> unique(y)
[1] 2 4 6 9
> duplicated(y)
[1] FALSE FALSE FALSE  TRUE  TRUE FALSE  TRUE
```

The any, all, and which functions operate on logical vectors. The first two return TRUE if the input vector contains at least one TRUE element, or if it contains only TRUE elements, respectively. which returns the indices of the elements that are TRUE.

```
> z1 <- c(TRUE, FALSE, FALSE)
> any(z1)
[1] TRUE
> all(z1)
[1] FALSE
> z2 <- c(TRUE, TRUE, TRUE)
> any(z2)
[1] TRUE
> all(z2)
[1] TRUE
> z3 <- c(FALSE, FALSE, FALSE)
> any(z3)
[1] FALSE
> all(z3)
[1] FALSE
> which(z1)
[1] 1
> which(z3)
integer(0)
```

Finally, is.na accepts a vector and returns a logical vector of the same length with elements that are TRUE if the corresponding element in the input is NA.

```
> x
 [1] 6 8 1 2 3 4 9 7 5 3
> is.na(x)
 [1] FALSE FALSE FALSE FALSE FALSE FALSE FALSE FALSE FALSE FALSE
> x[c(2,4)] <- NA # redefine x
> x
 [1]  6 NA  1 NA  3  4  9  7  5  3
> is.na(x)         # identify missing elements
 [1] FALSE  TRUE FALSE  TRUE FALSE FALSE FALSE FALSE FALSE FALSE
> is.numeric(x)    # Check if x is a numeric vector
[1] TRUE
> is.character(x) # Check if x is a character vector
[1] FALSE
```

See also: Problem 2.5 shows how to sort and order vectors.

2.3 WORK WITH CHARACTER VECTORS

Problem: You have a character vector and want to modify or get information from the character strings.

Solution: R has numerous functions that work on character vectors. Here we will give an example of some of the functions that operate on character strings.

The `toupper` and `tolower` functions translate each element of the input character vector to uppercase or lowercase, respectively.

`nchar` takes a character vector as input and returns a vector with the number of characters of each input element.

`grep` is a text search function. The first argument to `grep` is a regular expression containing the search pattern. The second argument is a character vector where matches are sought. `grep` returns a vector of indices where the regular expression pattern is found. The `ignore.case` option can be set to `TRUE` to make the search pattern case insensitive.

The `substr` function can be used to extract or replace substrings in a character vector. `substr` takes a character vector as first argument, and has two other required arguments: `start` and `stop` which should both be integer values and which define the starting and ending position of the substring extracted (the first character is position 1). If the selected substring is replaced, then the replacement string is inserted from the `start` position until the `stop` position.

`paste` accepts one or more R objects which are converted to character vectors and then concatenated together. The character strings are separated by the string given to the `sep` option which defaults to a single space, `sep=" "`. If the `collapse` option is specified, then the elements in the resulting vector are collapsed together to a single string where the individual elements are separated by the character string given by `collapse`.

```
> x <- c("Suki", "walk", "to", "the", "well", NA)
> toupper(x)                # convert to upper case
[1] "SUKI" "WALK" "TO"   "THE"  "WELL" NA
> tolower(x)                # convert to lower case
[1] "suki" "walk" "to"   "the"  "well" NA
> nchar(x)                  # no. of characters in each element of x
[1]  4  4  2  3  4 NA
> grep("a", x)              # elements that contain letter "a"
[1] 2
> grep("o", x)              # elements that contain letter "o"
[1] 3
```

```
> substr(x, 2, 3)       # extract 2nd to 3rd characters
[1] "uk" "al" "o"  "he" "el" NA
> substr(x, 2, 3) <- "X" # replace 2nd char with X
> x
[1] "SXki" "wXlk" "tX"   "tXe"  "wXll" NA
> substr(x, 2, 3) <- "XXXX"
> x
[1] "SXXi" "wXXk" "tX"   "tXX"  "wXXl" NA
> paste(x, "y")         # paste "y" to end of each element
[1] "SXXi y" "wXXk y" "tX y"   "tXX y"  "wXXl y" "NA y"
> paste(x, "y", sep="-")
[1] "SXXi-y" "wXXk-y" "tX-y"   "tXX-y"  "wXXl-y" "NA-y"
> paste(x, "y", sep="-", collapse=" ")
[1] "SXXi-y wXXk-y tX-y tXX-y wXXl-y NA-y"
```

See also: The `paste0` function is similar to `paste` except the separator argument is set to `sep=""` by default.

2.4 READ NON-ASCII CHARACTER VECTORS

Problem: You want to convert the encoding of an R character vector from one format to another format.

Solution: R automatically handles character strings encoded in ASCII, UTF-8, and the platform's native encoding. However, when R reads strings with another encoding then it may be necessary to either specify the encoding or to convert the character string from one encoding to another.

We can either set the encoding of a character vector by using the `Encoding` function or use `iconv` to convert between encodings and both functions take a character vector as first argument.

`Encoding` reads or sets the encoding of a character vector. We try that on a file created by Notepad under Windows XP and saved in the ANSI format.

```
> indata <- read.table("DKdata.txt", header=TRUE)
> indata
          name value
1       N\xe6h!     1
2 R\xf8dgr\xf8d     3
3           med     5
4      fl\xf8de     7
```

The `name` variable contains the character strings with the non-standard encoding. We need to extract the variable to use it as input for `Encoding` or `iconv` and also to convert it to a character string since `read.table` automatically converts characters strings to factors unless the `stringsAsFactors=FALSE` argument is included in the call to `read.table`.

The encoding for this ANSI file is called latin1.

```
> x <- as.character(indata$name)
> x
[1] "N\xe6h!"        "R\xf8dgr\xf8d" "med"
[4] "fl\xf8de"
> Encoding(x)                    # Read the encoding
[1] "unknown" "unknown" "unknown" "unknown"
> Encoding(x) <- "latin1"  # Set it to latin1
> x                              # Correct answer
[1] "Næh!"    "Rødgrød" "med"      "fløde"
```

The `iconvlist` function lists all the encodings that are available on the current system. Generally, it is not possible to detect the encoding of a character string with 100% certainty, and it is necessary to provide the information about the encoding just as shown above. The `iconv` converts a character vector from one encoding to another. The first argument is the character vector, while the `from` and `to` arguments set the old and new encoding, respectively. An empty string, `""`, for either `from` or `to` corresponds to encoding with the current locale. To convert a character vector from latin1 codepage to the codepage of the current system we can use the following code.

```
> x <- as.character(indata$name)
> x
[1] "N\xe6h!"        "R\xf8dgr\xf8d" "med"
[4] "fl\xf8de"
> iconv(x, from="latin1", to="")
[1] "Næh!"    "Rødgrød" "med"      "fløde"
```

See also: Many of the functions for reading text files data, `read.table`, `read.csv`, etc., have an argument `fileEncoding` that can be set to force the encoding used for parsing the text file at the time it is read.

2.5 SORT AND ORDER DATA

Problem: You want to sort or order a vector.

Solution: `sort` sorts a numeric, complex, character, or logical vector in ascending order. Set the argument `decreasing` to `TRUE` to sort it in decreasing order.

```
> x <- c(2, 4, 1, 5, 3, 4)
> x
[1] 2 4 1 5 3 4
> sort(x)
[1] 1 2 3 4 4 5
> sort(x, decreasing=TRUE)
[1] 5 4 4 3 2 1
> y <- c("b", "a", "A", "b", "B", "a")
> y
[1] "b" "a" "A" "b" "B" "a"
> sort(y)
[1] "A" "B" "a" "a" "b" "b"
```

If you want to find the order of the elements that sorts a vector, you can use the `order` function. The order arranges the argument in ascending order (descending when the `decreasing` option is `TRUE`).

```
> o <- order(x)
> o
[1] 3 1 5 2 6 4
> x[o[1]]  # Same as min(x)
[1] 1
> x[o]     # Same as sort(x)
[1] 1 2 3 4 4 5
```

Additional vectors can be given as input to `order` to break ties. Thus, the first input vector is ordered, and if any duplicates are present then the second vector is used to order those, and the third is used to order ties in the second vector and so forth.

```
> o1 <- order(y)
> o1
[1] 3 5 2 6 1 4
> o2 <- order(y, x)
> o2
[1] 3 5 2 6 1 4
```

Using multiple vectors as input to `order` is particularly useful for sorting data frames according to one or more variables.

```
> df <- data.frame(day=c(1, 2, 1, 2, 3),
+                  time=c("late", "late", "early", "late", "early"))
> df
  day  time
1   1  late
2   2  late
3   1 early
4   2  late
5   3 early
> o <- order(df$day, df$time)   # Order by day and then time
> df[o,]
  day  time
3   1 early
1   1  late
2   2  late
4   2  late
5   3 early
```

2.6 TRANSFORM A VARIABLE

Problem: You want to log transform (or use some other transformation) a vector or a variable in a data frame.

Solution: It is often necessary to create a new variable based on the existing variable(s) or to transform a variable, for example to change its scale.

In R this is done simply by applying the function that creates the new variable from the existing variables. If the new variable is to be part of a data frame, then the $-operator can be used to store the new variable inside a data frame.

```
> data(airquality)
> names(airquality)
[1] "Ozone"   "Solar.R" "Wind"    "Temp"    "Month"   "Day"
> celsius <- (airquality$Temp-32)/1.8            # New variable
> airquality$Celsius <- (airquality$Temp-32)/1.8 # Add to data frame
> head(airquality$Temp)
[1] 67 72 74 62 56 66
> head(airquality$Celsius)
[1] 19.44444 22.22222 23.33333 16.66667 13.33333 18.88889
```

Alternatively, the `transform` function can be used to create a new data frame from an existing data frame and at the same time define or redefine variables inside the new data frame. As its first argument `transform` takes the name of an existing data frame, and changed variables are put as additional arguments of the form tag=value as shown in the following.

```
> newdata <- transform(airquality, logOzone = log(Ozone))
> head(newdata)
  Ozone Solar.R Wind Temp Month Day  Celsius logOzone
1    41     190  7.4   67     5   1 19.44444 3.713572
2    36     118  8.0   72     5   2 22.22222 3.583519
3    12     149 12.6   74     5   3 23.33333 2.484907
4    18     313 11.5   62     5   4 16.66667 2.890372
5    NA      NA 14.3   56     5   5 13.33333       NA
6    28      NA 14.9   66     5   6 18.88889 3.332205
```

2.7 FIND THE VALUE OF X CORRESPONDING TO THE MAXIMUM OR MINIMUM OF Y

Problem: You have two numeric vectors, `x` and `y`, of the same length and want to find the value of `x` where `y` attains its maximum value.

Solution: Problem 2.2 showed how to use `which.max` and `which.min` functions to find the index of a vector where the first occurrence of the maximum or minimum value is found, respectively. These results can be trivially combined with another vector to return the value of a vector `x` corresponding to the location of the maximum of a vector `y` of the same length.

In the code below we wish to find the location of the highest peak of a function evaluated at different positions.

```
> x <- seq(0, pi/2, .2)
> x
[1] 0.0 0.2 0.4 0.6 0.8 1.0 1.2 1.4
> y <- sin(2*x)
> y
[1] 0.0000000 0.3894183 0.7173561 0.9320391 0.9995736 0.9092974
[7] 0.6754632 0.3349882
> x[which.max(y)]
[1] 0.8
```

2.8 CHECK IF ELEMENTS IN ONE OBJECT ARE PRESENT IN ANOTHER OBJECT

Problem: You want to check which of the elements in a vector, matrix, list, or data frame are present in another vector, matrix, list, or data frame.

Solution: The `%in%` operator checks if each of the elements in the object on the left-hand side has a match among the elements in the object on the right-hand side. The operator converts factors to character vectors before matching, and returns a logical vector that states the match for each element.

```
> x <- c(1, 2, 3, 1, NA, 4)
> y <- c(1, 2, 5, NA)
> x %in% y
[1]  TRUE  TRUE FALSE  TRUE  TRUE FALSE
> y %in% x
[1]  TRUE  TRUE FALSE  TRUE
> ! x %in% y   # Invert result: find objects NOT present
[1] FALSE FALSE  TRUE FALSE FALSE  TRUE
```

The `%in%` operator also works for matrices, data frames, and lists. Matrices are converted to vectors before matching. Lists and data frames are matched based on their individual elements.

```
> z <- matrix(2:7, ncol=2)
> z
     [,1] [,2]
[1,]    2    5
[2,]    3    6
[3,]    4    7
> x %in% z
[1] FALSE  TRUE  TRUE FALSE FALSE  TRUE
```

Lists (and therefore also data frames) as expanded and each of the list elements (variables for data frames) are checked if a similar object with the same values is found regardless of position and/or label.

```
> # c(1,2) is found in both
> data.frame(a=c(1,2), b=c(3,4)) %in% data.frame(f=c(5,6),
+                                                d=c(1,2))
[1]  TRUE FALSE
> list(1, 2, 3) %in% list(a=c(1,2), b=3)
[1] FALSE FALSE  TRUE
```

2.9 APPLY A FUNCTION TO SUBSETS OF A VECTOR

Problem: You want to apply a function to subsets of a vector.

Solution: It is often necessary to apply a function to subsets of a vector and then combine the results. R has several functions that enable the user to handle the split-apply-combine paradigm, and `tapply` works with vectors. It accepts a vector as first argument, and then a factor or list of factors that define the splits into subsets as second argument, and then the `FUN` argument sets the function that is to be applied to each subset. Any extra arguments after `FUN` are passed as arguments to `FUN`.

For the `airquality` dataset we wish to compute the mean temperature for each month. Thus we need to split the data into subsets defined by the month, and then compute the mean.

```
> data(airquality)
> tapply(airquality$Temp, airquality$Month, mean)
       5        6        7        8        9
65.54839 79.10000 83.90323 83.96774 76.90000
```

`tapply` returns a list unless the `FUN` function returns a scalar, in which case the result — as seen above — is simplified to a vector or matrix.

If we wish to split each month up into early, mid, and late month and look at the temperature for each period of each month then the `cut` function partitions the days and we can then feed that to a list.

```
> when <- cut(airquality$Day, 3, labels=c("Early", "Mid", "Late"))
> tapply(airquality$Temp, list(airquality$Month, when), mean)
     Early  Mid     Late
5 65.90909 63.7 67.00000
6 82.36364 77.3 77.11111
7 85.72727 81.8 84.00000
8 86.27273 79.2 86.20000
9 84.27273 73.0 72.22222
```

See also: See Problem 2.27 for information on how `cut` can be used to cut a numeric vector into a factor. If `tapply` returns a list and you want the result in a simpler form, then the `unlist` function may prove useful depending on the type of outcome returned by the function applied.

2.10 FILL IN MISSING VALUES WITH PREVIOUS VALUES

Problem: You have a vector of observations that contain missing values and you wish to replace it with the most recent non-`NA` prior to it.

Solution: To prevent typing, some values may only have been typed in once although they should essentially occur multiple times. For example, the id representing the unique identifier that defines an individual may only be typed the first time that the individual is registered.

The `filldown` function from the `MESS` package replaces `NA`'s with the most recent non-`NA` that appears before it in a vector.

```
> library(MESS)
> id <- c(1, NA, NA, NA, 2, NA, NA, NA, 3, NA, NA, NA)
> date <- c("2016-02-05", NA, NA, NA,
+           NA, NA, NA, NA,
+           "2016-02-06", NA, NA, NA)
> newid <- filldown(id)
> newid
 [1] 1 1 1 1 2 2 2 2 3 3 3 3
```

If we only wish to fill down non-`NA` for, say, each individual then we can use the approach outlined in Problem 2.9 to apply the `filldown` function to each subset of the vector.

```
> tapply(date, newid, filldown)   # Returns a list
$`1`
[1] "2016-02-05" "2016-02-05" "2016-02-05" "2016-02-05"

$`2`
[1] NA NA NA NA

$`3`
[1] "2016-02-06" "2016-02-06" "2016-02-06" "2016-02-06"
```

The result here is returned as a list so we use the `unlist` to combine the results into a vector.

```
> unlist(tapply(date, newid, filldown))
          11           12           13           14
"2016-02-05" "2016-02-05" "2016-02-05" "2016-02-05"
          21           22           23           24
          NA           NA           NA           NA
          31           32           33           34
"2016-02-06" "2016-02-06" "2016-02-06" "2016-02-06"
```

Here we have used the `filldown` function to fill out "fixed" values in the dataset; i.e., values that were missing in the sense that we had the actual values, they were just not entered in the data frame. The fill down approach described above can also be used to fill down observed measurement values from one time point to the next — the last observation carried forward approach. The last observation carried forward generally results in biased estimates and is virtually never the right approach to use for missing data in statistical analyses.

2.11 CONVERT COMMA AS DECIMAL MARK TO PERIOD

Problem: You have a vector of observations where a comma is used as the decimal mark. R sees the values of the vector as character strings and you want to convert it to numeric values.

Solution: R uses the period as the decimal mark but in many countries the comma is used as the decimal mark. When data are imported into R the decimal point can often be specified, for example with the `dec` option to `read.table` or `read.csv`.

If you have a vector of observations that are numbers using a comma as the decimal mark, then R will view the vector either as a factor or as a vector of character strings. In order to convert the values to numeric, we use the `sub` function to substitute the period decimal point for the comma and then call `as.numeric` to convert the resulting vector to numeric values. `sub` takes three arguments: the string pattern that should be matched, the character string that should be substituted for it, and the character vector on which it should be applied.

```
> x <- c("1,2", "1,5", "1,8", "1.9", "2")   # Example data
> as.numeric(sub(",", ".", x))  # Substitute , with . and convert
[1] 1.2 1.5 1.8 1.9 2.0
> f <- factor(x)                # It also works on factors
> as.numeric(sub(",", ".", f))  # Substitute , with . and convert
[1] 1.2 1.5 1.8 1.9 2.0
> as.numeric(x)          # Does not work - need to substitute first

Warning: NAs introduced by coercion

[1]  NA  NA  NA 1.9 2.0
```

Since `sub` returns a character vector and not a factor we can use the function `as.numeric` directly on the result from `sub` unlike the situation covered in Problem 2.21.

2.12 LAG OR SHIFT A VECTOR

Problem: You want to shift a vector up or down to create a lagged version.

Solution: Lagged variables are often used in connection with time series, but in many situations it may be relevant to created a "shifted" version of a variable — either by shifting the variable up or down. The built-in function `lag` works fine with time series objects but yields some peculiarities with vectors in general. In particular, it returns a time series object which may not be relevant.

The `data.table` package provides the `shift` function which is able to create shifted versions of a vector, list, or data frame. Besides the object to be shifted it accepts optional arguments `n`, and `fill`, which determine the number of positions to shift, the value to insert for the shifted positions, respectively. Furthermore, the argument `type` accepts one of the strings `lag` or `lead`, which determines if the vector is shifted up or down. Default values for the three options are 1, `NA`, and `lag`.

```
> library(data.table)
> x <- 1:8
> shift(x)
[1] NA  1  2  3  4  5  6  7
> shift(x, n=3)
[1] NA NA NA  1  2  3  4  5
> shift(x, n=2, fill=0)
[1] 0 0 1 2 3 4 5 6
> shift(LETTERS[1:5], type="lead")
[1] "B" "C" "D" "E" NA
```

If the input is a list or data frame then each element in the list/data frame is shifted and the result is returned as a list.

```
> df <- data.frame(1:5, LETTERS[1:5])
> shift(df, n=1, type="lag")
[[1]]
[1] NA  1  2  3  4

[[2]]
[1] <NA> A    B    C    D
Levels: A B C D E
```

Likewise, if `n` is a vector then each of the different shifting position values will be applied to the input object.

```
> shift(x, n=c(1,2))
[[1]]
[1] NA  1  2  3  4  5  6  7

[[2]]
[1] NA NA  1  2  3  4  5  6
```

See also: The time series packages `xts` and `zoo` have very fast and flexible lag functions built in, so for these objects it makes more sense to use those.

2.13 CALCULATE THE AREA UNDER A CURVE

Problem: You have a set of repeated measurements for each individual and wish to calculate the area under the curve defined by the set of measurements over time.

Solution: Repeated measurements occur for example when measurements are taken several times over time on a single individual and the resulting set of observations is essentially a response profile. In some situations it is desirable to summarize the set of repeated measurements by a single value like the mean, the maximum value, average increase, or the area under the response curve.

The area under the curve corresponds to the integral of the curve but since we only have information for the observed time points we need to approximate the underlying profile curve. If we assume that we can approximate the profile with straight lines between time points, then we can use the trapezoid rule to calculate the integral/area under the curve:

$$\int_a^b f(x)dx \approx \frac{1}{2}\sum_{i=2}^{n}(x_i - x_{i-1})(y_i + y_{i-1})$$

The `auc` function in the `MESS` package provides a way to compute the area under the curve. It expected two arguments as default: the first is a vector of x-values while the other is the vector (of the same length) of y-values. The example data and the fitted third-degree polynomial can be seen in Figure 2.1.

```
> library(MESS)
> x <- 1:5
> y <- c(1, 0, 1, 1, 5)
> auc(x, y)
[1] 5
> x2 <- c(4, 5, 1, 3, 2)  # Reorder the same data
> y2 <- c(1, 5, 1, 1, 0)
> auc(x2, y2)             # Reordering is fine. Same result
[1] 5
```

There are three optional arguments to `auc` that change the range and extrapolation of the `auc` function. `from` and `to` accept numeric values

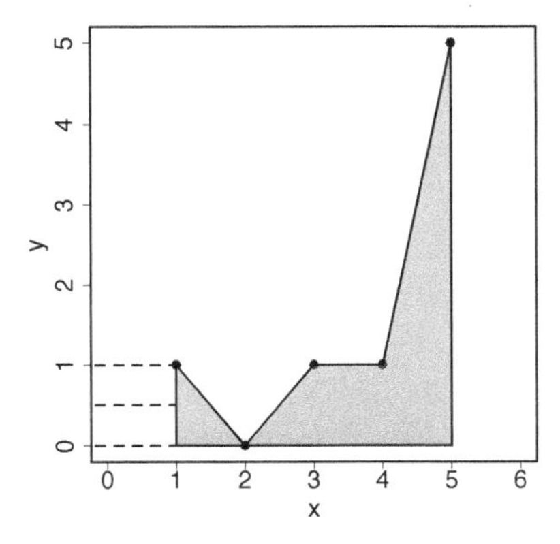

Figure 2.1: Computing the area under the curve.

and set the lower and upper limit for the range that should be considered, when computing the area under the curve. The `rule` argument specifies how interpolation outside the observation range takes place. The default value for `rule` is 1 which makes `auc` return `NA` outside the observational range. If `rule=2` then a combined "last observation carried forward" and "first observation carried backward" approach is used. If different extrapolations are needed for the left and right side of the observational range, then `rule` should be set to an integer vector of length 2 with elements corresponding to the extrapolation rule to the left and right, respectively.

The `yleft` and `yright` options to `approxfun` set the y-values that are used at the `from` and `to` values outside the observation range when `rule=2` instead of the "last observation carried forward" and "first observation carried backward" values.

```
> auc(x, y, from=0)                      # AUC from 0 to max(x)
[1] NA
> auc(x, y, from=0, rule=2)              # Allow extrapolation
[1] 6
> auc(x, y, from=0, rule=2, yleft=0)     # Use value 0 to the left
[1] 5.5
> auc(x, y, from=0, rule=2, yleft=.5)    # Use 1/2 to the left
[1] 5.75
```

If we wish to compute the area under the curve for each individual in a data frame we can use the `auc` function together with the `by` function. For example, to calculate the area under each growth curve for the `ChickWeight` data we can use the following code:

```
> data(ChickWeight)
> result <- by(ChickWeight, ChickWeight$Chick,
+              function(df) { auc(df$Time, df$weight) } )
> head(result)
ChickWeight$Chick
    18     16     15     13      9     20
  74.0  601.0  853.0 1397.5 1709.0 1608.0
```

We should make sure that the areas are comparable by ensuring that each area is based on the same interval from 0 to 21 days. Carry first and last observation backward and forward, respectively.

```
> result2 <- by(ChickWeight, ChickWeight$Chick, function(df) {
+               auc(df$Time, df$weight, from=0, to=21, rule=2)})
> head(result2)
ChickWeight$Chick
    18     16     15     13      9     20
 739.0 1087.0 1329.0 1397.5 1709.0 1608.0
```

See also: Use Problem 2.38 if you need to transform the data to long format before using `auc` if they are originally in wide format.

MATRICES, ARRAYS, AND TABLES

2.14 APPLY FUNCTION TO MARGINS OF A MATRIX/ARRAY

Problem: You want to apply a function to the rows or columns of a matrix or to the margin of an array.

Solution: R has a range of functions that can be used to execute a function repetitively by applying the function to each row or column of a matrix or to the margin of an array. This prevents us from having to write loops to perform some operation repeatedly, although the same result could be obtained with a loop.

The `apply` function is very versatile, and it requires three arguments: an array, a vector of the margin(s) over which the desired function should be applied (the `MARGIN` argument), and the function that should be executed repetitively (the `FUN` argument).

The `MARGIN` argument is used to specify which margin we want to apply the function over and which margin we wish to keep. For example, if the array we are using is two-dimensional then we can specify the margin to be either 1 (apply the function to each of the rows) or 2 (apply the function to the columns).

The function defined by the option FUN can be a built-in or user-defined function. The vectors of values defined by the MARGIN option is passed as the first argument to FUN and if additional parameters are included in the call to apply then they are passed on to the FUN function. As a consequence, the first argument of the function FUN should accommodate the correct input for apply to work.

```
> m <- matrix(c(NA, 2, 3, 4, 5, 6), ncol=2)
> m
     [,1] [,2]
[1,]   NA    4
[2,]    2    5
[3,]    3    6
> apply(m, 1, sum)                 # Compute the sum for each row
[1] NA  7  9
> apply(m, 2, mean)                # Compute the mean for each column
[1] NA  5
```

Additional arguments to the function that is applied can be given in the call to apply. For example, to disregard missing observations when computing the column means we add na.rm=TRUE to the call.

```
> apply(m, 2, mean, na.rm=TRUE)  # Compute the mean for each column
[1] 2.5 5.0
```

See also: See Problem 2.15 for an example of how to compute a matrix or array of proportions.

2.15 COMPUTE A MATRIX/ARRAY OF PROPORTIONS

Problem: You want to compute a matrix or array of proportions.

Solution: It is often relevant to convert numbers in a table to proportions — either relative to the total sum of the values in the matrix/array, or relative to a margin such as the row or column. This could be accomplished by the use of apply, but R has a function that simplifies this.

prop.table converts a matrix or array to the fraction relative to the specified margin. When no margin is specified then it is relative to the total sum of the elements.

```
> m <- matrix(c(1, 1, 2, 2, 7, 7), 2)
> m
     [,1] [,2] [,3]
```

```
[1,]    1    2    7
[2,]    1    2    7
> prop.table(m)
     [,1] [,2] [,3]
[1,] 0.05  0.1 0.35
[2,] 0.05  0.1 0.35
```

Here we just divide each of the elements in the matrix by the total sum (20). If we are interested in row probabilities then we divide the elements of each row by the sum of each row. Each row in our matrix sums to 10 so the row-wise probabilities are easy to verify. The rows are the first margin in a matrix so we set `margin=1`.

```
> prop.table(m, margin=1)
     [,1] [,2] [,3]
[1,]  0.1  0.2  0.7
[2,]  0.1  0.2  0.7
```

If we are interested in the column probabilities then we use `margin=2` as the columns are the second margin in a matrix/array. Since the two rows are identical in the example matrix we get that all of the column-wide proportions are exactly 0.5.

```
> prop.table(m, 2)
     [,1] [,2] [,3]
[1,]  0.5  0.5  0.5
[2,]  0.5  0.5  0.5
```

2.16 TRANSPOSE A MATRIX (OR DATA FRAME)

Problem: You want to transpose a matrix to swap columns and rows around.

Solution: The transpose function, `t`, transposes a matrix by interchanging the rows and columns of the matrix. The result is returned as a matrix.

```
> m <- matrix(1:6, ncol=2)      # Create matrix
> m
     [,1] [,2]
[1,]    1    4
[2,]    2    5
[3,]    3    6
```

```
> t(m)                                  # Transpose
     [,1] [,2] [,3]
[1,]    1    2    3
[2,]    4    5    6
```

The transpose function also works on data frames but it is generally a bad idea to transpose data frames because R converts the data frame to a matrix before transposing. As a consequence the result will be a character matrix if there are any non-numeric columns in the original data frame.

The following example shows how the transpose function swaps rows and columns around on an artificial data frame.

```
> mpdat <- data.frame(source=c("KU", "Spain", "Secret", "Attic"),
+                       rats=c(97, 119, 210, 358))
> mpdat
  source rats
1     KU   97
2  Spain  119
3 Secret  210
4  Attic  358
> t(mpdat)
       [,1]  [,2]    [,3]     [,4]
source "KU"  "Spain" "Secret" "Attic"
rats   " 97" "119"   "210"    "358"
```

The data frame has now been switched around but note that the values are enclosed in quotes as if they were text.

The rows of a matrix cannot have different types in R so it is necessary to remove character and factor columns from the original data frame before transposition if the numeric values of the transposed data frame are to be used.

```
> t(mpdat[,2])
     [,1] [,2] [,3] [,4]
[1,]   97  119  210  358
```

2.17 CREATE A TABLE OF COUNTS

Problem: You wish to create a contingency table of counts by cross-classifying one or more factors.

Solution: The `table` function cross-classifies and counts the number of observations for each classification category. As input, the `table` function takes one or more objects that can be interpreted as factors and then calculates the counts for each combination of factor levels.

We can count the number of occurrences for each value of a vector.

```
> x <- rbinom(20, size=4, p=.5)
> x
 [1] 1 2 2 3 1 3 4 2 2 0 1 1 2 2 3 2 3 4 2 3
> table(x)
x
0 1 2 3 4
1 4 8 5 2
```

When multiple vectors of the same length are given to `table` it cross-classifies them.

```
> sex <- sample(c("Male", "Female"), 20, replace=TRUE)
> sex
 [1] "Female" "Male"   "Female" "Male"   "Male"   "Male"
 [7] "Male"   "Male"   "Female" "Male"   "Male"   "Female"
[13] "Male"   "Male"   "Female" "Female" "Female" "Male"
[19] "Female" "Male"
> table(sex, x)
        x
sex      0 1 2 3 4
  Female 0 2 4 2 0
  Male   1 2 4 3 2
> treatment <- rep(c("Placebo", "Active"), 10)
> treatment
 [1] "Placebo" "Active"  "Placebo" "Active"  "Placebo" "Active"
 [7] "Placebo" "Active"  "Placebo" "Active"  "Placebo" "Active"
[13] "Placebo" "Active"  "Placebo" "Active"  "Placebo" "Active"
[19] "Placebo" "Active"
> table(sex, x, treatment)
, , treatment = Active

        x
sex      0 1 2 3 4
  Female 0 1 1 0 0
  Male   1 0 3 3 1

, , treatment = Placebo

        x
sex      0 1 2 3 4
  Female 0 1 3 2 0
  Male   0 2 1 0 1
```

R outputs several two-dimensional tables for multi-way tables with more than two dimensions. In the last example above, we have a three-dimensional table so the relevant two-dimensional table, `sex` by `x`, is printed for every other combination of values of the remaining dimensions.

Multi-way tables are more easily printed using the `ftable` function. The `Titanic` dataset contains information on survival of 2201 passengers on board the Titanic and the data are classified according to economic status (class), sex, and age group (child or adult). This four-dimensional array can be very compactly printed using `ftable`.

```
> data(Titanic)
> ftable(Titanic)
                   Survived  No Yes
Class Sex    Age
1st   Male   Child            0   5
             Adult          118  57
      Female Child            0   1
             Adult            4 140
2nd   Male   Child            0  11
             Adult          154  14
      Female Child            0  13
             Adult           13  80
3rd   Male   Child           35  13
             Adult          387  75
      Female Child           17  14
             Adult           89  76
Crew  Male   Child            0   0
             Adult          670 192
      Female Child            0   0
             Adult            3  20
```

See also: The `margin.table` function can be used to compute the sum of table entries (e.g., row or column sums), while `prop.table` can be used to compute entries in a contingency table as fractions of a table margin. The `xtabs` function can create contingency tables from factors using a formula interface.

2.18 CONVERT A TABLE OF COUNTS TO A DATA FRAME

Problem: You have a contingency table of counts and wish to convert the counts to a data frame.

Solution: Sometimes data are provided as a table of counts and it may be necessary to convert the counts to a data frame. The as.data.frame function converts a table representing the cross-tabulation of two or more categorical variables to a data frame where each row in the data frame represents a particular combination of categories (shown by a column for each dimension in the table) together with the frequency of that particular combination.

The HairEyeColor dataset consists of a cross-classification table of hair color, eye color, and sex for 592 U.S. statistics students.

```
> # Only look at males to keep output short
> maletable <- as.table(HairEyeColor[,,1])
> maletable
       Eye
Hair    Brown Blue Hazel Green
  Black    32   11    10     3
  Brown    53   50    25    15
  Red      10   10     7     7
  Blond     3   30     5     8
> freq <- as.data.frame(maletable)
> freq
    Hair   Eye Freq
1  Black Brown   32
2  Brown Brown   53
3    Red Brown   10
4  Blond Brown    3
5  Black  Blue   11
6  Brown  Blue   50
7    Red  Blue   10
8  Blond  Blue   30
9  Black Hazel   10
10 Brown Hazel   25
11   Red Hazel    7
12 Blond Hazel    5
13 Black Green    3
14 Brown Green   15
15   Red Green    7
16 Blond Green    8
```

The expand.table from the epitools package expands the contingency table into a data frame where each line corresponds to a single observation. As input, expand.table requires a table or array, x, which has both names(dimnames(x)) and dimnames(x) since these are converted to field names and factor levels, respectively.

```
> library(epitools)
> fulldata <- expand.table(HairEyeColor)
```

```
> head(fulldata)
   Hair   Eye  Sex
1 Black Brown Male
2 Black Brown Male
3 Black Brown Male
4 Black Brown Male
5 Black Brown Male
6 Black Brown Male
> summary(fulldata)
    Hair          Eye            Sex
 Black:108   Brown:220   Male  :279
 Brown:286   Blue :215   Female:313
 Red  : 71   Hazel: 93
 Blond:127   Green: 64
```

DATES AND TIMES

2.19 PARSING DATES AND TIMES

Problem: You want R to interpret character strings representing dates and times as dates and times and not as character vectors.

Solution: Date and time data come in many different formats, which can make it complicated to parse. For example, the ordering of month, day, and year, how months are represented (as numbers (1–12), abbreviated strings ("Feb"), or unabbreviated strings ("February")), whether or not leading zeroes are used for numbers less than 10, and the number of digits used to denote years, etc.

R has three built-in date/time classes (Date, POSIXct, and POSIXlt). Here we will use the `lubridate` package which essentially is a wrapper for POSIXct but with a more intuitive syntax that makes working with dates and times a lot easier and much more flexible.

The `lubridate` package provides six basic functions to parse strings for dates: `dmy`, `dym`, `mdy`, `myd`, `ydm`, and `ymd` where the letters represent the order in which day, month, and year appear in the string.

```
> library(lubridate)
> string1 <- "1971-08-20"
> string2 <- "11-04-01"   # Year with only two digits
> date1 <- ymd(string1)
> date2 <- ymd(string2)   # 11 becomes 2011
> date1
[1] "1971-08-20"
> date2
```

```
[1] "2011-04-01"
> string3 <- c("02/28/00", "Feb 29th, 2000", "March 01 00", "3-31-00")
> date3 <- mdy(string3)   # Super flexible
> date3
[1] "2000-02-28" "2000-02-29" "2000-03-01" "2000-03-31"
> date2 - date1            # Difference between dates
Time difference of 14469 days
```

Note that the time zone is not automatically added to the dates but can be set with the `tz` argument.

```
> date1 <- ymd(string1, tz="Europe/Copenhagen")
> date1
[1] "1971-08-20 CET"
> ymd(date2, tz="UTC") - date1 # Difference changed a bit
Time difference of 14469.04 days
```

`lubridate` uses the same approach to handle time: combine `h`, `m`, and `s` to indicate the order that the hours, minutes, and seconds appear in, and this can be combined with year, month, and day.

```
> time1 <- hms(c("08:34:12", "15-32-41"))
> time1
[1] "8H 34M 12S"    "15H -32M -41S"
> time2 <- hm(c("12:11", "5,30"))
> time2
[1] "12H 11M 0S" "5H 30M 0S"
> ydm_hms("71-20-08 09:05:21")  # Combine date and time
[1] "1971-08-20 09:05:21 UTC"
> ydm_hm("71-20-08 09:05")      # Combine date and time no seconds
[1] "1971-08-20 09:05:00 UTC"
> ydm_hm("71-20-08 09:05 PM")   # AM/PM specification allowed
[1] "1971-08-20 21:05:00 UTC"
```

The date/time parsing functions are very flexible and even filter out unnecessary text.

```
> ydm_hm("Trial started on 71-20-08 at 09:05 PM and was successful")
[1] "1971-08-20 21:05:00 UTC"
```

Arithmetic operations on dates/times are vectorized and work as expected. We can add/subtract numbers (which are considered seconds by default) or use one of the functions `minutes`, `hours`, `days`, `months`, or `years` to quickly specify a period.

Table 2.3: Some of the codes available for extracting strings from dates

Symbol	Interpretation	Example
%d	day as a number	21
%a	abbreviated weekday	Fri
%A	unabbreviated weekday	Thursday
%m	month as a number	09
%b	abbreviated month	Jan
%B	unabbreviated month	February
%y	2-digit year	10
%Y	4-digit year	2011
%H	hour on 24-hour clock	23
%I	hour on 12-hour clock	11
%M	minutes	34
%S	seconds	51
%p	AM/PM string	PM

```
> date4 <- mdy(c("Mar 31 2011", "Apr 1 2011"))
> date4 + 30          # Add 30 seconds
[1] "2011-04-30" "2011-05-01"
> date4 + days(30)  # Add 30 days
[1] "2011-04-30" "2011-05-01"
> date4 + months(1) # Add 1 month
[1] NA           "2011-05-01"
```

Note that adding a month might result in an illegal date (April 31st does not exist) which produces an NA.

See also: The `today` and `now` functions return the current date and current date-time.

2.20 EXTRACT AND FORMAT DATE/TIME INFORMATION

Problem: You wish to extract and format dates/times from an R date/-time object.

Solution: R has three built-in date/time classes (Date, POSIXct, and POSIXlt). The primary focus here will be on the `lubridate` package which is a wrapper for POSIXct that makes date and time extraction a lot more flexible.

By default, R prints dates in the format "4-digit year–2-digit month–2-digit day," but that can be changed with the format function. The format function can output a date in any desired format by supplying the format function with an output format given by the codes in Table 2.3. The lubridate uses the stamp function to format dates/times, and it also accepts the codes from Table 2.3.

The stamp function generates a function that can be used to format date/time values. It expects a character vector as input and uses the character vector to infer the relevant template(s) to format the date/time values.

```
> library(lubridate)
> string <- c("02/28/00", "Feb 29th, 2000", "March 01 00", "3-31-00")
> D <- mdy(string)          # Parse as dates
> D
[1] "2000-02-28" "2000-02-29" "2000-03-01" "2000-03-31"
> stamp("March 1, 1999")(D) # Make and apply stamp like "March 1, 1999"

Multiple formats matched: "%B %d, %Y"(1), "March %m, %Y"(1)
Using: "%B %d, %Y"

[1] "February 28, 2000" "February 29, 2000" "March 01, 2000"
[4] "March 31, 2000"
```

Note that stamp finds two formats that match "March 1, 1999": one which uses the unabbreviated month, day, and year, and another which has a prefix string "March" then month and year (see Table 2.3), and it uses the first. The second match is clearly not correct as March is the actual month and not a general string, and the first number represents the day.

The orders argument can be used to specify the order of the date/-time in the character vector, and it accepts a vector of formatting strings corresponding to the date/time functions found in the lubridate package. For example, to force the string above to interpret "March" as a string and not as part of the date we could write

```
> stamp("March 1, 1999", orders="dy")(D) # Only expect day and year

Using: "March %d, %Y"

[1] "March 28, 2000" "March 29, 2000" "March 01, 2000"
[4] "March 31, 2000"
```

in which case all the output has the string prefix March. stamp works analogously when times are present, and the `quiet=TRUE` argument can be set to prevent `stamp` from writing the output format that is used. Note that the necessary order string for times is supposed to be in upper case as in the example for hours and minutes below.

```
> D2 <- ydm_hm(c("71-20-08 09:05"), c("71-18-08 09:05"),
+              c("00-28-02 09:15"), c("01-20-12 13:10"))
> mystamp <- stamp("Date of birth: Sunday, Jan 1, 1999 3:34 pm",
+                  quiet=TRUE)
> mystamp(D2)
[1] "Date of birth: Friday, Aug 20, 1971 09:05 AM"
[2] "Date of birth: Wednesday, Aug 18, 1971 09:05 AM"
[3] "Date of birth: Monday, Feb 28, 2000 09:15 AM"
[4] "Date of birth: Thursday, Dec 20, 2001 13:10 PM"
> mystamp <- stamp("Date of birth: Sunday, Jan 1, 1999 3:34",
+                  orders="mdyHM")

Multiple formats matched: "Date of birth: %A, %b %d, %Y %H:%M"(1),
"Date of birth: Sunday, %b %d, %Y %H:%M"(1)
Using: "Date of birth: %A, %b %d, %Y %H:%M"

> mystamp(D2)
[1] "Date of birth: Friday, Aug 20, 1971 09:05"
[2] "Date of birth: Wednesday, Aug 18, 1971 09:05"
[3] "Date of birth: Monday, Feb 28, 2000 09:15"
[4] "Date of birth: Thursday, Dec 20, 2001 13:10"
```

See also: The help page for `strptime` to see all possible conversion symbols.

FACTORS

2.21 CONVERT A FACTOR TO NUMERIC

Problem: You want to convert a factor to a vector of numeric values based on the factor labels.

Solution: At first glance `as.numeric(myfactor)` might appear to be a viable solution to convert the factor `myfactor` to a numeric vector. However, `as.numeric(myfactor)` returns a numeric vector of *the index* of the levels and not the actual values. Instead we need to use the `levels` function to return the actual labels and then convert these. This also ensures that R automatically converts non-numeric values to `NA`'s and prints a warning message if any of the levels cannot be converted to a numeric value.

```
> f <- factor(c(1:4, "A", 8, NA, "B"))
> f
[1] 1    2    3    4    A    8    <NA> B
Levels: 1 2 3 4 8 A B
> vect <- as.numeric(levels(f))[f]

Warning: NAs introduced by coercion

> vect
[1]  1  2  3  4 NA  8 NA NA
> as.numeric(f)                         # Not correct
[1]  1  2  3  4  6  5 NA  7
```

2.22 CONVERT A FACTOR TO CHARACTER STRINGS

Problem: You want to convert a factor to a vector of character strings based on the factor labels.

Solution: Extracting the factor labels as character string is done automatically by the `as.character` function. It returns each label as a string and keeps `NA`.

```
> f <- factor(c(1:4, "A", 8, NA, "B"))
> f
[1] 1    2    3    4    A    8    <NA> B
Levels: 1 2 3 4 8 A B
> vect <- as.character(f)
> vect
[1] "1" "2" "3" "4" "A" "8" NA  "B"
```

2.23 ADD A NEW LEVEL TO AN EXISTING FACTOR

Problem: You want to add a new factor level to an existing factor.

Solution: Sometimes it is necessary to add a new level to an existing factor; for example, if an existing level is split up or because new observations that contain levels not present in the existing factor are introduced.

It is not possible to add a new category directly just by assigning the new value to the relevant observations. Instead it is necessary to first redefine the factor with the additional category and then assign the new values to the relevant observations afterwards.

In the example below we wish to convert the first two observations to a new level called `A`.

```
> f <- factor(c(1, 2, 3, 2, 3, 1, 3, 1, 2, 1))
> f
 [1] 1 2 3 2 3 1 3 1 2 1
Levels: 1 2 3
> f[c(1,2)] <- "A"    # Not working
> f
 [1] <NA> <NA> 3    2    3    1    3    1    2    1
Levels: 1 2 3
```

Instead we need to add the new levels to the factor before we change or add observations that have this new category. This is done by redefining the factor and forcing the new category to be part of the levels.

```
> f <- factor(f, levels=c(levels(f), "A"))
> f[c(1,2)] <- "A"    # Can now assign the value "new level"
> f
 [1] A A 3 2 3 1 3 1 2 1
Levels: 1 2 3 A
```

2.24 COMBINE THE LEVELS OF A FACTOR

Problem: You want to group some of the levels of an existing factor to create a new factor with fewer levels.

Solution: Use the `levels` function to set the new levels of the original factor.

```
> f <- factor(c(1, 1, 2, 3, 2, 1, "A", 3, 3, 3, "A"))
> levels(f)    # Print existing levels
[1] "1" "2" "3" "A"
```

The `levels` function prints the existing levels for the factor `f`. If we wish to combine, say, levels 1 and 3 then we assign new levels to the factor with the same number of elements but where the levels that should be combined have the same labels.

```
> levels(f) <- c("13", "2", "13", "A")
> f
 [1] 13 13 2  13 2  13 A  13 13 13 A
Levels: 13 2 A
```

An alternative approach that some users find easier to read is to first select the elements from the factor levels using R's indexing possibilities and then assigning them the new factor name

```
> f <- factor(c(1, 1, 2, 3, 2, 1, "A", 3, 3, 3, "A"))
> f
 [1] 1 1 2 3 2 1 A 3 3 3 A
Levels: 1 2 3 A
> levels(f)[c(1,3)] <- "1&3"
> f
 [1] 1&3 1&3 2   1&3 2   1&3 A   1&3 1&3 1&3 A
Levels: 1&3 2 A
```

Finally a third approach is to use a named list, where the names correspond to the new values of the factor and the elements define which levels are to be combined.

```
> f <- factor(c(1, 1, 2, 3, 2, 1, "A", 3, 3, 3, "A"))
> f
 [1] 1 1 2 3 2 1 A 3 3 3 A
Levels: 1 2 3 A
> levels(f) <- list("1+3" = c(1, 3), "2"=2, "A"="A")
> f
 [1] 1+3 1+3 2   1+3 2   1+3 A   1+3 1+3 1+3 A
Levels: 1+3 2 A
```

2.25 REMOVE UNUSED LEVELS OF A FACTOR

Problem: You want to remove the levels of a factor that do not occur in the dataset.

Solution: Use the `droplevels` function on an existing factor to create a new factor that only contains the levels that are present in the original factor.

```
> f <- factor(c(1, 2, 3, 2, 3, 2, 1), levels=c(0, 1, 2, 3))
> f
[1] 1 2 3 2 3 2 1
Levels: 0 1 2 3
> ff <- droplevels(f)   # Drop unused levels from f
> ff
[1] 1 2 3 2 3 2 1
Levels: 1 2 3
```

The `droplevels` function can also be used on data frames, in which case the function is applied to all factors in the data frame. The `except` option takes a vector of indices representing columns where unused factor levels are *not* to be dropped. In the code below we drop the levels for all factors in a data frame except for variable 2.

```
> df <- data.frame(f1=f, f2=f)
> newdf <- droplevels(df, except=2)
> newdf$f1
[1] 1 2 3 2 3 2 1
Levels: 1 2 3
> newdf$f2
[1] 1 2 3 2 3 2 1
Levels: 0 1 2 3
```

See also: Alternatively, the `factor` function can be applied to an existing factor to create a new factor that only contains the levels that are present in the original factor.

2.26 CHANGE THE REFERENCE LEVEL

Problem: You want to reorder the levels of a factor to ensure that a specific level comes first and becomes the reference level.

Solution: By default, R orders factor levels according to their alphabetical/lexicographic order. This is especially useful for models, when the first factor level is taken as the reference level.

The `relevel` function sets the first level of an unordered factor. It takes the factor as first argument, and the `ref` argument specifies which factor level is to be the first. `ref` expects a string with the label of the new reference level.

```
> f <- factor(LETTERS[1:5])
> f
[1] A B C D E
Levels: A B C D E
> f <- relevel(f, ref="D")      # Set level D as reference
> f
[1] A B C D E
Levels: D A B C E
```

See also: The `Relevel` function from the `Epi` package can set the order of each of the factor levels and not just the first level.

2.27 CUT A NUMERIC VECTOR INTO A FACTOR

Problem: You want to divide a numeric vector into intervals and then convert the result into a factor.

Solution: `cut` divides the range of a numeric vector into intervals which can be converted to a factor. If we want to divide the range of a numeric vector into n intervals of the same length, we just include n as the second argument to `cut`.

```
> x <- (1:15)**2
> x
 [1]   1   4   9  16  25  36  49  64  81 100 121 144 169 196 225
> cut(x, 3)  # Cut the range of x into 3 interval of same length
 [1] (0.776,75.7] (0.776,75.7] (0.776,75.7] (0.776,75.7]
 [5] (0.776,75.7] (0.776,75.7] (0.776,75.7] (0.776,75.7]
 [9] (75.7,150]   (75.7,150]   (75.7,150]   (75.7,150]
[13] (150,225]    (150,225]    (150,225]
Levels: (0.776,75.7] (75.7,150] (150,225]
```

If we wish to make categories of roughly the same size, we should divide the numeric vector based on its quantiles. In that case, we can use the `quantile` function, as shown:

```
> # Cut the range into 3 intervals of roughly the same size
> cut(x, quantile(x, probs=seq(0, 1, length=4)),
+     include.lowest=TRUE)
 [1] [1,32.3]    [1,32.3]    [1,32.3]    [1,32.3]    [1,32.3]
 [6] (32.3,107] (32.3,107] (32.3,107] (32.3,107] (32.3,107]
[11] (107,225]  (107,225]  (107,225]  (107,225]  (107,225]
Levels: [1,32.3] (32.3,107] (107,225]
```

DATA FRAMES AND LISTS

2.28 SELECT A SUBSET OF A DATA FRAME

Problem: You want to select a subset of cases of a data frame based on some criteria.

Solution: Specific cases of a data frame can be selected using the built-in index functions in R. Alternatively, the `subset` function can be used to define the criteria to select cases from a data frame. The first argument to `subset` is the name of the original data frame and the `subset` argument is a logical statement that identifies the desired cases.

The logical statement can contain variables from the original data frame. For example, if we wish to select the cases from the `earthquakes` data where the magnitude of the earthquake is greater than 7.5 then we do as follows:

```
> library(MESS)
> data(earthquakes)
> subset(earthquakes, mag>7.5)
                          time latitude longitude  depth mag
1817  2015-11-24T22:50:54.460Z -10.0614  -71.0180 622.09 7.6
1818  2015-11-24T22:45:38.850Z -10.5360  -70.9645 605.72 7.6
6056  2015-09-16T22:54:32.860Z -31.5729  -71.6744  22.44 8.3
12023 2015-05-30T11:23:02.110Z  27.8386  140.4931 664.00 7.8
14186 2015-04-25T06:11:25.950Z  28.2305   84.7314   8.22 7.8
                                     place       type
1817        211km S of Tarauaca, Brazil earthquake
1818          175km WNW of Iberia, Peru earthquake
6056           48km W of Illapel, Chile earthquake
12023 189km WNW of Chichi-shima, Japan earthquake
14186              36km E of Khudi, Nepal earthquake
```

Several logical statements may be combined to achieve a more complex subsetting. If we want to select tremors from a mining explosion with magnitude greater than 3.6 then we use the command:

```
> subset(earthquakes, type=="mining explosion" & mag>3.6)
                          time latitude longitude depth mag
7200  2015-08-28T19:37:33.390Z  43.6025 -105.2930     0 3.8
11621 2015-06-07T13:20:24.080Z  22.6379  -12.4410     0 4.1
15869 2015-03-19T21:10:08.630Z  43.8335 -105.2754     0 3.7
18416 2015-01-28T17:06:44.170Z  44.0068 -105.2311     0 3.7
                                place             type
7200                          Wyoming mining explosion
11621 9km S of Zouerate, Mauritania mining explosion
15869                         Wyoming mining explosion
18416                         Wyoming mining explosion
```

The `select` argument to the `subset` function enables you to extract specific variables from the original data frame. To select only the `time` and `place` variables from the `earthquakes` data frame when we are only interested in observations with a magnitude greater than 7 and a depth more than 100 km we type:

```
> subset(earthquakes, mag>7 & depth>100,
+        select=c("time", "place"))
                          time                              place
```

```
1817  2015-11-24T22:50:54.460Z       211km S of Tarauaca, Brazil
1818  2015-11-24T22:45:38.850Z          175km WNW of Iberia, Peru
3508  2015-10-26T09:09:42.560Z   45km E of Farkhar, Afghanistan
3788  2015-10-20T21:52:02.560Z     34km NE of Port-Olry, Vanuatu
12023 2015-05-30T11:23:02.110Z 189km WNW of Chichi-shima, Japan
```

Note that when a factor is subsetted, it will not automatically drop levels that are no longer present. That can result in tables with empty and irrelevant cells and figures with empty bars, etc.

See also: The `filter` function from the `dplyr` package works similarly to `subset` but is much faster. Use Problem 2.25 to remove unused factor levels.

2.29 SELECT THE COMPLETE CASES OF A DATA FRAME

Problem: You want to select a subset of cases of a data frame for which there are no missing values.

Solution: The `complete.cases` function returns a logical vector indicating which cases are complete. The function takes a sequence of vectors, matrices and data frames as arguments.

The New York daily air quality measurements from the `airquality` dataset contains missing values for the ozone and solar radiation variables.

```
> data(airquality)
> head(airquality)
  Ozone Solar.R Wind Temp Month Day
1    41     190  7.4   67     5   1
2    36     118  8.0   72     5   2
3    12     149 12.6   74     5   3
4    18     313 11.5   62     5   4
5    NA      NA 14.3   56     5   5
6    28      NA 14.9   66     5   6
> complete <- complete.cases(airquality)
> head(complete, 20)
 [1]  TRUE  TRUE  TRUE  TRUE FALSE FALSE  TRUE  TRUE  TRUE FALSE
[11] FALSE  TRUE  TRUE  TRUE  TRUE  TRUE  TRUE  TRUE  TRUE  TRUE
> ccdata <- airquality[complete,]
> head(ccdata)
  Ozone Solar.R Wind Temp Month Day
1    41     190  7.4   67     5   1
2    36     118  8.0   72     5   2
3    12     149 12.6   74     5   3
```

```
4    18     313 11.5   62     5   4
7    23     299  8.6   65     5   7
8    19      99 13.8   59     5   8
```

If we know that we will only use some of the variables from a data frame, then the complete cases selection should be based on those variables only. For example, if we are only interested in the ozone and temperature values, then we would specify these variables as arguments to `complete.cases` function instead of the whole data frame.

```
> complete2 <- complete.cases(airquality$Ozone, airquality$Temp)
> oztemp <- airquality[complete2,]
> head(oztemp)
  Ozone Solar.R Wind Temp Month Day
1    41     190  7.4   67     5   1
2    36     118  8.0   72     5   2
3    12     149 12.6   74     5   3
4    18     313 11.5   62     5   4
6    28      NA 14.9   66     5   6
7    23     299  8.6   65     5   7
```

The `complete.cases` function is particularly useful with respect to model reduction when two statistical models are compared since the fit of the two models should be based on the same data.

2.30 DELETE A VARIABLE FROM A DATA FRAME

Problem: You want to delete a variable from a data frame.

Solution: There exists several functions in various packages to R that allow you to rename and delete variables in a data frame. However, you can remove a variable in a data frame directly simply by setting it to `NULL`.

```
> mydat <- data.frame(x=1:5, y=factor(c("A", "B", "A", "B", NA)))
> mydat
  x    y
1 1    A
2 2    B
3 3    A
4 4    B
5 5 <NA>
> mydat$y <- NULL   # Remove the y variable from mydat
> mydat
  x
```

```
1 1
2 2
3 3
4 4
5 5
```

2.31 APPLY FUNCTION TO EACH VARIABLE IN A DATA FRAME

Problem: You have a data frame or list and want to apply a function to each variable or element.

Solution: It is often necessary to iterate over a data frame to extract information or apply a function to each variable and R has a range of functions that can be used to execute a function repetitively. This prevents us from having to write loops to perform some operation repeatedly, although the same result could be obtained with a loop.

The `lapply` function is the work horse for applying a function to each element of a list. `lapply` applies a function over a list so it will automatically work on every variable in the data frame and returns the result as a list. `sapply` behaves exactly as `lapply` except it tries to simplify the result. `lapply` and `sapply` need just two arguments: the data frame or list and the function to apply (given by the `FUN` argument).

```
> library(MESS)
> data(earthquakes)
> lapply(earthquakes, FUN=class) # Apply the class function
$time
[1] "factor"

$latitude
[1] "numeric"

$longitude
[1] "numeric"

$depth
[1] "numeric"

$mag
[1] "numeric"

$place
[1] "factor"
```

```
$type
[1] "factor"
> sapply(earthquakes, class)     # Simplify output
     time  latitude longitude     depth         mag     place
 "factor" "numeric" "numeric" "numeric" "numeric"  "factor"
     type
 "factor"
```

We might encounter problems when we apply functions to all variables in a data frame because the variable types may differ. For example, if we compute the mean for each variable in the earthquakes dataset we get a result that contains NA because some of the variables are factors and not numeric.

```
> sapply(earthquakes, mean)   # Compute the mean of each variable
      time   latitude  longitude      depth        mag
        NA  8.8687875  0.6886241 75.3692862  4.2234975
     place       type
        NA         NA
```

Additional options to the applied function can be specified in the call to lapply, or sapply. For example, if we wish to disregard missing observations and trim away the lower and upper 10% of observations for each variable, then we write the extra arguments to the mean function when calling sapply.

```
> sapply(earthquakes, mean, trim=.1)
     time  latitude longitude     depth       mag     place
       NA  9.091773  2.369823 44.275283  4.236905        NA
     type
       NA
```

It is possible to apply user-defined functions in lapply. For example, if you want to convert all factor variables in a data frame to character strings because they were imported as factors you can use lapply, but you only want to make the conversion if the variable was a factor. We use the is.factor to return TRUE if the input is a factor, and idea from Problem 2.22 for the actual conversion.

```
> newlist <- lapply(earthquakes, function(i) {
+                       if (is.factor(i)) {  # Check if it is a factor
+                           as.character(i)  # Convert
+                       } else {             # Otherwise do nothing
+                           i
```

```
+                        }
+                    })
> sapply(newlist, class)     # New classes
       time    latitude   longitude       depth         mag
"character"   "numeric"   "numeric"   "numeric"   "numeric"
      place        type
"character" "character"
```

2.32 SPLIT DATA FRAME INTO SUBSETS AND APPLY FUNCTION TO EACH PART

Problem: You want to apply the same function repetitively on every row or group of rows in a data frame.

Solution: R has a range of functions that can be used to execute a function repetitively by applying the function to rows or group of rows of a data frame. Note that this is different from Problem 2.31 where we iterated over variables/columns in the data frame.

The `by` function applies a function to subsets of a data frame, and it requires three arguments. The first argument should be a data frame, the second argument should be a factor (or list of factors) with the same length(s) as the number of rows in the data frame, and the last required argument is the function that should be applied to each subset of the data frame defined by the factor.

For example, to calculate the mean of each variable for each month in the `airquality` dataset we use the `colMeans` function together with `by` and use the `airquality$Month` variable to split the data frame into groups. Additional argument to the applied function can be specified in the call to `by` as extra arguments. Below we add the extra argument `na.rm=TRUE` to disregard missing values when using `colMeans`.

```
> data(airquality)
> by(airquality, airquality$Month, colMeans, na.rm=TRUE)
airquality$Month: 5
    Ozone   Solar.R      Wind      Temp     Month       Day
 23.61538 181.29630  11.62258  65.54839   5.00000  16.00000
-------------------------------------------------------
airquality$Month: 6
    Ozone   Solar.R      Wind      Temp     Month       Day
 29.44444 190.16667  10.26667  79.10000   6.00000  15.50000
-------------------------------------------------------
airquality$Month: 7
```

```
     Ozone    Solar.R       Wind       Temp      Month
 59.115385 216.483871   8.941935  83.903226   7.000000
       Day
 16.000000
------------------------------------------------
airquality$Month: 8
     Ozone    Solar.R       Wind       Temp      Month
 59.961538 171.857143   8.793548  83.967742   8.000000
       Day
 16.000000
------------------------------------------------
airquality$Month: 9
    Ozone   Solar.R      Wind      Temp     Month       Day
 31.44828 167.43333  10.18000  76.90000   9.00000  15.50000
```

Here the mean of each variable is calculated for each month.

See also: Alternatively, the `group_by` function from the `dplyr` package can be used to split the data, apply a function to each split part, and combine the results.

2.33 APPLY FUNCTION TO EACH ROW OF A DATA FRAME

Problem: You want to apply a function repetitively to every row of a data frame and use values from each row as arguments in the function call.

Solution: Sometimes you want to apply a function repeatedly using information from the variables at each row as input to the function.

When the function of interest is vectorized (i.e., it can handle arguments that are vectors and apply the function element-wise) then the relevant columns of the data frame can be used directly as input to the function. Alternatively, the `do.call` function can execute a function call using a list of named arguments as the corresponding named arguments in the function call. Since a data frame is also a list we can use `do.call` directly to apply the function on each row of the data frame. The first argument to `do.call` should be the function to call, while the `args` argument is a named list of input values that are passed to the function. Recall that `do.call` uses the names of the variables in the data frame to match the corresponding argument in the function call, so they need to match.

For example, to compute the power for a two-sample comparison of means, we simply use the columns from the data frame as input in the call to `power.t.test`.

```
> df1 <- data.frame(delta=seq(0.7, 1, .1),
+                   n=seq(20, 50, 10))
> do.call("power.t.test", df1)

     Two-sample t test power calculation

              n = 20, 30, 40, 50
          delta = 0.7, 0.8, 0.9, 1.0
             sd = 1
      sig.level = 0.05
          power = 0.5782714, 0.8614222, 0.9780440, 0.9986074
    alternative = two.sided

NOTE: n is number in *each* group
```

If the input data frame contains more variables than are needed in the function call then we get an error unless the function accepts extra input through the ... arguments because the names of the additional variables cannot be matched.

```
> df2 <- data.frame(delta=seq(0.7, 1, .1),
+                   n=seq(20, 50, 10),
+                   extra=1:4)
> df2
  delta  n extra
1   0.7 20     1
2   0.8 30     2
3   0.9 40     3
4   1.0 50     4
> do.call("power.t.test", df2)

Error in power.t.test(delta = c(0.7, 0.8, 0.9, 1), n = c(20, 30, 40, 50:
unused argument (extra = 1:4)
```

In those situations we can create a wrapper function with a dots argument, ..., so it accepts extra arguments which are not used (output not shown).

```
> do.call(function(n, delta, ...) { power.t.test(n=n, delta=delta) },
+         df2)
```

The `power.t.test` function is only partially vectorized: it is vectorized when the power for a two-sample comparison of means is computed, but not if you want the inverse problem of computing the sample size given the power and smallest significant difference. For non-vectorized functions, the `mapply` function can be used, but then the ordering of the arguments should match the order of the arguments in the function *or* the name of the arguments should be explicitly stated. `mapply` returns a list if the argument `SIMPLIFY=FALSE` is supplied and otherwise (the default) the result will try to be simplified to a vector, array or list.

```
> df3 <- data.frame(delta=seq(0.7, 1, .1),
+                   power=c(0.58, 0.86, 0.98, 0.998))
> res <- mapply(power.t.test, delta=df3$delta, power=df3$power)
> res
            [,1]
n           20.0778
delta       0.7
sd          1
sig.level   0.05
power       0.58
alternative "two.sided"
note        "n is number in *each* group"
method      "Two-sample t test power calculation"
            [,2]
n           29.87815
delta       0.8
sd          1
sig.level   0.05
power       0.86
alternative "two.sided"
note        "n is number in *each* group"
method      "Two-sample t test power calculation"
            [,3]
n           40.76533
delta       0.9
sd          1
sig.level   0.05
power       0.98
alternative "two.sided"
note        "n is number in *each* group"
method      "Two-sample t test power calculation"
            [,4]
n           47.80142
delta       1
sd          1
sig.level   0.05
power       0.998
alternative "two.sided"
note        "n is number in *each* group"
method      "Two-sample t test power calculation"
```

To extract the number sample size per group needed for the given combination of delta and power we extract the first result from each function call and use `unlist` to convert the result to a vector.

```
> unlist(res[1,])
[1] 20.07780 29.87815 40.76533 47.80142
```

See also: See Problem 3.49 for more information on computing power using `power.t.test`.

2.34 COMBINE TWO DATASETS

Problem: You want to combine two data frames that contain the same variables but consist of different cases.

Solution: The `rbind` function merges two data frames "vertically" by binding the rows from the two datasets. The two data frames must have the same variables (and these variables should be of the same types), but the variables do not have to appear in the same order.

```
> df1 <- data.frame(id=1:3, x=11:13, y=factor(c("A", "B", "C")))
> df1
  id  x y
1  1 11 A
2  2 12 B
3  3 13 C
> df2 <- data.frame(id=6:10, y=factor(1:5), x=1:5)
> df2
  id y x
1  6 1 1
2  7 2 2
3  8 3 3
4  9 4 4
5 10 5 5
> combined <- rbind(df1, df2)
> combined
  id  x y
1  1 11 A
2  2 12 B
3  3 13 C
4  6  1 1
5  7  2 2
6  8  3 3
7  9  4 4
8 10  5 5
```

If one of the data frames contains variables that the other does not, then there are two possible options before joining them: Either delete the extra variables that only exist in one of the two data frames (see Problem 2.30), or create the missing variable(s) and set them to NA (missing) for the data frame where they are not present. The two possible solutions are shown below.

```
> df1 <- data.frame(id=1:3, x=11:13, y=factor(c("A", "B", "C")),
+                   z=1:3)
> df2 <- data.frame(id=6:10, y=factor(1:5), x=1:5)
> df1$z <- NULL    # Remove the z variable from data frame df1
> rbind(df1,df2)  # Now we can join the data frames
  id  x y
1  1 11 A
2  2 12 B
3  3 13 C
4  6  1 1
5  7  2 2
6  8  3 3
7  9  4 4
8 10  5 5
> df1 <- data.frame(id=1:3, x=11:13, y=factor(c("A", "B", "C")),
+                   z=1:3)
> df2 <- data.frame(id=6:10, y=factor(1:5), x=1:5)
> df2$z <- NA      # Add a vector z of NAs to data frame df2
> rbind(df1,df2)  # Now we can join the data frames
  id  x y  z
1  1 11 A  1
2  2 12 B  2
3  3 13 C  3
4  6  1 1 NA
5  7  2 2 NA
6  8  3 3 NA
7  9  4 4 NA
8 10  5 5 NA
```

2.35 MERGE DATASETS

Problem: You want to merge two data frames that have the same cases but contain different variables.

Solution: You have two sets of data frames on the same set of individuals and want to merge them "horizontally" by combining the columns of the two datasets. This can be done with the merge function. In most cases, you want to merge the two data frames by one or more common key variables to make sure that the rows are matched correctly. The

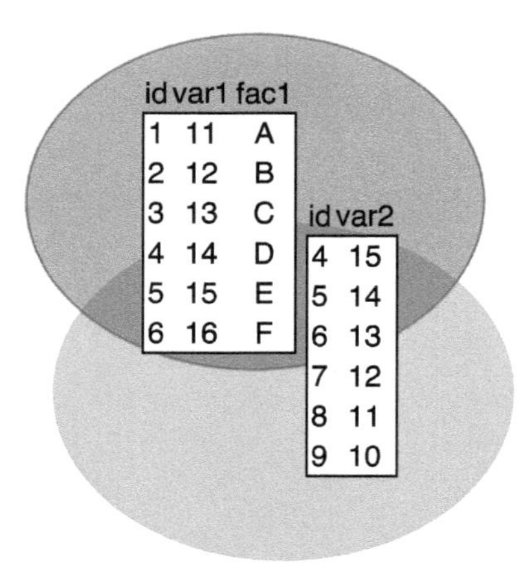

Figure 2.2: Examples of merging two data frames in R. The blue area corresponds to a (left) outer join, the gray to a (right) outer join, while the intersection or union gives an inner or outer join, respectively.

matching is controlled by the `by` optional argument which defaults to columns with names that are present in both data frames.

We start by creating two example data frames (see Figure 2.2 for a graphical representation).

```
> dfx <- data.frame(id=factor(1:6), var1=11:16,
+                   fac1=factor(c("A", "B", "C", "D", "E", "F")))
> dfy <- data.frame(id=factor(c(4,5,6,7,8,9)), var2=15:10)
> dfx
  id var1 fac1
1  1   11    A
2  2   12    B
3  3   13    C
4  4   14    D
5  5   15    E
6  6   16    F
> dfy
  id var2
1  4   15
2  5   14
3  6   13
4  7   12
5  8   11
6  9   10
```

There are four ways to merge two datasets.

Inner join. Only keep the observations that are present in both datasets which corresponds to the intersection of the blue and gray sets in Figure 2.2.

```
> merge(dfx, dfy, by="id")
  id var1 fac1 var2
1  4   14    D   15
2  5   15    E   14
3  6   16    F   13
```

Note how the unmatched observations from dfx and dfy are discarded from the merged data frame since the id does not appear in both data frames.

Outer join. Keep all the observations from both datasets (corresponding to the union of the blue and gray sets in Figure 2.2). Setting the all=TRUE argument includes all observations from both datasets.

```
> merge(dfx, dfy, by="id", all=TRUE)
  id var1 fac1 var2
1  1   11    A   NA
2  2   12    B   NA
3  3   13    C   NA
4  4   14    D   15
5  5   15    E   14
6  6   16    F   13
7  7   NA <NA>   12
8  8   NA <NA>   11
9  9   NA <NA>   10
```

Left outer join. Keep all observations and rows from the "left" dataset and add the variables from the "right" dataset *if* they have a corresponding id match in the left dataset. In Figure 2.2 this would correspond to the blue set. A left outer join is undertaken with the all.x=TRUE argument.

```
> merge(dfx, dfy, by="id", all.x=TRUE)
  id var1 fac1 var2
1  1   11    A   NA
2  2   12    B   NA
3  3   13    C   NA
4  4   14    D   15
5  5   15    E   14
6  6   16    F   13
```

Right outer join. A right outer join is the opposite of a left outer join and ensures that all the rows from the second data frame are always kept in the merged dataset. Observations from the left dataset are only included if they have a corresponding match in the right dataset (the gray area in Figure 2.2). Setting the `all.y=TRUE` argument forces a right outer join.

```
> merge(dfx, dfy, by="id", all.y=TRUE)
  id var1 fac1 var2
1  4   14    D   15
2  5   15    E   14
3  6   16    F   13
4  7   NA <NA>   12
5  8   NA <NA>   11
6  9   NA <NA>   10
```

Note that `NA` is inserted for all the outer joins variables when a row or id only appear in one of the two data frames.

See also: The `dplyr` package has a number of specialized functions for merging datasets.

2.36 ADD NEW OBSERVATIONS TO A DATA FRAME

Problem: You want to add extra new observations to an existing data frame.

Solution: New observations can be added to an existing data frame by first inserting the new observations in a data frame that contains the same variables and then using Problem 2.34 to join the observations from the two data frames together.

Alternatively, the `rbind` function can be used to merge a list to a data frame by adding the elements of the list to the data frame. The data frame and the list must have the same number of variables/elements and if the list is unnamed then the order of the elements in the list should match the order of the variables in the data frame.

```
> df1 <- data.frame(id=1:3, x=11:13, y=factor(c("A", "B", "C")))
> df1
  id  x y
1  1 11 A
2  2 12 B
3  3 13 C
```

```
> newobs <- list(4, 10, "B")    # The new observation to add
> df2 <- rbind(df1, newobs)
> df2
  id  x y
1  1 11 A
2  2 12 B
3  3 13 C
4  4 10 B
```

Note that if the order of the elements in the list of the new observation does not match the order of the variables in the data frame then the variable types in the resulting data frame might change to accommodate the new observation. In particular, if the data frame contains a factor and the new observation contains a factor level that does not already exist in the data frame then the factor variable in the new observation will be set to missing. Problem 2.34 could be used in situations with new factor levels.

```
> newobs2 <- list(4, "B", 10)  # Order different from data frame
> df3 <- rbind(df1, newobs2)   # Add the new observation

Warning in `[<-.factor`(`*tmp*`, ri, value = structure(c(1L, 2L, 3L, NA),
.Label = c("A", : invalid factor level, NA generated

> df3
  id  x    y
1  1 11    A
2  2 12    B
3  3 13    C
4  4  B <NA>
```

Note that the factor level for the new observation becomes `NA` instead of 10 since '10' is not part of the possible levels that were found in the original `df1` data frame. Also, since we added a character string to the `x` variable the whole variable is converted from numeric to character.

```
> df3$x        # The x variable is converted to character
[1] "11" "12" "13" "B"
```

See also: The `bind_rows` function from the `dplyr` package binds rows and is faster than the built-in `rbind` function.

2.37 STACK THE COLUMNS OF A DATA FRAME TOGETHER

Problem: You have a data frame and want to stack the individual columns together to form a new data frame.

Solution: Observations from different groups, situations, or conditions are sometimes stored as separate variables in a data frame.

The `reshape2` package is ideal for modifying data frames from wide to long format and vice versa, and it is discussed in more detail in Problem 2.38. The `melt` function takes the individual columns and transforms them into a new data frame. By default, `melt` stacks all the numeric variables into a new variable called `value` and includes a column, `variable`, that contains the column name corresponding to the column in the original data frame that provided the value.

```
> library(reshape2)
> df1 <- data.frame(time1=c(1, 2, 3, 4),
+                   time2=c(3, NA, 6, 7))
> df1
  time1 time2
1     1     3
2     2    NA
3     3     6
4     4     7
> melt(df1)

No id variables; using all as measure variables

  variable value
1    time1     1
2    time1     2
3    time1     3
4    time1     4
5    time2     3
6    time2    NA
7    time2     6
8    time2     7
```

Non-numeric variables are considered id variables while the numeric variables are considered measured variables. Id variable values are not added to the stacked vector but are kept (and replicated) as separate identifiers for each corresponding value.

For example, `melt` gives the same number of observations in our stack of values if we add a factor to the data frame, but we get an additional column giving the corresponding id.

```
> df1$id <- c("Reed", "Sue", "Ben", "Johnny")
> df1
  time1 time2     id
1     1     3   Reed
```

```
2      2    NA     Sue
3      3     6     Ben
4      4     7 Johnny
> melt(df1)

Using id as id variables

      id variable value
1   Reed    time1     1
2    Sue    time1     2
3    Ben    time1     3
4 Johnny    time1     4
5   Reed    time2     3
6    Sue    time2    NA
7    Ben    time2     6
8 Johnny    time2     7
```

melt has two arguments, id.vars and measure.vars, that can be used to specify which variables are to be considered id variables and measured variables, respectively. Both arguments accept either a vector of integers or character strings giving the indices or names of the variables to be included as id variables and measured variables. If only one of the two is specified then the remaining variables are automatically used for the other argument.

For example, if we should use the x variable as id and use the id variable as a measured variable then we would write

```
> melt(df1, id.vars=c("time1"))
  time1 variable  value
1     1    time2      3
2     2    time2   <NA>
3     3    time2      6
4     4    time2      7
5     1       id   Reed
6     2       id    Sue
7     3       id    Ben
8     4       id Johnny
```

See also: The stack function from base R has the same functionality as melt (with default arguments) but disregards non-numeric variables in the data frame. unstack reverses the stack operation.

ID	time1	time2	time3
1	10	20	30
2	11	21	31

ID	time	value
1	1	10
2	1	11
1	2	20
2	2	21
1	3	30
2	3	31

Figure 2.3: Examples of a data frame in "wide" format (left) and in "long" format (right).

2.38 RESHAPE A DATA FRAME FROM WIDE TO LONG FORMAT OR VICE VERSA

Problem: You have a data frame that is in "wide" format and want to transform it to the "long" format usually required by R functions or vice versa.

Solution: Data frames can be input and stored in "wide" or "long" formats. The terms wide and long loosely refer to the shape of the dataset when it is entered in a spreadsheet.

The wide format generally has each variable saved as a separate column while each observation is stored in its own row in the long format (see Figure 2.3). Another way to think about wide and long formats when repeated measurements are available on each subject is that, for the wide format, a subject's repeated responses can all be found in a single row with the measurements over time found in different columns, while in the long format, each row is one time point per subject. Thus, in the long format each individual may result in multiple records/rows and time-constant variables will be constant across these records, while the time-varying variables will change across the records.

Many of R's statistical functions require data frames to be in the "long" format where there is only one observation/record on each row.

Consider the table below with repeated measurements (over four time points) for five individuals. These data are in the wide format since we have all of the measurements from each individual on a single line.

```
person age gender time1 time2 time3 time4
     I  20   Male    10    12    15    16
    II  32   Male    10    12    15    16
   III  20 Female    10    NA    15    16
    IV  20   Male    10    12    15    16
     V  20   Male    10    12    15    16
```

We use the `reshape2` package to change the format of the data. There are two main concepts in the `reshape2` package: melt and cast. Melt referes to the situation of shaping the data in long format (corresponding to molten material that becomes elongated) and is handled by the `melt` function while the `dcast` function casts the molten data in a different data frame mold. `melt` and `dcast` classify variables in two types: identifier variables and measurement variables. Identifier variables are used for columns that identify each individual subject and for variables that are to be kept constant over time, while measured variables represent what is measured on the subject.

`melt` and `dcast` have two arguments, `id.vars` and `measure.vars`, that can be used to specify which variables are to be considered id variables and measured variables, respectively. Both arguments accept either a vector of integers or character strings giving the indices or names of the variables to be included as id variables and measured variables. If only one of the two is specified then the remaining variables are automatically used for the other argument.

We read the data shown above and transform it from wide format to long. The identifier variables are `person`, `age`, and `gender`. `person` identifies the individual and should be included in `id.vars` while `age` and `gender` should be included in `id.vars` because they are constant over time. Variables not included in `id.vars` are automatically assumed to be part of `measure.vars`.

```
> library(reshape2)
> indata <- read.table("wide.txt", header=TRUE)
> indata
  person age gender time1 time2 time3 time4
1      I  20   Male    10    12    15    16
2     II  32   Male    10    12    15    16
3    III  20 Female    10    NA    15    16
4     IV  20   Male    10    12    15    16
5      V  20   Male    10    12    15    16
> long <- melt(indata, id.vars=c("person", "age", "gender"))
> long
```

```
   person age gender variable value
1       I  20   Male    time1    10
2      II  32   Male    time1    10
3     III  20 Female    time1    10
4      IV  20   Male    time1    10
5       V  20   Male    time1    10
6       I  20   Male    time2    12
7      II  32   Male    time2    12
8     III  20 Female    time2    NA
9      IV  20   Male    time2    12
10      V  20   Male    time2    12
11      I  20   Male    time3    15
12     II  32   Male    time3    15
13    III  20 Female    time3    15
14     IV  20   Male    time3    15
15      V  20   Male    time3    15
16      I  20   Male    time4    16
17     II  32   Male    time4    16
18    III  20 Female    time4    16
19     IV  20   Male    time4    16
20      V  20   Male    time4    16
```

When melting we get two new columns besides the variables included in `id.vars`. By default they are named `variable` and `value` corresponding to the column name and the corresponding measurement.

The `dcast` function takes a data frame in long format and a formula where the variables on the left-hand side specify the variables that define the rows of the output and the variables on the right-hand side of the formula specify the variables that define the columns. `dcast` tries to guess the column that contains the values insert in the data frame or it can be specified with the `value.var` argument. In the code below we cast to wide format with rows defined by the person variable and columns corresponding to the `variable` variable. R correctly guesses that the `value` columns is to be used to fill out elements in the data frame.

```
> dcast(long, person ~ variable)
  person time1 time2 time3 time4
1      I    10    12    15    16
2     II    10    12    15    16
3    III    10    NA    15    16
4     IV    10    12    15    16
5      V    10    12    15    16
```

The `bdstat` dataset from the `MESS` package contains the number of Danish births per month and year for the period from 1901 to mid 2013.

The data frame is in long format but we would like to cast it to a form where the years are given as rows and the months are the columns/-variables. The data frame should contain the births so we specify that argument manually.

```
> library(MESS)
> data(bdstat)
> head(bdstat)
  year month births dead
1 1901     1   5734 3485
2 1901     2   5546 3250
3 1901     3   6760 3868
4 1901     4   6469 3840
5 1901     5   6465 3318
6 1901     6   6129 2987
> wide <- dcast(bdstat, year ~ month, value.var="births")
> head(wide)
  year    1    2    3    4    5    6    7    8    9   10   11
1 1901 5734 5546 6760 6469 6465 6129 6131 6270 6195 5926 5726
2 1902 6014 5724 6462 6199 6225 5897 6013 6210 6287 6000 5911
3 1903 6175 5755 6536 6405 6612 5984 5865 6060 6020 5876 5368
4 1904 6110 6154 6778 6305 6332 6055 6050 6375 6070 5908 5610
5 1905 6250 5873 6599 6610 6443 5999 5947 6190 6112 5912 5469
6 1906 6025 5874 6687 6388 6356 6060 6142 6371 6344 6092 5874
    12
1 5868
2 5897
3 5695
4 5945
5 5678
6 6004
```

After casting we now have the months as columns and the years as rows.

melt and dcast essentially simplify the use of the reshape function from R. However, when there are multiple time-varying observations then we need to use the reshape function.

Consider the following dataset in "wide" format where we have observations on five individuals (indicated by the person column) time-constant covariates age and gender and time-varying observations y1, ..., y4 that represent our variable of interest measured at 4 different time points. t1, ..., t4 indicate the actual times (days) when the individuals were measured, and we can see that this differs between individuals. We would like to convert this dataset to the "long" format.

```
person age gender t1 t2 t3 t4 y1 y2 y3 y4
```

```
    I  20   Male  0  1  2  3 10 12 15 16
   II  32   Male  0  2  4  5 10 12 15 16
  III  20 Female  0  1  2  3 10 NA 15 16
   IV  20   Male  0  1  2  3 10 12 15 16
    V  20   Male  0  1  NA 3 10 12 15 16
```

The `reshape` function transforms data frames between the "wide" and "long" formats and vice versa. In order to transform a data frame from "wide" to "long" format, we should at least provide the `varying` and `direction="long"` options besides the original data frame when calling `reshape`. `varying` names the set(s) of time-varying variables that correspond to a single variable in the long format. The `varying` option should either be a single vector of column names (or indices) if there is only one time-varying variable or a list of vectors of variable names or a matrix of names if there are multiple time-varying variables.

Two other often-used options when converting from "wide" to "long" format are `idvar` and `v.names`. `idvar` sets the name of the covariate that identifies which observations belong to the same individual in the "long" format, and the default name is `id` unless something else is specified. `v.names` can be used to provide a vector of variable names in the long format that correspond to multiple columns in the "wide" format. If nothing is specified, then R will try to guess variable names in the long format.

In the code below we assume that the data frame listed above is stored in a text file named `widedata.txt`. We wish to transform the "wide" data to the "long" format and make sure that `t1`, ..., `t4` are converted to a time-varying covariate named `day` and that `y1`, ..., `y4` are converted to a time-varying covariate entitled `response`.

```
> indata <- read.table("widedata.txt", header=TRUE)
> indata
  person age gender t1 t2 t3 t4 y1 y2 y3 y4
1      I  20   Male  0  1  2  3 10 12 15 16
2     II  32   Male  0  2  4  5 10 12 15 16
3    III  20 Female  0  1  2  3 10 NA 15 16
4     IV  20   Male  0  1  2  3 10 12 15 16
5      V  20   Male  0  1 NA  3 10 12 15 16
> long <- reshape(indata,
+                 varying=list(c("y1", "y2", "y3", "y4"),
+                              c("t1", "t2", "t3", "t4")),
+                 v.names=c("response", "day"),
+                 direction="long")
> long
```

```
    person age gender time response day id
1.1      I  20   Male    1       10   0  1
2.1     II  32   Male    1       10   0  2
3.1    III  20 Female    1       10   0  3
4.1     IV  20   Male    1       10   0  4
5.1      V  20   Male    1       10   0  5
1.2      I  20   Male    2       12   1  1
2.2     II  32   Male    2       12   2  2
3.2    III  20 Female    2       NA   1  3
4.2     IV  20   Male    2       12   1  4
5.2      V  20   Male    2       12   1  5
1.3      I  20   Male    3       15   2  1
2.3     II  32   Male    3       15   4  2
3.3    III  20 Female    3       15   2  3
4.3     IV  20   Male    3       15   2  4
5.3      V  20   Male    3       15  NA  5
1.4      I  20   Male    4       16   3  1
2.4     II  32   Male    4       16   5  2
3.4    III  20 Female    4       16   3  3
4.4     IV  20   Male    4       16   3  4
5.4      V  20   Male    4       16   3  5
```

The `long` data frame now contains the same data in the "long" format and it can be used directly with most of R's modeling functions. Apart from the time-constant covariates from the original dataset and the new time-varying variables `reshape` has created two additional covariates: `id` and `time`.

We could have set `idvar="person"` to prevent `reshape` from creating a new variable, `id`, since we already had a variable in the dataset that could be used to identify individuals. In the `long` data frame we see that the ys from the original dataset are combined in the vector `response` and the ts are combined in the variable `time`.

The `direction="wide"` option should be set if we wish to convert a dataset from "long" to "wide" format. When converting to the "wide" format, it is important to specify both the `idvar` and the `timevar` options so R knows how to identify observations from a single individual and how to order the time-varying measurements for each individual. Unless time-varying variables are specified with the `v.names` option, `reshape` assumes that all variables — except those used with `idvar` and `timevar` — are time-varying.

```
> reshape(long, idvar="person", timevar="time",
+         v.names=c("day", "response"),
+         direction="wide")
```

```
    person age gender id day.1 response.1 day.2 response.2
1.1      I  20   Male  1     0         10     1         12
2.1     II  32   Male  2     0         10     2         12
3.1    III  20 Female  3     0         10     1         NA
4.1     IV  20   Male  4     0         10     1         12
5.1      V  20   Male  5     0         10     1         12
    day.3 response.3 day.4 response.4
1.1     2         15     3         16
2.1     4         15     5         16
3.1     2         15     3         16
4.1     2         15     3         16
5.1    NA         15     3         16
```

Column names are generated by adding a number to the variable names specified in `v.names`. If we want to manually specify the column names then we need to supply the `varying` argument as shown in the code below.

```
> reshape(long, idvar="person", timevar="time",
+         v.names=c("day", "response"),
+         varying=list(c("y1", "y2", "y3", "y4"),
+                      c("t1", "t2", "t3", "t4")),
+         direction="wide")
    person age gender id y1 t1 y2 t2 y3 t3 y4 t4
1.1      I  20   Male  1  0 10  1 12  2 15  3 16
2.1     II  32   Male  2  0 10  2 12  4 15  5 16
3.1    III  20 Female  3  0 10  1 NA  2 15  3 16
4.1     IV  20   Male  4  0 10  1 12  2 15  3 16
5.1      V  20   Male  5  0 10  1 12 NA 15  3 16
```

The output above shows that the `timevar` variable used to define the time-varying measurements from each individual is removed from the dataset since measurements (and their order) are now all gathered on a single line for each individual.

2.39 CONVERT A DATA FRAME TO A VECTOR

Problem: You have a data frame or matrix of values and want to convert all the values into a single vector.

Solution: You have data frame or a matrix of observations and wish to combine all the observations into a single vector. Data type conversion from data frames to vectors can be accomplished using `as.matrix` which first converts the data frame to a matrix and then use `as.vector` subsequently to convert the matrix into a vector.

```
> df <- data.frame(a=rep(1,5), b=rep(2,5), c=rep(3,5))
> df
  a b c
1 1 2 3
2 1 2 3
3 1 2 3
4 1 2 3
5 1 2 3
> vector <- as.vector(as.matrix(df))  # Convert to vector
> vector
 [1] 1 1 1 1 1 2 2 2 2 2 3 3 3 3 3
```

If all variables in the original data frame are numeric as in the example above, then the resulting vector will be numeric. Otherwise, the result will be a character vector as shown in the following example.

```
> df <- data.frame(a=rep(1,5),
+                  b=factor(c("A", "A", "A", "B", "B")))
> df          # Show data frame with both numeric and factor
  a b
1 1 A
2 1 A
3 1 A
4 1 B
5 1 B
> as.vector(as.matrix(df))    # Result is vector of strings
 [1] "1" "1" "1" "1" "1" "A" "A" "A" "B" "B"
```

## CHAPTER 3

Statistical analyses

Model formulas play a central role when specifying statistical models in many of the R functions. The basic format for a formula is of the form `response ~ predictor(s)`. The ~ is read "is modeled by" so we interpret the expression as a specification that the response is modeled by a combination of the predictors. The predictor variable(s) can be either numeric variables, factors, or a combination of both.

Generally, terms are included in the model with the + operator which indicates that the explanatory variables have an additive effect on the response variable (possible through a transformation). An intercept term is implicitly included in R formulas and it can be dropped from the R code by including the term `-1`. Likewise, other terms can be removed by the `-` operator — not just the intercept.

The code below uses the `ChickWeight` dataset to show examples of model formulas in combination with the linear model function, `lm` (the output from the calls below is not shown). The `ChickWeight` data frame contains information on the effect of diets on early growth of chickens.

```
> data(ChickWeight)
> attach(ChickWeight)
> lm(weight ~ Time)           # Linear regression
> lm(weight ~ Time - 1)       # Linear regression through origin
> lm(weight ~ Diet)           # One-way ANOVA
> lm(weight ~ Diet - 1)       # One-way ANOVA without ref. group
> lm(weight ~ Diet + Chick)   # Two-way additive ANOVA
> # Analysis of covariance, common slope, one intercept per diet
> lm(weight ~ Time + Diet)
```

Interactions can be specified in different ways in the model formula. The `*`, `:`, and `^` operators all introduce interaction terms in

Table 3.1: Operators for model formulas

Operator	Effect in model formula
+	add the term to the model
-	leave out the term from the model
:	introduce interaction between the terms
*	introduce interaction and lower-order effect of the terms
^	add all terms including interactions up to a certain order
/	nest the terms on the left within those on the right

the model, but in different ways. `var1:var2` adds an interaction between the two terms. `var1*var2` includes not only the interaction but also the main effects, so `var1*var2` is identical to `var1 + var2 + var1:var2`. The formula `(var1 + var2 + var3)^k` includes all the main effects and all possible interactions up to order k. For example, the formula `(var1 + var2 + var3)^3` corresponds to `var1*var2*var3` while the command `(var1 + var2 + var3)^2` corresponds to the formula `var1*var2*var3 - var1:var2:var3` which we also can write as `var1*var2 + var1*var3 + var2*var3`. The nested operator is written as `var1/var2` which means that we fit a model with a main effect of `var1` plus `var2` within `var1`. This is equivalent to the model `var1 + var1:var2`.

```
> lm(weight ~ Diet*Chick)     # Two-way ANOVA with interaction
> lm(weight ~ Diet + Chick + Diet:Chick) # Same as above
> # Analysis of covariance, one slope and intercept for each diet
> lm(weight ~ Time*Diet)
> lm(weight ~ Diet/Chick)                # Nested model
> lm(weight ~ Diet + Diet:Chick)         # Same as above
> lm(weight ~ Diet*Chick - Chick)        # Same as above
```

A summary of operators for model formulas can be seen in Table 3.1.

The inhibit function, `I`, inhibits the interpretation of formula operators, `+`, `-`, etc., so they can be used as arithmetical operators in the formulas. Normal arithmetic expressions, for example logarithms or square-roots can be included in the formulas for both the response variable and explanatory variables.

The `pressure` dataset contains data on the relation between temperature in degrees Celsius and vapor pressure of mercury (in millimeters of mercury). If we wish to model the logarithm of the mercury pressure as a quadratic function of temperature measured in Fahrenheit, then we write

```
> data(pressure)
> lm(log(pressure) ~ I(temperature*9/5+32) +
+                    I((temperature*9/5+32)^2), data=pressure)
```

As shown in the code above, the response variable can have a function applied to it — in this case the natural logarithm.

The `interaction` function computes the factor identical to what is obtained by specifying an interaction term in a model. We can use the `interaction` function outside model formulas to see how the resulting interaction factor looks. In the following code, two factors are used to create a new factor that is the interaction between the two original factors. We can set the argument `drop=TRUE` to `interaction` to remove any unused factor levels for the interaction.

```
> f1 <- factor(c("A", "B", "C", "B", "A", "C"))
> f2 <- factor(c(1, 2, 1, 2, 1, 2))
> interaction(f1,f2)
[1] A.1 B.2 C.1 B.2 A.1 C.2
Levels: A.1 B.1 C.1 A.2 B.2 C.2
> interaction(f1,f2, drop=TRUE) # Drop unused factor levels
[1] A.1 B.2 C.1 B.2 A.1 C.2
Levels: A.1 C.1 B.2 C.2
```

The `offset` function is used inside model formulas to fix the estimated coefficient of a numeric explanatory variable at the value 1.

```
> lm(weight ~ offset(Time))     # Fix the slope of time at 1
> newy <- weight - Time         # Subtract the effect of time
> lm(newy ~ 1)                  # Same result as above
```

If we wish to fix the regression parameter at a value other than 1, then we simply multiply by the desired constant inside `offset`.

```
> lm(weight ~ offset(2.7*Time))    # Fix slope at 2.7

Call:
lm(formula = weight ~ offset(2.7 * Time))

Coefficients:
(Intercept)
      92.88
```

Table 3.2: Functions for common summary statistics

Symbol/function	Description
`mean(x)`	mean
`median(x)`	median
`sd(x)`	standard deviation
`var(x)`	variance
`IQR(x)`	inter-quartile range
`cor(x,y)`	correlation between `x` and `y`
`sum(x)`	sum
`cumsum(x)`	cumulative or aggregated sum
`min(x)` and `max(x)`	minimum and maximum value
`range(x)`	range
`quantile(x)`	quantile

The vertical bar, `|`, is used in some model functions (e.g., `lme` and `lmer`) to indicate conditioning. See Problems 3.16 and 3.17 for examples.

DESCRIPTIVE STATISTICS

3.1 COMPUTE SUMMARY STATISTICS

Problem: You wish to compute a common summary statistic.

Solution: The distribution of a numeric variable is often summarized by presenting summary statistics of the data. Typical summary statistics for the central tendency include mean and median, while summary statistics for the dispersion often include the standard deviation, variation, range, or inter-quartile range. Table 3.2 lists some functions for computing summary statistics from a vector.

The code below shows examples where these summary functions are used.

```
> x <- c(6, 8, 1:4)
> x
[1] 6 8 1 2 3 4
> mean(x)          # mean of x
[1] 4
> median(x)        # median
[1] 3.5
> sd(x)            # standard deviation
```

```
[1] 2.607681
> IQR(x)          # inter-quartile range
[1] 3.25
> y <- (1:6)**2   # New vector
> y
[1]  1  4  9 16 25 36
> cor(x,y)        # Pearson correlation
[1] -0.3783846
> cor(x, y, method="spearman") # Spearman correlation
[1] -0.3714286
> cor(x, y, method="kendall")  # Kendall correlation
[1] -0.06666667
> sum(x)          # sum of elements in x
[1] 24
> cumsum(x)       # cumulative sum of all elements of x
[1]  6 14 15 17 20 24
> min(x)          # minimum value
[1] 1
> max(x)          # maximum value
[1] 8
> range(x)        # range. Same as the values above
[1] 1 8
> quantile(x)     # get quartiles
  0%  25%  50%  75% 100%
1.00 2.25 3.50 5.50 8.00
> quantile(x, probs=c(.15, .25, .99))  # specific quantiles
 15%  25%  99%
1.75 2.25 7.90
```

Several of the functions used above take optional arguments. In particular the `na.rm` option can be set to `TRUE` to ensure that missing observations are disregarded in the computations as shown in the following.

```
> x[c(2,4)] <- NA        # redefine x
> x                      # vector with missing values
[1]  6 NA  1 NA  3  4
> sum(x)                 # sum returns missing
[1] NA
> sum(x, na.rm=TRUE)     # unless we remove the NA's
[1] 14
> mean(x)                # mean also returns missing
[1] NA
> mean(x, na.rm=TRUE)    # unless we remove the NA's
[1] 3.5
```

3.2 CREATE DESCRIPTIVE TABLE

Problem: You want to create a descriptive table to present the data in a publication.

Solution: Descriptive tables that present the data are often part of a scientific publication. The `univariateTable` function from the `Publish` can generate a publication-ready table presenting the data. By default, `univariateTable` computes the mean and standard deviation for numeric variables and the distribution of observations in categories for factors. These values can be computed and compared across groups.

The main input to `univariateTable` is a formula where the right-hand side lists the variables to be included in the table while the left-hand side — if present — specifies a grouping variable. In the example below we look at a dataset on forced expiratory volume in children, where the age (in years), height (in inches), and lung capacity (in liters) are registered for boys and girls some of which are exposed to smoking. We compute the descriptive statistics grouped by smoking exposure.

```
> library(isdals)   # Get the data
> data(fev)
> library(Publish)
> univariateTable(Smoke ~ Age + Ht + FEV + Gender, data=fev)
  Variable      Level Smoke = 0 (n=589) Smoke = 1 (n=65)
1      Age mean (sd)          9.5 (2.7)       13.5 (2.3)
2       Ht mean (sd)         60.6 (5.7)       66.0 (3.2)
3      FEV mean (sd)          2.6 (0.9)        3.3 (0.7)
4   Gender          0        279 (47.4)        39 (60.0)
5                   1        310 (52.6)        26 (40.0)
  Total (n=654) p-value
1     9.9 (3.0)  <1e-04
2    61.1 (5.7)  <1e-04
3     2.6 (0.9)  <1e-04
4    318 (48.6)
5    336 (51.4)  0.0714
```

The mean age for children exposed to smoke is 13.5 years (with a standard deviation of 2.3 years) while the corresponding mean and standard deviation for the children not exposed to smoking is 9.5 and 2.7, respectively. The number of individuals in each group (589 and 65) is listed and for categorical variables the number of observations in each category is listed with the relative percentages for each group. Finally, the groups are compared with a one-way analysis-of-variance for numeric variables, and a chi-square test for categorical variables. Mean age is clearly different in the two groups ($p < 0.0001$) while gender is not ($p = 0.0714$).

`univariateTable` prints the mean and standard deviation for numeric variables but for non-normal variables it may be more appropriate to present the median and inter-quartile range. This is forced on a variable-by-variable basis by encompassing the relevant variables in `Q()`. The `showTotals` argument can be set to `FALSE` to remove the column with the totals.

```
> univariateTable(Smoke ~ Age + Q(Ht) + FEV + Gender,
+                 showTotals=FALSE, data=fev)
  Variable          Level Smoke = 0 (n=589)  Smoke = 1 (n=65)
1      Age      mean (sd)          9.5 (2.7)        13.5 (2.3)
2       Ht median [iqr] 61.0 [57.0, 64.5] 66.0 [63.5, 68.0]
3      FEV      mean (sd)          2.6 (0.9)         3.3 (0.7)
4   Gender              0        279 (47.4)        39 (60.0)
5                       1        310 (52.6)        26 (40.0)
  p-value
1  <1e-04
2  <1e-04
3  <1e-04
4
5  0.0714
```

The median heights for the two groups are 61 and 66 inches, respectively, and the inter-quartile ranges are $[57.0, 64.5]$ and $[63.5, 68.0]$. The Krushkal–Wallis' test is used to compare the distributions for non-normal variables. Here, the height is clearly different for the exposure and non-exposure groups.

`univariateTable` accepts three optional arguments, `summary.format`, `Q.format`, and `freq.format` that can be used to change the output for numeric, modified numeric, and categorical variables, respectively.

See also: The `stat.desc` function from the `pastecs` package can also create parts of a descriptive table.

LINEAR MODELS

3.3 FIT A LINEAR REGRESSION MODEL

Problem: You want to fit a linear regression model to describe the linear relationship between two quantitative variables.

Solution: Linear models (of which linear regression is a special case with just a single quantitative explanatory variable) are fitted in R using the function lm. lm expects a model formula of the form y ~ x where y is the quantitative response variable and x is the quantitative explanatory variable.

In the trees dataset, we seek to model the volume of the trees as a linear function of the height. We can use the data option to specify a data frame that contains the variables or just make sure that the variables are accessible.

```
> data(trees)
> result <- lm(Volume ~ Height, data=trees)
> result

Call:
lm(formula = Volume ~ Height, data = trees)

Coefficients:
(Intercept)       Height
    -87.124        1.543
```

The fitted linear regression has two parameters, the intercept and slope, and they are listed as (Intercept) and Height in the output, respectively. The summary function can be used to get additional information on the estimated parameters.

```
> summary(result)

Call:
lm(formula = Volume ~ Height, data = trees)

Residuals:
    Min      1Q  Median      3Q     Max
-21.274  -9.894  -2.894  12.068  29.852

Coefficients:
            Estimate Std. Error t value Pr(>|t|)
(Intercept) -87.1236    29.2731  -2.976 0.005835 **
Height        1.5433     0.3839   4.021 0.000378 ***
---
Signif. codes:  0 '***' 0.001 '**' 0.01 '*' 0.05 '.' 0.1 ' ' 1

Residual standard error: 13.4 on 29 degrees of freedom
Multiple R-squared:  0.3579,Adjusted R-squared:  0.3358
F-statistic: 16.16 on 1 and 29 DF,  p-value: 0.0003784
```

From the summary output we find the estimates, their corresponding standard errors and t-tests for the hypothesis that the parameter

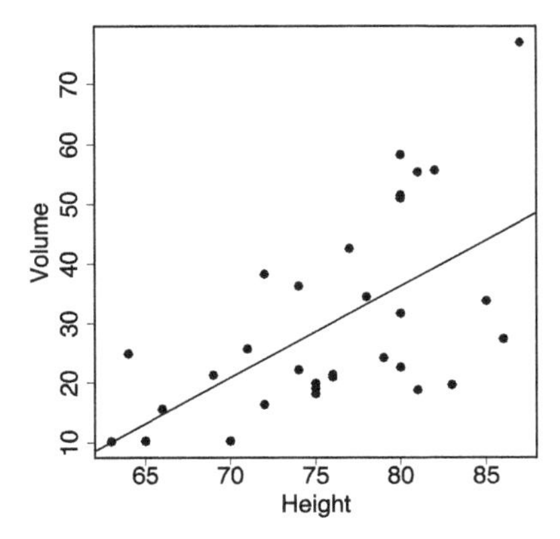

Figure 3.1: Fit of a linear regression model to the cherry tree data.

equals zero. The slope is significantly different from zero ($p = 0.000378$), and the volume increases on average by 1.5433 for each unit increase of height. The residual standard error of 13.4 is the estimated standard deviation of the residuals.

We can use the `abline` function to add the fitted regression line to an existing plot by using the fitted model as argument. The output is shown in Figure 3.1.

```
> plot(trees$Height, trees$Volume,      # Make scatter plot
+      xlab="Height", ylab="Volume")
> abline(result)                        # Add fitted line
```

See also: Problems 3.33 and 4.21 show how to make model validation for linear models.

3.4 FIT A MULTIPLE LINEAR REGRESSION MODEL

Problem: You want to fit a multiple linear regression model to describe the linear relationship between a quantitative response variable and several explanatory quantitative variables.

Solution: Multiple linear regression is an extension of linear regression (see Problem 3.3) where we model a quantitative response as a function of two or more quantitative explanatory variables.

The cherry tree dataset `trees` contains information not only on the tree volume and height, but also on the tree girth (diameter at breast height). We want to use both the height and girth, which are measures that are easily acquired to model the volume of the tree.

```
> data(trees)
> result <- lm(Volume ~ Height + Girth, data=trees)
> result

Call:
lm(formula = Volume ~ Height + Girth, data = trees)

Coefficients:
(Intercept)       Height        Girth
   -57.9877       0.3393       4.7082
```

The summary function is used to extract estimates and their corresponding standard errors.

```
> summary(result)

Call:
lm(formula = Volume ~ Height + Girth, data = trees)

Residuals:
    Min      1Q  Median      3Q     Max
-6.4065 -2.6493 -0.2876  2.2003  8.4847

Coefficients:
            Estimate Std. Error t value Pr(>|t|)
(Intercept) -57.9877     8.6382  -6.713 2.75e-07 ***
Height        0.3393     0.1302   2.607   0.0145 *
Girth         4.7082     0.2643  17.816  < 2e-16 ***
---
Signif. codes:  0 '***' 0.001 '**' 0.01 '*' 0.05 '.' 0.1 ' ' 1

Residual standard error: 3.882 on 28 degrees of freedom
Multiple R-squared:  0.948,Adjusted R-squared:  0.9442
F-statistic:   255 on 2 and 28 DF,  p-value: < 2.2e-16
```

The three parameters for the model are the intercept and two partial regression slopes, and their estimates can all be found in the output from summary: 0.3393 for height and 4.7082 for girth. We see that both height and girth are statistically significant with p-values of 0.0145 and 0, respectively. The residual standard error of 3.882 is the estimated standard deviation of the residuals.

See also: Problem 3.33 and 4.21 show how to make model validation for linear models.

3.5 FIT A POLYNOMIAL REGRESSION MODEL

Problem: You wish to fit a polynomial regression model where the relationship between the response variable and an explanatory variable is a higher-order polynomial.

Solution: In some situations the functional relationship between the response variable and a quantitative explanatory variable should be modeled as a kth order polynomial.

Polynomial regression is a special case of multiple linear regression. Thus we can use Problem 3.4 and the `lm` function to fit a polynomial regression model by treating the different exponents of the explanatory variable as distinct independent variables in a multiple regression model.

In the example we wish to model the number of deaths due to AIDS for a 19-year period. We use the inhibit function, `I`, to ensure that the `^` operator is interpreted as exponentiation and not as a formula interaction. Also, we subtract 1980 from the years to prevent numerical instability; 1999^3 is a very large number and if we analyze the data on the original scale then R might run into numerical difficulties and will be unable to estimate all the model parameters.

```
> year <- seq(1981, 1999)
> deaths <- c(339, 1201, 3153, 6368, 12044, 19404, 29105,
+             36126, 43499, 49546, 60573, 79657, 79879,
+             73086, 69984, 61124, 49379, 43225, 41356)
>
> newyear <- year - 1980
> model <- lm(deaths ~ newyear + I(newyear^2) + I(newyear^3))
> summary(model)

Call:
lm(formula = deaths ~ newyear + I(newyear^2) + I(newyear^3))

Residuals:
    Min      1Q  Median      3Q     Max
-9983.7 -2102.0  -782.8   635.6 12592.4

Coefficients:
             Estimate Std. Error t value Pr(>|t|)
(Intercept)   3199.14    6783.11   0.472 0.643975
newyear      -4754.21    2862.05  -1.661 0.117442
I(newyear^2)  1717.37     328.10   5.234 0.000101 ***
I(newyear^3)   -73.14      10.80  -6.771 6.29e-06 ***
---
Signif. codes:  0 '***' 0.001 '**' 0.01 '*' 0.05 '.' 0.1 ' ' 1
```

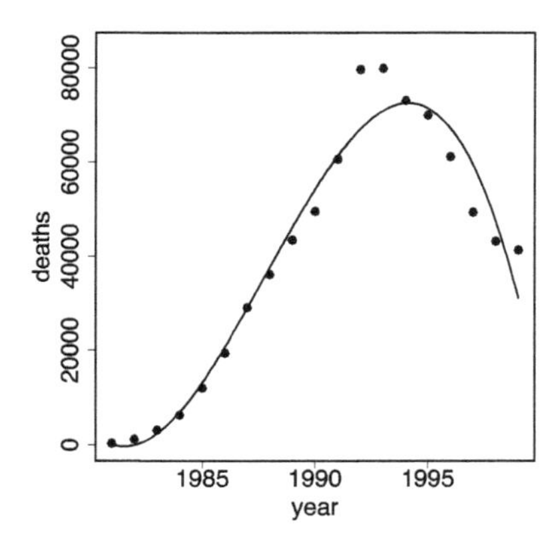

Figure 3.2: Fit of a third-degree polynomial to AIDS deaths data.

```
Residual standard error: 5985 on 15 degrees of freedom
Multiple R-squared:  0.9591,Adjusted R-squared:  0.9509
F-statistic: 117.3 on 3 and 15 DF,  p-value: 1.232e-10
```

Thus we model the average number of AIDS-related deaths as the third-degree polynomial

$$y = -73.14x^3 + 1717.37x^2 - 4754.21x + 3199.14,$$

where y is the number of deaths and x is the number of years since 1980. The `summary` output also shows that the third-order coefficient is highly significant ($p < 0.0001$) so a second-order polynomial would provide a significantly worse fit.

We can plot the data and the predicted relationship using the `predict` function as shown below.

```
> # Plot the data and add the fitted cubic line
> plot(year, deaths)
> lines(seq(1981, 1999, .1),
+       predict(model, data.frame(newyear=seq(1, 19, .1))))
```

The data and the fitted third-degree polynomial can be seen in Figure 3.2.

3.6 FIT A ONE-WAY ANALYSIS OF VARIANCE

Problem: You wish to perform a one-way analysis of variance to test if the means from several groups are equal.

Solution: One-way analysis of variance is used to test if the means of two or more groups are equal. Since one-way analysis of variance is a special case of a linear model where the explanatory variable is a single categorical factor, we can use the `lm` function for the analysis — we just need to make sure that the explanatory variable is coded as a factor.

The `antibio` dataset from the `isdals` package provides data from an experiment concerned with exploring the influence of antibiotics on decomposition of dung organic material from heifers. There are five types of antibiotics and a control group and the outcome measures the amount of organic material.

```
> library(isdals)
> data(antibio)
> head(antibio)
      type  org
1 Ivermect 3.03
2 Ivermect 2.81
3 Ivermect 3.06
4 Ivermect 3.11
5 Ivermect 2.94
6 Ivermect 3.06
> model <- lm(org ~ type, data=antibio)
> model

Call:
lm(formula = org ~ type, data = antibio)

Coefficients:
 (Intercept)   typeControl  typeEnroflox  typeFenbenda
     2.89500      -0.29167      -0.18500      -0.06167
typeIvermect  typeSpiramyc
     0.10667      -0.04000
> summary(model)

Call:
lm(formula = org ~ type, data = antibio)

Residuals:
     Min       1Q   Median       3Q      Max
-0.29000 -0.06000  0.01833  0.07250  0.18667

Coefficients:
             Estimate Std. Error t value Pr(>|t|)
(Intercept)   2.89500    0.04970  58.248  < 2e-16 ***
typeControl  -0.29167    0.07029  -4.150 0.000281 ***
typeEnroflox -0.18500    0.07029  -2.632 0.013653 *
typeFenbenda -0.06167    0.07029  -0.877 0.387770
```

```
typeIvermect  0.10667    0.07029   1.518 0.140338
typeSpiramyc -0.04000    0.07858  -0.509 0.614738
---
Signif. codes:  0 '***' 0.001 '**' 0.01 '*' 0.05 '.' 0.1 ' ' 1

Residual standard error: 0.1217 on 28 degrees of freedom
Multiple R-squared:  0.5874,Adjusted R-squared:  0.5137
F-statistic: 7.973 on 5 and 28 DF,  p-value: 8.953e-05
```

Categorical explanatory variables are parameterized differently from quantitative explanatory variables in R output. For categorical variables, one of the levels is chosen to be the reference level and the other categories/levels are parameterized as differences or contrasts relative to this reference level.

The `(Intercept)` term corresponds to the reference level of the treatment factor and here that is the α-Cypermethrin antibiotics, so the average level of organic material for α-Cypermethrin is 2.90. The estimate listed for `typeControl` is the contrast relative to the reference level. Thus, the average decrease for treatment B is $2.90 + -0.292 = 2.60$. Likewise the standard error found on the `typeControl` line corresponds to the standard error of *the difference* between type Control and α-Cypermethrin. The t-value and p-value printed on each line are the values found by testing the hypothesis that the parameter is equal to zero. Hence, for the contrast `typeControl` we get a t-value of -4.15 and a corresponding p-value of 0.00028. We clearly reject the hypothesis that the difference between the reference level (α-Cypermethrin) and the control group is zero. The test for the `(Intercept)` parameter corresponds to the hypothesis that the overall level of the reference group (α-Cypermethrin) equals zero, which we reject with a p-value of 0.

The `drop1` function tests each explanatory variable by removing the terms from the model formula and comparing the fit to the original model. `drop1` can be used for model reduction and for one-way analysis of variance we should set the option `test="F"` to ensure that `drop1` computes test statistics (and corresponding p-values) based on the F distribution.

```
> drop1(model, test="F")
Single term deletions

Model:
org ~ type
       Df Sum of Sq    RSS    AIC F value    Pr(>F)
```

```
<none>              0.4150 -137.8
type    5   0.59082 1.0058 -117.7  7.9726 8.953e-05 ***
---
Signif. codes:  0 '***' 0.001 '**' 0.01 '*' 0.05 '.' 0.1 ' ' 1
```

From the output it is clear that there is a highly significant effect of type (F statistic of 7.97 and a p-value that is zero) so we reject the hypothesis that the organic material is the same for all antibiotic types. Since we only have one explanatory variable for one-way analysis of variance, we get the exact same result as we got from the F statistic in the `summary` output above.

See also: If you only wish to compare the means of two groups, then you should use the `t.test` function which does not assume that the variance of each group is identical (Problem 3.22). For more than two groups the `oneway.test` function can test equality of means without assuming equal variances of each group. See Problem 3.52 for the non-parametric Kruskal–Wallis test. Problem 4.8 shows how to make boxplots to visually compare the groups.

3.7 MAKE POST-HOC PAIRWISE COMPARISONS

Problem: Use post-hoc pairwise comparisons to determine which pairs of groups that differ from each other after a categorical explanatory variable was found to be significant.

Solution: Pairwise comparison tests or multiple comparison tests make multiple pairwise comparisons of individual group means. Post-hoc pairwise comparison tests refer to the situation where pairwise comparisons are used *after* an analysis of variance has found that a categorical explanatory variable with multiple categories is significant (see for example Problems 3.6 or 3.8) which is the situation we consider here. Once a categorical variable is found to be significant we know that the means differ significantly across the variable and we would like to conclude which of the groups are actually different.

We will start by using the example with organic material and antibiotics from Problem 3.6. As we can see below we have 6 categories and their means are clearly not the same ($p = 0$) so it is meaningful to conduct pairwise comparisons among the six different groups.

```
> library(isdals)
> data(antibio)
> model <- lm(org ~ type, data=antibio)
> drop1(model, test="F")
Single term deletions

Model:
org ~ type
       Df Sum of Sq    RSS    AIC F value    Pr(>F)
<none>               0.4150 -137.8
type    5   0.59082 1.0058 -117.7  7.9726 8.953e-05 ***
---
Signif. codes:  0 '***' 0.001 '**' 0.01 '*' 0.05 '.' 0.1 ' ' 1
```

Based on the overall F test we do not know which pairs of antibiotics that are significantly different from each other.

Tukey's Honest Significant Differences (Tukey's HSD) is an often-used approach for post-hoc pairwise comparisons after significant effects since it compares the difference between each pair of means with appropriate adjustment for multiple testing and provides confidence intervals rather than just p values.

The `TukeyHSD` function computes Tukey's HSD and corresponding confidence intervals. It requires an `aov` object as its primary input so we convert our model fit from `lm` to aov when calling `TukeyHSD`. There are two important arguments to `TukeyHSD`: `which` and `conf.level`. `which` should be a character vector that lists the terms from the fitted model for which the Tukey intervals should be calculated (default is all terms) while `conf.level` sets the desired confidence level (default is 95%).

```
> result <- TukeyHSD(aov(model))
> result
  Tukey multiple comparisons of means
    95% family-wise confidence level

Fit: aov(formula = model)

$type
                        diff         lwr         upr     p adj
Control-Alfacyp  -0.29166667 -0.50645972 -0.07687362 0.0034604
Enroflox-Alfacyp -0.18500000 -0.39979305  0.02979305 0.1225956
Fenbenda-Alfacyp -0.06166667 -0.27645972  0.15312638 0.9488454
Ivermect-Alfacyp  0.10666667 -0.10812638  0.32145972 0.6563131
Spiramyc-Alfacyp -0.04000000 -0.28014593  0.20014593 0.9953987
Enroflox-Control  0.10666667 -0.10812638  0.32145972 0.6563131
Fenbenda-Control  0.23000000  0.01520695  0.44479305 0.0304908
```

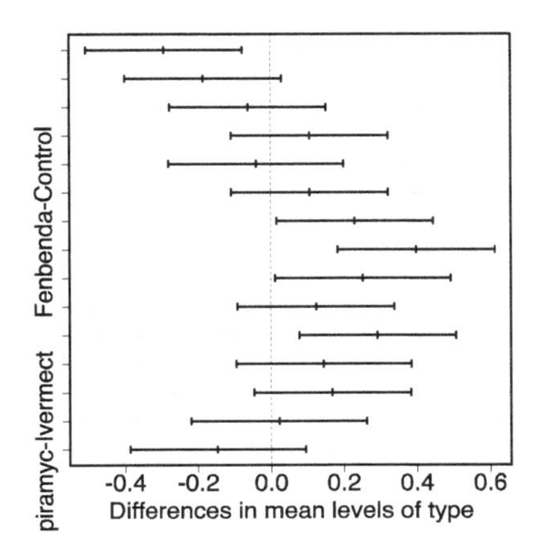

Figure 3.3: Pairwise confidence intervals after Tukey HSD correction for multiple testing. The pairwise differences are plotted in the same order as the output from `TukeyHSD`.

```
Ivermect-Control   0.39833333  0.18354028  0.61312638 0.0000612
Spiramyc-Control   0.25166667  0.01152074  0.49181260 0.0358454
Fenbenda-Enroflox  0.12333333 -0.09145972  0.33812638 0.5093714
Ivermect-Enroflox  0.29166667  0.07687362  0.50645972 0.0034604
Spiramyc-Enroflox  0.14500000 -0.09514593  0.38514593 0.4549043
Ivermect-Fenbenda  0.16833333 -0.04645972  0.38312638 0.1923280
Spiramyc-Fenbenda  0.02166667 -0.21847926  0.26181260 0.9997587
Spiramyc-Ivermect -0.14666667 -0.38681260  0.09347926 0.4424433
```

Here we see the lower and upper limits of the pairwise multiple testing corrected confidence intervals and the corresponding adjusted p-values. We see that all of the treatments but Enroflox are (barely) statistically different from the control. We can also plot the results to see a visual representation of the pairwise differences and confidence intervals (see Figure 3.3 for the output).

```
> plot(result)
```

If there are multiple categorical explanatory variables then the `which` argument can set which variables are used to compute the adjusted pairwise confidence intervals.

See also: The `glht` function from the `multcomp` package can be used to compute Tukey's Honest Significant Differences and it has extra options for tweaking the analysis.

3.8 FIT A TWO-WAY ANALYSIS OF VARIANCE

Problem: You wish to perform a two-way analysis of variance to examine how the means are influenced by two categorical explanatory variables.

Solution: Two-way analysis of variance is used to test how two categorical explanatory variables influence the mean level of the response variable. Two-way analysis of variance is a special case of a linear model where there are two categorical explanatory variables so we can use the `lm` function for the analysis — we just need to make sure that the explanatory variables are coded in R as factors. We can include (and test for) an interaction between the two explanatory variables in the model if there are multiple observations per combination of the two explanatory variables. Otherwise we can only consider the additive model, where the two explanatory variables influence the response independently of each other.

The `interaction.plot` function plots the means for different combinations of the two explanatory factors, which might illustrate possible interactions. `interaction.plot` takes three arguments: `x.factor` which is the explanatory factor plotted along the x axis, `trace.factor` is the other explanatory variable that determines the individual traces, and `response` which is the response variable.

In the example below, we use the `fev` data from the `isdals` package to examine if and how gender (0=female, 1=male) and exposure to smoking (0=no, 1=yes) influence the forced expiratory volume (FEV) in children.

```
> library(isdals)
> data(fev)
> # Convert variables to factors and get meaningful labels
> fev$Gender <- factor(fev$Gender, labels=c("Female", "Male"))
> fev$Smoke <- factor(fev$Smoke, labels=c("No", "Yes"))
> interaction.plot(fev$Gender, fev$Smoke, fev$FEV,
+                  lwd=3, col=c("black", "blue"))
> model <- lm(FEV ~ Gender + Smoke + Gender*Smoke, data=fev)
> model

Call:
lm(formula = FEV ~ Gender + Smoke + Gender * Smoke, data = fev)

Coefficients:
       (Intercept)           GenderMale            SmokeYes
            2.3792               0.3552              0.5867
```

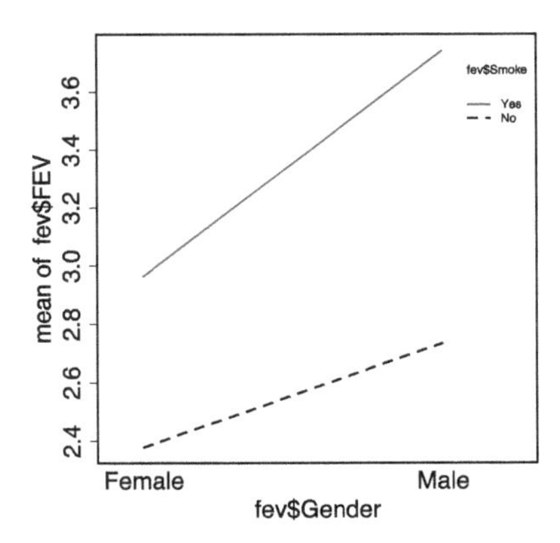

Figure 3.4: Interaction plot for the forced expiratory volume data.

```
GenderMale:SmokeYes
             0.4221
```

Figure 3.4 shows the interaction plot for the `fev` dataset. If there is no interaction, then the traces should be roughly parallel. The figure does not suggest a large interaction between the two variables but we include an interaction between gender and smoking status in our initial model in order to test if there is an interaction. The output from the starting model shows that non-smoking females are the reference group, and that they have an average forced expiratory volume of 2.379. non-smoking males have an average FEV of $2.379 + 0.355 = 2.734$. Similarly, smoking females have an average FEV of $2.379 + 0.587 = 2.966$ while smoking males have an average level of $2.379 + 0.355 + 0.587 + 0.422 = 3.743$.

The `drop1` function tests each explanatory variable by removing the term from the model formula while preserving the hierarchical principle. `drop1` can be used for model reduction and the option `test="F"` ensures that `drop1` computes test statistics (and corresponding p-values) based on the F distribution.

```
> drop1(model, test="F")
Single term deletions

Model:
FEV ~ Gender + Smoke + Gender * Smoke
             Df Sum of Sq    RSS     AIC F value  Pr(>F)
<none>                    433.40 -261.08
Gender:Smoke  1    2.5127 435.91 -259.30  3.7684 0.05266 .
```

```
---
Signif. codes:  0 '***' 0.001 '**' 0.01 '*' 0.05 '.' 0.1 ' ' 1
```

The results from drop1 show that the interaction is almost significant ($p = 0.05266$) when testing at a significance level of 5%. Here we choose to remove the interaction and refit the model without the interaction term.

```
> model2 <- lm(FEV ~ Gender + Smoke, data=fev)
> drop1(model2, test="F")
Single term deletions

Model:
FEV ~ Gender + Smoke
       Df Sum of Sq    RSS     AIC F value    Pr(>F)
<none>              435.91 -259.30
Gender  1    25.436 461.35 -224.21  37.986 1.249e-09 ***
Smoke   1    33.681 469.60 -212.63  50.300 3.450e-12 ***
---
Signif. codes:  0 '***' 0.001 '**' 0.01 '*' 0.05 '.' 0.1 ' ' 1
> summary(model2)

Call:
lm(formula = FEV ~ Gender + Smoke, data = fev)

Residuals:
     Min       1Q   Median       3Q      Max
-1.95758 -0.63880 -0.03123  0.52159  3.03942

Coefficients:
            Estimate Std. Error t value Pr(>|t|)
(Intercept)  2.35788    0.04774  49.394  < 2e-16 ***
GenderMale   0.39571    0.06420   6.163 1.25e-09 ***
SmokeYes     0.76070    0.10726   7.092 3.45e-12 ***
---
Signif. codes:  0 '***' 0.001 '**' 0.01 '*' 0.05 '.' 0.1 ' ' 1

Residual standard error: 0.8183 on 651 degrees of freedom
Multiple R-squared:  0.112,Adjusted R-squared:  0.1093
F-statistic: 41.07 on 2 and 651 DF,  p-value: < 2.2e-16
```

Both of the main effects (gender and smoking exposure) are statistically significant as shown by drop1, so we conclude that there is an additive effect of both gender and smoking status on forced expiratory volume. From the summary output we see that boys on average have a forced expiratory volume that is 0.3957 liters larger than girls and that persons exposed to smoking have a larger expiratory volume than non-smokers! (This somewhat surprising result is caused by the fact that age is a confounder that is associated with both the smoking status and FEV, and generally the smokers are older than the non-smokers.)

See also: Problem 3.9 analyzes the full dataset with all explanatory variables.

3.9 FIT A LINEAR NORMAL MODEL

Problem: You wish to analyze a dataset using a linear normal model where multiple explanatory variables (both categorical and quantitative) and possibly their interactions may be present.

Solution: The class of linear normal models is extremely versatile and can be used to analyze data in many situations, and all of the situations mentioned in this section are special cases of the linear normal model. Linear normal models are fitted in R by the `lm` function, with the appropriate model formula which can include both quantitative and categorical explanatory variables as well as interactions between them.

We use the `fev` data from the `isdals` package to illustrate how forced expiratory volume — a surrogate for lung capacity — is influenced by gender (0=female, 1=male), exposure to smoking (0=no, 1=yes), age, and height for 654 children aged 3–19 years. The effect of height on forced expiratory volume is modeled as a quadratic polynomial and we allow for interactions between both gender and smoking status as well as between age and smoking status. Thus, we allow the effect of smoking on lung capacity to be different for boys and girls, and we also allow for the effect of age to depend on smoking status. A scatter plot of the data is shown in Figure 3.5.

```
> library(isdals)
> data(fev)
> # Convert variables to factors and get meaningful labels
> fev$Gender <- factor(fev$Gender, labels=c("Female", "Male"))
> fev$Smoke <- factor(fev$Smoke, labels=c("No", "Yes"))
> summary(fev)
      Age              FEV               Ht             Gender
 Min.   : 3.000   Min.   :0.791   Min.   :46.00   Female:318
 1st Qu.: 8.000   1st Qu.:1.981   1st Qu.:57.00   Male  :336
 Median :10.000   Median :2.547   Median :61.50
 Mean   : 9.931   Mean   :2.637   Mean   :61.14
 3rd Qu.:12.000   3rd Qu.:3.119   3rd Qu.:65.50
```

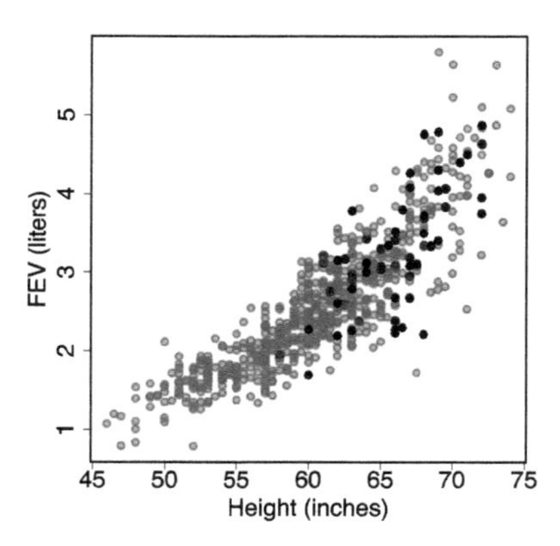

Figure 3.5: Scatter plot of the `fev` dataset. The smokers are colored black, non-smokers are blue.

```
 Max.   :19.000   Max.   :5.793   Max.   :74.00
 Smoke
 No :589
 Yes: 65



> plot(FEV ~ Ht, pch=20, col=c("Blue", "Black")[Smoke],
+    xlab="Height (inches)", ylab="FEV (liters)",  data=fev)
```

We start by fitting the initial model. Recall that interactions like `Smoke*Age` inherently include the main effects of `Smoke` and `Age` so we do not have to specify those directly. The functions `drop1` (with option `test="F"` to compute the F statistic and get proper p-values based on the F distribution) and `summary` can be used for model reductions and extracting parameter estimates, respectively.

```
> model <- lm(FEV ~ Ht + I(Ht^2) + Smoke*Gender + Smoke*Age,
+             data=fev)
> drop1(model, test="F")
Single term deletions

Model:
FEV ~ Ht + I(Ht^2) + Smoke * Gender + Smoke * Age
             Df Sum of Sq    RSS     AIC F value    Pr(>F)
<none>                    100.50 -1208.9
Ht            1    4.6001 105.10 -1181.6 29.5684 7.663e-08 ***
```

```
I(Ht^2)        1    8.7753 109.28 -1156.2 56.4056 1.968e-13 ***
Smoke:Gender   1    0.3906 100.89 -1208.4  2.5108    0.1136
Smoke:Age      1    0.3525 100.85 -1208.6  2.2656    0.1328
---
Signif. codes:  0 '***' 0.001 '**' 0.01 '*' 0.05 '.' 0.1 ' ' 1
```

Both the linear and quadratic terms of height are highly significant, but neither of the two interactions is significant in this initial model. We choose to remove the interaction between smoke and gender and refit the reduced model.

```
> model2 <- lm(FEV ~ Ht + I(Ht^2) + Gender + Smoke*Age,
+               data=fev)
> drop1(model2, test="F")
Single term deletions

Model:
FEV ~ Ht + I(Ht^2) + Gender + Smoke * Age
          Df Sum of Sq    RSS     AIC F value    Pr(>F)
<none>                 100.89 -1208.4
Ht         1    4.9069 105.80 -1179.3 31.4668 3.009e-08 ***
I(Ht^2)    1    9.2725 110.16 -1152.9 59.4627 4.720e-14 ***
Gender     1    1.3763 102.27 -1201.5  8.8258   0.00308 **
Smoke:Age  1    0.2557 101.15 -1208.7  1.6399   0.20080
---
Signif. codes:  0 '***' 0.001 '**' 0.01 '*' 0.05 '.' 0.1 ' ' 1
> # Remove the insignificant interaction and refit
> model3 <- lm(FEV ~ Ht + I(Ht^2) + Gender + Smoke + Age,
+               data=fev)
> drop1(model3, test="F")
Single term deletions

Model:
FEV ~ Ht + I(Ht^2) + Gender + Smoke + Age
        Df Sum of Sq    RSS     AIC F value    Pr(>F)
<none>               101.15 -1208.7
Ht       1    4.7552 105.90 -1180.7 30.4644 4.922e-08 ***
I(Ht^2)  1    9.1323 110.28 -1154.2 58.5063 7.354e-14 ***
Gender   1    1.2918 102.44 -1202.4  8.2759   0.00415 **
Smoke    1    0.8493 102.00 -1205.2  5.4411   0.01997 *
Age      1    9.0777 110.22 -1154.5 58.1563 8.658e-14 ***
---
Signif. codes:  0 '***' 0.001 '**' 0.01 '*' 0.05 '.' 0.1 ' ' 1
```

Since none of the terms in this model are insignificant, `model3` becomes our final model, and we conclude that there is a significant effect of height, gender, smoke, and age on forced expiratory volume. These effects work additively as there are no interactions in the final model. We can get the parameter estimates from the final model using `summary` on the final model object.

```
> summary(model3)

Call:
lm(formula = FEV ~ Ht + I(Ht^2) + Gender + Smoke + Age, data = fev)

Residuals:
     Min       1Q   Median       3Q      Max 
-1.61190 -0.22716  0.00619  0.22418  1.80565 

Coefficients:
              Estimate Std. Error t value Pr(>|t|)    
(Intercept)  6.8945787  1.4993579   4.598 5.12e-06 ***
Ht          -0.2742341  0.0496850  -5.519 4.92e-08 ***
I(Ht^2)      0.0031251  0.0004086   7.649 7.35e-14 ***
GenderMale   0.0945352  0.0328613   2.877  0.00415 ** 
SmokeYes    -0.1332112  0.0571079  -2.333  0.01997 *  
Age          0.0694646  0.0091089   7.626 8.66e-14 ***
---
Signif. codes:  0 '***' 0.001 '**' 0.01 '*' 0.05 '.' 0.1 ' ' 1

Residual standard error: 0.3951 on 648 degrees of freedom
Multiple R-squared:  0.794,Adjusted R-squared:  0.7924 
F-statistic: 499.4 on 5 and 648 DF,  p-value: < 2.2e-16
```

The parameter estimates show that, on average, the forced expiratory volume increases 0.0695 liters per year, boys have a larger lung capacity than girls (on average 0.0945 liters higher), and that smoking reduces the lung capacity by 0.1332 liters. The parabola defined by the effect of height opens upward so not surprisingly the lung capacity increases with height (given the values of height available in the data frame).

Note that while it is possible to define interactions between quantitative explanatory variables in the model formulas in R, it may be quite difficult to interpret their meaning. Basically, the values of the two vectors are just multiplied together element-wise so it is indeed possible to combine "apples and oranges." In most situations, however, it makes more sense to model interactions between quantitative explanatory variables by categorizing the quantitative variables first and then to include the interaction between the categorized variables in the model (see Problem 2.27 for how to categorize a quantitative variable).

See also: Problems 3.33 and 4.21 cover model validation for linear normal models. The `biglm` function in package `biglm` can be used to fit linear models to large datasets with many cases.

3.10 FIT A PENALIZED REGRESSION MODEL

Problem: You want to fit a penalized regression model such as lasso, ridge regression, or elastic net to your data for simultaneous variable selection and shrinkage.

Solution: Penalized regression methods such as the lasso (least absolute shrinkage and selection operator), ridge regression, or elastic net algorithm, for example, are useful approaches for simultaneous variable selection and shrinkage estimation.

The lasso, ridge regression, and elastic net are shrinkage and/or selection methods which minimize the sum of squared errors (for linear models) or the normalized negative log-likelihood (for general models) with a penalty on the size of the estimated coefficients. In practice, the penalty has the form

$$\lambda\left[(1-\alpha)||\beta||_2^2 + \alpha||\beta||_1\right],$$

where $\alpha \in [0, 1]$ is the elastic net mixing parameter that mixes between the lasso ($\alpha = 1$) and ridge regression ($\alpha = 0$), and $\lambda \geq 0$ is the penalty parameter.

Penalized regression methods are particularly — but not only — useful in situations where the number of variables is much larger than the number of observations in the dataset (popularly referred to as the $n \ll p$ situation) since the variable selection and shrinkage allows all variables to be included and considered in the model simultaneously.

The `glmnet` package implements various variants of the lasso method in the `glmnet` function. The first two arguments, `x` and `y`, are the (numeric) design matrix and the outcome vector, respectively. By default, the columns in the design matrix `x` are standardized (i.e., have mean zero and variance 1), but this default behavior can be changed by setting the `standardize` argument to `FALSE`. The coefficients are always returned on the original scale.

The `family` argument determines the family of models as for `glm` and defaults to a model with Gaussian errors. `alpha` is the elastic net mixture parameter and sets the penalty mixture between the two different penalties and should be between 0 and 1 (0 is ridge and 1 is lasso), and it defaults to 1.

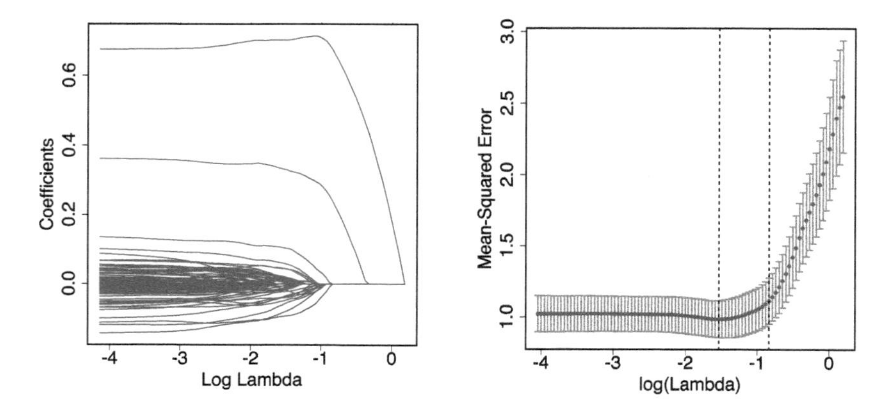

Figure 3.6: Left panel: each curve represents a variable and its corresponding coefficient for different values of the penalty parameter (no penalty towards the left and high penalty towards the right of the graph). Right panel: the mean-squared cross-validation error for different values of the penalty parameter.

In the example below we simulate 100 observation with 100000 predictors each. Predictors 1 and 2 are the only ones to have a true effect on the outcome. The results can be plotted using `plot`, where the `xvar` argument can be used to determine what is shown on the x axis. The default is the value of the L1 Norm (the $\|\beta\|_1$ value above), while `xvar="lambda"` uses the $\log(\lambda)$ value, and `xvar="dev"` plots the percentage of deviance explained. The plot using the $\log(\lambda)$ values is shown in the left-hand plot of Figure 3.6, which shows that there are two variables that have non-zero coefficients even for relatively large lambda values whereas the remaining 99998 predictors are only included with a non-zero coefficient in the model when the penalty becomes small. Thus an interesting $\log(\lambda)$ might be somewhere around -1.

```
> library(glmnet)
> set.seed(12345)     # Set seed to keep results reproducible
> x <- matrix(rnorm(100*100000), ncol=100000)
> y <- rnorm(100, mean=x[,1] + .5*x[,2])
> result <- glmnet(x, y)
> plot(result, xvar="lambda", col="blue")
```

Printing the object returned from `glmnet` shows a matrix that lists the percentage of the deviance explained for different values of λ (output not shown here). The estimated coefficients for a given value of λ are extracted with the `coef` function, where the `s` argument specifies the desired value of the penalty λ. The `coef` function also accepts the `exact` argument which — when set to `TRUE` — ensures that exact values of λ are used when the estimated coefficients are returned. Otherwise an approximation based on linear interpolation is made.

Below we return the estimated coefficients for $\lambda = \exp(-1)$ based on the plot in Figure 3.6. Because of the large number of predictors we only return the first handful.

```
> head(coef(result, s=exp(-1)))
6 x 1 sparse Matrix of class "dgCMatrix"
                     1
(Intercept) 0.05222911
V1          0.70778762
V2          0.28505182
V3          .
V4          .
V5          .
```

The first two estimated coefficients and the intercept are all non-zero when the penalty is $\lambda = \exp(-1)$ but the vast majority of coefficients are set to zero (represented by a '.' in the output). An intercept is always included in the fitted model unless the argument `intercept=FALSE` is set.

The choice of λ can be decided using cross-validation instead of being decided subjectively from the plot. `cv.glmnet` automates the process of N-fold cross-validation of the penalized regression model. The first two arguments to `cv.glmnet` should be the design matrix `x` and a response vector `y` just as in `glmnet`. Other arguments from `glmnet` can also be used, and the `nfold` argument controls the number of folds, N, used for cross-validation.

`cv.glmnet` returns information from the cross-validation based on the fitted models of the N different subsets of the data. We plot the result to see the cross-validation error as a function of the penalty parameter, λ. The output is seen in the right-hand plot of Figure 3.6 and it shows the cross-validation mean-squared error curve (the blue dotted line) as well as the upper and lower standard deviation bars for different values of λ.

```
> cvres <- cv.glmnet(x, y)
> plot(cvres)
```

The minimum mean-squared error is achieved with a λ around $\exp(-1.5) = 0.22$. The two vertical dashed lines represent the minimum cross-validation error (the left-most vertical line) and the largest λ with a cross-validation error that is within one standard error of the minimum error (the right-most vertical line). The values for λ corresponding to the two lines shown in the plot can be extracted directly from the cross-validated result.

```
> cvres$lambda.min
[1] 0.2154956
> cvres$lambda.1se
[1] 0.4329808
```

Since the smallest mean-squared error is found for $\lambda = 0.215$ we can extract the coefficients from the original model fit when using this penalty by supplying the value to the `s` argument when calling `coef`. When using the `coef` function on the cross-validated object we can also use the string `lambda.min` or `lambda.1se` as values for `s` to use the corresponding values.

```
> head(coef(cvres, s = "lambda.min"))
6 x 1 sparse Matrix of class "dgCMatrix"
                     1
(Intercept) 0.04396796
V1          0.70149526
V2          0.33180747
V3          .
V4          .
V5          .
```

The `predict` function can be used to predict the expected outcomes based on the estimated coefficients from a given penalty. For example, to compute the expected outcome for individuals that have the same set of variables as the first three observations in the dataset we call

```
> head(predict(cvres, newx = x[1:3,], s = "lambda.min"))
              1
[1,]  0.5532949
[2,] -0.2506483
[3,]  0.6237499
```

where the predictors used are supplied in the argument `newx`.

GENERALIZED LINEAR MODELS

3.11 FIT A LOGISTIC REGRESSION MODEL

Problem: You want to fit a logistic regression model to describe the relationship between a binary response variable and a set of explanatory variables.

Solution: Logistic regression models are appropriate for binary outcomes; i.e., when the response variable is categorical with exactly two categories (typically denoted "success" and "failure"). Logistic regression models the probability associated with one of the response categories as a function of one or more explanatory variables.

Logistic regression models are part of the class of generalized linear models which can be fitted in R using the `glm` function. `glm` takes a model formula as first argument and needs the `family=binomial` argument to ensure that the logistic regression model is used. The model formula for logistic regression models can include the response variable in two different ways: either as a factor or numeric vector with exactly two categories, or as a matrix with two columns, where the first column denotes the number of successes and the second column is the number of failures. We will focus on the first situation here.

The logistic regression model formula in `glm` can be specified in the same manner as for `lm` when we have a dataset with exactly one observation per row. The response variable should be a factor, and the first level of this factor will denote failures for the binomial distribution while all other levels will denote successes. If the response variable is numeric, it will be converted to a factor, so a numeric vector of zeros and ones will automatically work as a response variable where "1" is considered success.

Below, we wish to use the `birthwt` data from the `MASS` package to examine if low birth weight is influenced by the mother's race, mother's weight (in pounds), and/or the mother's smoking status. The `summary` function prints the parameter estimates.

```
> library(MASS)
> data(birthwt)
> model <- glm(low ~ factor(race) + smoke + lwt, family=binomial,
+              data=birthwt)
> summary(model)

Call:
glm(formula = low ~ factor(race) + smoke + lwt, family = binomial,
```

```
    data = birthwt)

Deviance Residuals:
    Min       1Q   Median       3Q      Max
-1.5278  -0.9053  -0.5863   1.2878   2.0364

Coefficients:
               Estimate Std. Error z value Pr(>|z|)
(Intercept)    -0.10922    0.88211  -0.124  0.90146
factor(race)2   1.29009    0.51087   2.525  0.01156 *
factor(race)3   0.97052    0.41224   2.354  0.01856 *
smoke           1.06001    0.37832   2.802  0.00508 **
lwt            -0.01326    0.00631  -2.101  0.03562 *
---
Signif. codes:  0 '***' 0.001 '**' 0.01 '*' 0.05 '.' 0.1 ' ' 1

(Dispersion parameter for binomial family taken to be 1)

    Null deviance: 234.67  on 188  degrees of freedom
Residual deviance: 215.01  on 184  degrees of freedom
AIC: 225.01

Number of Fisher Scoring iterations: 4
```

The parameter estimates listed with the `summary` function show that the effect of smoking status is 1.06, which is — since `glm` uses the logit link function by default — the estimated log odds ratio of low birth weight for smokers (`smoke=1`) compared to non-smokers (`smoke=0`). Likewise, race 2 has an estimated log odds ratio of 1.29 for low birth weight compared to the reference race (which is race 1). The exponential function, `exp`, can be used to transform the log odds ratio estimates back to odds ratios. Thus the odds ratio of low birth weight for smokers compared to non-smokers is $\exp(1.06) = 2.886$ with a corresponding 95% confidence interval for the odds ratio given by

```
> exp(1.060 +  c(-1,1)*1.96*0.3783)     # 95% CI for smoking
[1] 1.375108 6.058535
```

The last two columns in the output table from `summary` show the Wald test statistic and corresponding p-value for the hypothesis of testing each parameter equal to zero.

The `drop1` function (with argument `test="Chisq"` to use the χ^2-distribution for likelihood ratio model reductions) can be used to obtain the model reduction tests while still preserving the hierarchical principle.

```
> drop1(model, test="Chisq")
Single term deletions

Model:
low ~ factor(race) + smoke + lwt
             Df Deviance    AIC    LRT Pr(>Chi)
<none>           215.01 225.01
factor(race)  2   224.34 230.34 9.3260 0.009438 **
smoke         1   223.26 231.26 8.2444 0.004088 **
lwt           1   219.97 227.97 4.9601 0.025939 *
---
Signif. codes:  0 '***' 0.001 '**' 0.01 '*' 0.05 '.' 0.1 ' ' 1
```

The results from the `drop1` function show that both race, mother's weight, and smoking status are significant (with p-value of 0.009438, 0.02594 and 0.004088, respectively).

Note that the manner in which the response variable is coded is important when interpreting the output. `glm` models the probability that successes occur so if we want to model the probability of failures we should change the sign of all parameter estimates.

By default, `glm` uses the logit link function for binomial data; i.e., it models the probability of success through the logarithm of the odds of success. Another possibility is to use the probit link function which is useful when it is reasonable to assume that each result of success or failure is actually the discretely observed outcome of a continuous underlying normal distribution. The probit link function can replace the default logit function by adding the link option to the family argument; i.e., `family=binomial(link="probit")`.

There is only one parameter in the binomial distribution — the probability of success — so once that parameter is estimated, then the variance is fully given. Sometimes, however, the mean may be correctly modeled but the observed variance is larger than the expected variance so there is overdispersion. Overdispersion can be caused by omission of important explanatory variables, correlation between binary responses, or a misspecified link function. The quasi-binomial family can be used with `glm` to allow for overdispersion, simply by using specifying `family=quasibinomial`.

In the example below we refit the birth weight data using the probit link function and allowing for overdispersion.

```
> model2 <- glm(low ~ factor(race) + smoke + lwt,
+               family=quasibinomial(link="probit"),
+               data=birthwt)
> summary(model2)

Call:
glm(formula = low ~ factor(race) + smoke + lwt,
    family = quasibinomial(link = "probit"), data = birthwt)

Deviance Residuals:
    Min       1Q   Median       3Q      Max
-1.5216  -0.9116  -0.5759   1.2838   2.0680

Coefficients:
               Estimate Std. Error t value Pr(>|t|)
(Intercept)   -0.072490   0.515706  -0.141   0.8884
factor(race)2  0.791295   0.304201   2.601   0.0100 *
factor(race)3  0.589826   0.241147   2.446   0.0154 *
smoke          0.652215   0.221789   2.941   0.0037 **
lwt           -0.008067   0.003643  -2.214   0.0280 *
---
Signif. codes:  0 '***' 0.001 '**' 0.01 '*' 0.05 '.' 0.1 ' ' 1

(Dispersion parameter for quasibinomial family taken to be 0.9897768)

    Null deviance: 234.67  on 188  degrees of freedom
Residual deviance: 214.53  on 184  degrees of freedom
AIC: NA

Number of Fisher Scoring iterations: 4
```

The overdispersion parameter of 0.9897 is very close to 1 which suggests no overdispersion in this case. While there is no difference in the structure of the final model (both race and smoke are still significant) there is still a discrepancy between the parameter estimates from the model with the logit link function (`model` above) and the model with the probit link function (`model2`). The estimates from the probit model represent changes in the z score for the cumulative standard normal distribution. When the estimates are transformed to probabilities using the inverse logit or the cumulative standard normal distribution, $\Phi(z)$, for the logit and probit link, respectively, then the probabilities of success are generally very similar except for probabilities very close to zero or one. The probability of low birth weight for a child born to a mother who smokes, has race 1, and weighs 120 pounds is calculated for the two models as:

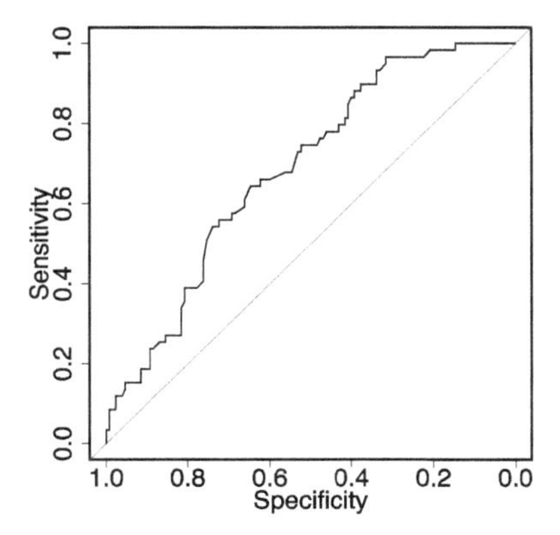

Figure 3.7: ROC curve based on a fit of a logistic regression model for low birth weight. The area under the ROC curve is 0.6855.

```
> exp(-0.10922 + 1.06001 + 120*-0.01326) /
+     (1 + exp(-0.10922 + 1.06001 + 120*-0.01326)) # Logit
[1] 0.3451539
> pnorm(-0.072490 + 0.652215 + 120*-0.008067)      # Probit
[1] 0.3488915
```

Both models provide virtually the same result.

Receiver-operating characteristic (ROC) curves are an excellent way to compare diagnostic tests. The ROC curve plots the true positive rate (sensitivity) against the false positive rate (1 - specificity). Each point on the ROC curve represents a sensitivity/specificity pair corresponding to a particular decision threshold, and a test has better accuracy the closer it is to the upper left corner of the plot. The area under the ROC curve quantifies the overall ability of the test to discriminate between the individuals who have and do not have the disease. A perfect test with no false positives and no false negatives has an area of 1.00, while a useless test has an area of 0.5.

There are several packages to plot ROC curves in R and here we will focus on the `pROC` package. The `roc` function accepts a simple model formula where there is just one predictor on the right-hand side (if multiple predictors are passed then the `roc` function is run once for each predictor). Below we extract the prediction probabilities from the full model and pass those to the predictor in the `roc` function. The resulting ROC curve is shown in Figure 3.7 and the area under the ROC curve is 0.6855.

```
> library(pROC)
> pred <- predict(model, type=c("response"))
> roccurve <- roc(low ~ pred, data=birthwt)
> plot(roccurve)

Call:
roc.formula(formula = low ~ pred, data = birthwt)

Data: pred in 130 controls (low 0) < 59 cases (low 1).
Area under the curve: 0.6855
```

See also: The `elrm` package can be used for exact logistic regression modeling, which might be applicable when the sample size is small or when some categorical predictors have levels that result in only one type of outcome. The help page for `help(family)` lists other available link functions for logistic regression, including the complementary log-log.

3.12 FIT A CONDITIONAL LOGISTIC REGRESSION MODEL

Problem: You want to fit a conditional logistic regression model to describe the relationship between a binary response variable and a set of explanatory variables.

Solution: Conditional logistic regression is an alternative to ordinary logistic regression (see Problem 3.11) when there are matched or clustered data so it is possible to use subjects (or their matches) as their own controls. Just like ordinary logistic regression, conditional logistic regression models are appropriate for binary outcomes; i.e., when the response variable is categorical with exactly two categories (typically denoted "success" and "failure"), and the probability associated with one of the response categories is modeled as a function of one or more explanatory variables.

Conditional logistic regression models can be fitted with the `clogit` function from the `survival` package. `clogit` takes a model formula as first argument where the response variable is a logical vector, and it uses the `strata` function inside the model formula to identify the observations that are linked. By default, `clogit` uses exact calculations when the parameters are estimated but that may be time-consuming for larger datasets. The `method` argument can be set to either `efron` or `breslow` to employ various approximations.

Below, we wish to use the `matched` data from the `MESS` package to analyze if flu vaccination prevents hospitalization for flu (i.e., reduces the number of severe cases). One-hundred and fifty cases — all aged 65 and older — were sampled and each were matched with two controls on sex and age. Medical records were used to determine whether cases and controls had received a flu vaccine shot and whether they suffered from underlying lung disease since lung disease could be a confounder. In the formula below we test if the patient is a case to obtain a logical vector as outcome and include `id` in `strata` to account for each matched trio.

```
> library(survival)
> library(MESS)
> data(matched)
> model <- clogit(iscase=="Case" ~ vaccine + lung + strata(id),
+                 data=matched)
> summary(model)
Call:
coxph(formula = Surv(rep(1, 450L), iscase == "Case") ~ vaccine +
    lung + strata(id), data = matched, method = "exact")

  n= 450, number of events= 150

                     coef exp(coef) se(coef)      z Pr(>|z|)
vaccineVaccinated -0.4008    0.6698   0.2233 -1.795   0.0726
lungDisease        1.3053    3.6889   0.2348  5.558 2.72e-08

vaccineVaccinated .
lungDisease       ***
---
Signif. codes:  0 '***' 0.001 '**' 0.01 '*' 0.05 '.' 0.1 ' ' 1

                  exp(coef) exp(-coef) lower .95 upper .95
vaccineVaccinated    0.6698     1.4930    0.4324     1.038
lungDisease          3.6889     0.2711    2.3281     5.845

Rsquare= 0.077   (max possible= 0.519 )
Likelihood ratio test= 36.28  on 2 df,   p=1.324e-08
Wald test            = 32.59  on 2 df,   p=8.36e-08
Score (logrank) test = 37.28  on 2 df,   p=8.026e-09
```

Conditional logistic regression uses the logit transformation so any estimated parameters will be log odds. The parameter estimates listed with the `summary` function show that the vaccine increases the odds of hospitalization with a factor 0.6698 (i.e., it becomes smaller) which means that the vaccine appears to be beneficial although the effect is not statistically significant ($p = 0.0726$). The 95% confidence interval for odds ratio of vaccine can be read off the output and is $(0.4324; 1.038)$. Lung disease is highly significant but we already knew that underlying lung disease was a risk factor so that result is not that surprising. We can also see that the odds for hospitalization are 3.69 times larger for patients with lung disease compared to patients without lung disease.

Note that unlike ordinary logistic regression there is no intercept in the summary output. This is a consequence of conditioning on the clusters — we are comparing effects within clusters and here the intercept cancels out and plays no role in the estimation (and thus cannot be estimated).

The drop1 function (with argument test="Chisq" to use the χ^2-distribution for likelihood ratio model reductions) can be used to obtain the model reduction tests while still preserving the hierarchical principle.

```
> drop1(model, test="Chisq")
Single term deletions

Model:
Surv(rep(1, 450L), iscase == "Case") ~ vaccine + lung + strata(id)
           Df    AIC    LRT  Pr(>Chi)
<none>        297.30
vaccine     1 298.61   3.31   0.06887 .
lung        1 329.05  33.75 6.268e-09 ***
strata(id)  0 533.14 235.84
---
Signif. codes:  0 '***' 0.001 '**' 0.01 '*' 0.05 '.' 0.1 ' ' 1
```

The results from the likelihood-ratio tests from the drop1 function are similar to the output from the Wald tests printed by summary, and we still have that vaccine is borderline significant and that lung disease is highly significant.

3.13 FIT AN ORDINAL LOGISTIC REGRESSION MODEL

Problem: You want to fit an ordinal logistic regression model to describe the relationship between an ordinal response variable and a set of explanatory variables.

Solution: Logistic regression models can be extended to handle response variables with more than two categories as shown in Problem 3.14. Here we assume that the response categories are ordered (i.e., when the response categories are ordinal) and we want to model how different explanatory variables influence the probability of observing the different polytomous outcomes.

Ordinal data occur when it is possible to order and rank the observations but when the distance between the categories is unknown. Examples of ordered variables include disease severity (least severe to most severe or disease stages), pain scales, survey responses (from strongly disagree to strongly agree) and classifications of continuous variables.

The proportional odds model is an example of a model for ordinal response data where the explanatory variables have a unified effect on all response categories. The ordinal nature of the response is preserved by considering the odds of a response category *or a lesser category* so we are essentially looking at cumulative probabilities for the ordered response categories. The major advantage of the proportional odds model lies in the interpretation of the effect of explanatory variables: The change in the cumulative odds for any response category is the same regardless of which response category we consider.

Several implementations of ordinal regression models exist in R. Here we will focus on the `clm` function (found in the `ordinal` package) which implements proportional odds logistic (and probit) regression modeling. `clm` expects a model formula with an ordered factor as response variable. A non-ordered response variable is converted to an ordered factor where the ordering is given by the ordering of the response factor levels.

In the example below we use the `icecreamads` dataset from the MESS package to model ice cream consumption as a function of advertising type, temperature, and price. The consumption is scored in three categories (low, medium, and high) and we first use the `ordered` function on the response variable to ensure that R uses the correct ordering.

```
> library(ordinal)
> library(MESS)
> data(icecreamads)
> resp <- ordered(icecreamads$Consumption)
> model <- clm(Consumption ~ Temperature + Price + Advertise,
+              data=icecreamads)
> summary(model)
formula: Consumption ~ Temperature + Price + Advertise
data:    icecreamads

 link  threshold nobs logLik AIC   niter max.grad cond.H
 logit flexible  30   -20.20 52.40 6(0)  4.98e-11 4.6e+07

Coefficients:
                    Estimate Std. Error z value Pr(>|z|)
Temperature          0.14727    0.04001   3.681 0.000232 ***
Price               -6.11008   51.27316  -0.119 0.905143
```

```
Advertiseradio         0.27664    1.01090   0.274 0.784347
Advertisetelevision  0.06982    1.05090   0.066 0.947028
---
Signif. codes:  0 '***' 0.001 '**' 0.01 '*' 0.05 '.' 0.1 ' ' 1

Threshold coefficients:
               Estimate Std. Error z value
1_low|2_medium    3.560     14.400   0.247
2_medium|3_high   6.941     14.419   0.481
```

The estimates from the `summary` output are split into two sections. The effect of the explanatory variables are listed in the "Coefficients" section and represent logarithms of odds ratios — just as for logistic regression. For any given consumption level, we have that each increase in temperature by 1 degree Fahrenheit will result in increased odds of a higher category (i.e., more consumption) since the estimate 0.1473 is positive.

Thus, for a given category the odds of being placed in a higher (more consumption) category increase by a factor $\exp(0.1473) = 1.159$ for each extra degree. A 95% confidence interval for, say, the odds ratio between the cumulative probabilities for two adverts that differ only in temperature by 10 degrees Fahrenheit is

```
> exp(10*0.1473 + 10*qnorm(.975)*c(-1,1)*0.04001)
[1] 1.991361 9.556118
```

Hence, we are 95% confident that the interval from $[1.991; 9.556]$ contains the true odds that the consumption will be larger than a given category if the temperature is increased by 10 degrees.

The section of results titled "Thresholds" gives the intercepts for all but the last response category. Thus, the estimated probability that, say, a poster advert will result in low consumption when the temperature is zero degrees and the standardized price is zero (the reference level here of having cheap ice cream and a freezing temperature is somewhat outside the normal range of values for these variables) is

```
> exp(3.56)/(1 + exp(3.56))
[1] 0.9723476
```

If we wish to calculate the probability that, say, an advertising campaign run during a period with 50-degree temperatures with a standardized ice cream price of 0.28 would result in medium consumption then we should keep in mind that we are modeling cumulative probabilities so we have to *subtract* the probability of the first category from the probability of "medium consumption or less" to obtain the probability of "medium consumption."

```
> exp(6.941-(0.147*50-6.11*0.28))/(1+exp(6.941-(0.147*50-6.11*0.28))) -
+ exp(3.560-(0.147*50-6.11*0.28))/(1+exp(3.560-(0.147*50-6.11*0.28)))
[1] 0.6750028
```

Note here, that it is not a typo that there is a negative sign before the coefficients for the explanatory variables. Recall that we are modeling cumulative probabilities, so an increase in cumulative probability means that it is more likely to observe a *lower* response category/score. The change of sign (which normally is positive for the other models we consider) for the explanatory variable(s) is due to the parameterization and because we prefer to have positive coefficients indicate an association with higher categories/scores. Thus, a positive estimate means that higher response categories are more likely and a negative estimate means that lower response categories are more likely. In this case the coefficient for pulse is negative, so individuals with a higher pulse are more likely to have more frequent exercise status.

The probabilities of the individual categories can also be obtained by the `predict` function with option `type="prob"`. It returns a list and the `fit` element shows the estimated prediction probabilities.

```
> head(predict(model, type="prob")$fit)
[1] 0.8086686 0.5979040 0.6317847 0.7612425 0.1418436 0.7991064
```

By default, `clm` uses the logit link function for binomial data; i.e., it models the probability of success through the logarithm of the odds of success. Another possibility is to use the probit link function which is useful when it is reasonable to assume that each result of success or failure is actually the discretely observed outcome of a continuous underlying normal distribution of probabilities. The probit link function can be used instead of the default logit function by setting the argument `method="probit"`. Alternatively, `clm` allows for a complementary log-log or a Cauchy latent variable links with the `method="cloglog"` or `method="cauchit"` arguments, respectively.

The functions `summary` and `drop1` (with argument `test="Chisq"` to use the χ^2-distribution for likelihood ratio model reductions) can be used to obtain the parameter estimates and model reduction tests. Above we see

```
> drop1(model, test="Chisq")
Single term deletions

Model:
Consumption ~ Temperature + Price + Advertise
            Df    AIC     LRT  Pr(>Chi)
<none>         52.403
Temperature  1 74.712 24.3096 8.203e-07 ***
Price        1 50.417  0.0142    0.9051
Advertise    2 48.484  0.0812    0.9602
---
Signif. codes:  0 '***' 0.001 '**' 0.01 '*' 0.05 '.' 0.1 ' ' 1
```

Only temperature is significant. The type of advertising campaign and price has no substantial effect on the consumption of ice cream. When the temperature increases the consumption goes up regardless.

One of the assumptions underlying ordinal logistic (and ordinal probit) regression is that the relationship between each pair of outcome groups is the same. Thus, the proportional odds assumption should be checked before the model is used. We check this assumption by first fitting a multinomial regression model (see Problem 3.14) and then we compare the deviance differences between the two models. Under the null hypothesis that the ordinal regression model is correct, the deviance difference approximately follows a χ^2-distribution with a number of degrees of freedom corresponding to the difference in number of parameters between the two models. Deviances are extracted from model fits using the `deviance` function (for results from the `multinom` function), and as twice the negative log-likelihood for the `clm` results, and the effective number of degrees of freedom are found for both models as the `edf` element.

```
> library(nnet)
> multi <- multinom(Consumption ~ Price + Advertise + Temperature,
+                   data=icecreamads)
# weights:  18 (10 variable)
initial  value 32.958369
iter  10 value 20.039731
iter  20 value 19.989620
iter  30 value 19.955271
iter  40 value 19.886545
iter  50 value 19.783157
iter  60 value 19.782853
iter  70 value 19.780696
iter  80 value 19.780481
iter  90 value 19.779109
```

```
iter 100 value 19.778951
final  value 19.778951
stopped after 100 iterations
> 1-pchisq(-2*as.numeric(logLik(model)) - deviance(multi),
+          df=multi$edf - model$edf)
[1] 0.9323418
```

Hence we fail to reject the proportional odds regression model ($p = 0.932$) which means it makes sense to use the proportional odds model to analyze the ordinal response variable.

See also: The `vglm` function from the `VGAM` package can fit ordinal logistic models (with option `family=propodds`) using vector generalized linear models. The `ordinal` package also handles mixed effect ordinal regression with a similar syntax to `lme4` (see Problem 3.16)

3.14 FIT A MULTINOMIAL LOGISTIC REGRESSION MODEL

Problem: You want to fit a multinomial logit regression model to describe the relationship between a polytomous categorical response variable and a set of explanatory variables.

Solution: Logistic regression can be extended to handle response variables that have more than two categories. Here we will assume that the response is nominal (consists of unordered categories) — see Rule 3.13 for ordered categories (i.e., when the response categories are ordinal) — and we want to model how a set of explanatory variables influences the probability of observing the different polytomous outcomes.

Multinomial logistic regression models can be fitted in R using neural networks which are implemented by the `nnet` package. The `multinom` function handles multinomial logistic regression models and the function takes a model formula as first argument. The model formula for multinomial logistic regression can include the response variable in two different ways: either as a factor with k categories (where the first category will be the reference level) or as a matrix with k columns where the elements in the matrix are interpreted as counts for the different categories. We will focus on the first situation here.

The `alligator` data frame from the `isdals` package contains information on the length and primary food choice ("fish", "invertebrates", or "other") of 59 alligators. It is of interest to examine how the primary food choice depends on the size of the alligators.

```
> library(nnet)
> library(isdals)
> data(alligator)
> head(alligator)
  length          food
1   1.24 Invertebrates
2   1.32          Fish
3   1.32          Fish
4   1.40          Fish
5   1.42 Invertebrates
6   1.42          Fish
> model <- multinom(food ~ length, data=alligator)
# weights:  9 (4 variable)
initial  value 64.818125
iter  10 value 49.170710
final  value 49.170622
converged
> summary(model)
Call:
multinom(formula = food ~ length, data = alligator)

Coefficients:
              (Intercept)     length
Invertebrates    4.079701 -2.3553303
Other           -1.617713  0.1101012

Std. Errors:
              (Intercept)    length
Invertebrates    1.468640 0.8032870
Other            1.307274 0.5170823

Residual Deviance: 98.34124
AIC: 106.3412
```

Presenting the results from a multinomial logistic regression model can be somewhat tricky since there are multiple equations and multiple comparisons to present. The estimates printed by the `summary` function are log-odds values relative to the reference outcome ("fish" in this situation), so the estimates refer to the change in the log odds of the outcome relative to the reference outcome associated with a change in the explanatory variable.

For example, the estimate -2.355 for length of invertebrates says that the change in log odds of eating invertebrates relative to eating fish changes by -2.355 for each increase in alligator length of 1 meter. The estimate is *not* the overall change in odds of eating invertebrates for two alligators who differ by a length of 1 meter. Likewise, the odds ratio that

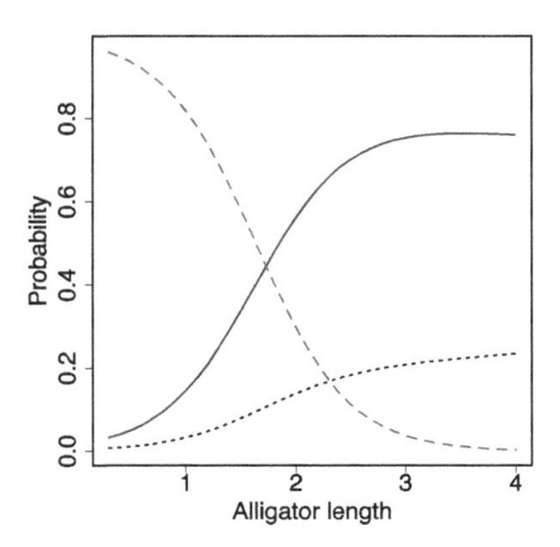

Figure 3.8: Probabilities of primary food preference for alligators of different lengths. Solid, dashed, and dotted lines correspond to "fish", "invertebrates", and "other", respectively.

an alligator of, say, length 2 meters prefers "invertebrates" to "other" is $\exp((4.080 - 2 \cdot 2.355) - (-1.618 + 2 \cdot 0.110)) = 2.152$.

The `predict` prints out the most probable class or the probabilities for the individuals classes (when argument `response="probs"`) based on the fitted model.

```
> head(predict(model, response="probs"))
[1] Invertebrates Invertebrates Invertebrates Invertebrates
[5] Invertebrates Invertebrates
Levels: Fish Invertebrates Other
```

The first couple of observations clearly have a substantial preference for eating invertebrates. Since we only have a single quantitative explanatory variable in this model, we can illustrate how the probabilities of the three response groups change by alligator length. This is done below where we again use `predict` to estimate the probabilities of the different responses under the model. The result is shown in Figure 3.8 where we see that the probabilities of the three responses sum to one for all lengths.

```
> len <- seq(0.3, 4, .1)
> matplot(len, predict(model, newdata=data.frame(length=len),
+                      type="probs"),
+         type="l", xlab="Alligator length", ylab="Probability")
```

The `anova` function can be used for model reduction for multinomial logistic regression models by comparing the fit of two nested models. If we wish to test the hypothesis that length has no influence on the preference of food then we compare the full model to a model where `length` has been removed.

```
> null <- multinom(food ~ 1, data=alligator)
# weights:  6 (2 variable)
initial  value 64.818125
iter  10 value 57.570928
iter  10 value 57.570928
iter  10 value 57.570928
final  value 57.570928
converged
> anova(null, model)                    # Compare the two models
Likelihood ratio tests of Multinomial Models

Response: food
   Model Resid. df Resid. Dev   Test    Df LR stat.
1      1       116  115.14186
2 length       114   98.34124 1 vs 2     2 16.80061
       Pr(Chi)
1
2 0.0002247985
```

The likelihood ratio test statistic for comparing the model where alligator length is included to the model where alligator length is excluded yields a p-value of 0.000224 so there is a clear effect of alligator size on primary food choice.

See also: The `vglm` function from the `VGAM` package can fit multinomial logistic models (with option `family=multinomial`) using vector generalized linear models. Also, the `mlogit` package can be used for multinomial logistic regression analysis, and it includes extra functionality for analyzing mixed-multinomial logistic regression models.

3.15 FIT A POISSON REGRESSION MODEL

Problem: You want to fit a Poisson regression model to describe the relationship between a count response variable and a set of explanatory variables.

Solution: Poisson regression models can be used to deal with situations where we wish to model count data; e.g., when the response variable is obtained by counting the number of occurrences of an event.

Poisson regression models are part of the class of generalized linear models which can be fitted using the `glm` function in R if the argument `family=poisson` is specified. The first argument to `glm` is the model formula, where the variable corresponding to the response should be a vector of non-negative integers. `summary` prints the estimates and standard errors from a fitted model.

The `bees` dataset from the `MESS` package contains information on the number of bees caught in different-colored cages at four different locations for two types of bees. It is of interest to see if either of the two types of bees has a preference for the color of the cages. For brevity of output, we model the number of caught bees with a main effect of location, and an interaction between color and bee type although the data allow for a full three-way interaction.

```
> library(MESS)
> data(bees)
> beedata <- subset(bees, Time=="july1") # Look at one day only
> head(beedata)
    Locality Replicate  Color  Time       Type Number id
1  Havreholm         A  White july1 Bumblebees      1  1
4  Havreholm         A Yellow july1 Bumblebees      2  1
7  Havreholm         A   Blue july1 Bumblebees      0  1
10 Havreholm         A  White july1   Solitary      1  1
13 Havreholm         A Yellow july1   Solitary      4  1
16 Havreholm         A   Blue july1   Solitary      3  1
> model <- glm(Number ~ Locality + Type*Color,
+              family=poisson, data=beedata)
> summary(model)

Call:
glm(formula = Number ~ Locality + Type * Color, family = poisson,
    data = beedata)

Deviance Residuals:
    Min       1Q   Median       3Q      Max
-2.1172  -0.9826  -0.6551   0.6573   2.3147

Coefficients:
                        Estimate Std. Error z value Pr(>|z|)
(Intercept)              -1.0647     0.7147  -1.490 0.136268
LocalityKragevig         -0.4745     0.2407  -1.971 0.048706
LocalitySaltrup          -1.8608     0.4063  -4.580 4.66e-06
LocalitySvaerdborg       -1.8608     0.4063  -4.580 4.66e-06
TypeSolitary              2.5649     0.7338   3.495 0.000473
ColorWhite                1.3863     0.7906   1.754 0.079509
ColorYellow               1.8718     0.7596   2.464 0.013726
TypeSolitary:ColorWhite  -1.7540     0.8479  -2.069 0.038589
```

```
TypeSolitary:ColorYellow  -2.1342     0.8157  -2.616 0.008888

(Intercept)
LocalityKragevig         *
LocalitySaltrup          ***
LocalitySvaerdborg       ***
TypeSolitary             ***
ColorWhite               .
ColorYellow              *
TypeSolitary:ColorWhite  *
TypeSolitary:ColorYellow **
---
Signif. codes:  0 '***' 0.001 '**' 0.01 '*' 0.05 '.' 0.1 ' ' 1

(Dispersion parameter for poisson family taken to be 1)

    Null deviance: 162.785  on 71  degrees of freedom
Residual deviance:  84.215  on 63  degrees of freedom
AIC: 195.42

Number of Fisher Scoring iterations: 6
```

The log link is the default link function for Poisson data in `glm` so it models the natural logarithm of the expected counts. The parameter estimates for a Poisson regression model with log link have relative risk interpretations: the estimated effect of an explanatory variable is multiplicative on the rate, and thus leads to a risk ratio or relative risk. The parameter estimates listed with the `summary` function are relative to the reference group which consists of bumblebees and blue cages and the location `Havreholm`. For example, the relative risk of observing a solitary bee in a yellow cage relative to a bumblebee in a blue cage at the same location is

```
> exp(2.5649 + 1.8718 - 2.1342)
[1] 9.999149
```

The estimates also show that, apparently, bumblebees prefer yellow and white cages to blue, while solitary bees appear to be slightly more indifferent to cage color.

The last two columns in the `summary` output show the Wald test statistic and corresponding p-value for the hypothesis of testing each parameter equal to zero. The `drop1` function (with argument `test="Chisq"` to use the χ^2-distribution for likelihood ratio model reductions) can be used for model reduction tests.

```
> drop1(model, test="Chisq")
Single term deletions

Model:
Number ~ Locality + Type * Color
           Df Deviance    AIC    LRT  Pr(>Chi)
<none>          84.215 195.43
Locality    3  132.051 237.26 47.836 2.308e-10 ***
Type:Color  2   93.956 201.16  9.740  0.007672 **
---
Signif. codes:  0 '***' 0.001 '**' 0.01 '*' 0.05 '.' 0.1 ' ' 1
```

The results from the `drop1` function show that there is an interaction between bee type and cage color (the likelihood ratio test yields a p-value of 0.007672) and that location is extremely significant (the p-value is virtually zero).

The log link is the default link function for Poisson data in `glm` so it models the natural logarithm of the expected counts. Two alternate link functions are available for Poisson models, namely `identity` and `sqrt`, and they are selected by adding the link option to the family argument, for example `family=poisson(link="identity")`.

Poisson regression models can also be used with rate data, where the rate is a count of events divided by the corresponding exposure. Different observational units may have different exposures (e.g., some counts are registered over time intervals of different lengths, or the counts could be based on groups of different sizes) so for rate data, an offset variable is included in the model to represent some measure of exposure. An offset is a variable that is forced to have a regression coefficient of 1, and for the log link function the offset contains the (natural) logarithm of the exposure. The offset is included in the model by adding an `offset(log(exposure))` term to the explanatory variable(s) in the model formula (where `exposure` is the name of the exposure variable).

For the Poisson distribution, the variance is fully determined by the mean so once that parameter is estimated then the variance is given. Sometimes, however, the data are more variable than the Poisson distribution predicts and we say that there is overdispersion. Overdispersion can be caused by omission of important explanatory variables, correlation between responses, or a misspecified link function. The quasi-Poisson family can be used with the `glm` function to allow for overdispersion, simply by using specifying `family=quasipoisson`.

Below we refit the Poisson model for the `bees` data allowing for overdispersion.

```
> model2 <- glm(Number ~ Locality + Type*Color,
+               family=quasipoisson, data=beedata)
> drop1(model2, test="Chisq")
Single term deletions

Model:
Number ~ Locality + Type * Color
           Df Deviance scaled dev.  Pr(>Chi)
<none>          84.215
Locality    3  132.051      33.388 2.668e-07 ***
Type:Color  2   93.956       6.798    0.0334 *
---
Signif. codes:  0 '***' 0.001 '**' 0.01 '*' 0.05 '.' 0.1 ' ' 1
> summary(model2)

Call:
glm(formula = Number ~ Locality + Type * Color, family = quasipoisson,
    data = beedata)

Deviance Residuals:
    Min       1Q   Median       3Q      Max
-2.1172  -0.9826  -0.6551   0.6573   2.3147

Coefficients:
                         Estimate Std. Error t value Pr(>|t|)
(Intercept)               -1.0647     0.8554  -1.245 0.217863
LocalityKragevig          -0.4745     0.2881  -1.647 0.104580
LocalitySaltrup           -1.8608     0.4863  -3.826 0.000302
LocalitySvaerdborg        -1.8608     0.4863  -3.826 0.000302
TypeSolitary               2.5649     0.8783   2.920 0.004848
ColorWhite                 1.3863     0.9463   1.465 0.147898
ColorYellow                1.8718     0.9092   2.059 0.043651
TypeSolitary:ColorWhite   -1.7540     1.0150  -1.728 0.088859
TypeSolitary:ColorYellow  -2.1342     0.9764  -2.186 0.032555

(Intercept)
LocalityKragevig
LocalitySaltrup          ***
LocalitySvaerdborg       ***
TypeSolitary             **
ColorWhite
ColorYellow              *
TypeSolitary:ColorWhite  .
TypeSolitary:ColorYellow *
---
Signif. codes:  0 '***' 0.001 '**' 0.01 '*' 0.05 '.' 0.1 ' ' 1

(Dispersion parameter for quasipoisson family taken to be 1.43273)
```

```
    Null deviance: 162.785  on 71  degrees of freedom
Residual deviance:  84.215  on 63  degrees of freedom
AIC: NA

Number of Fisher Scoring iterations: 6
```

The overdispersion parameter is 1.433 which suggests some overdispersion in this situation. While there is no difference in the fixed effect parameters, we see from the output that the standard errors are larger now. This is also seen in the test for the hypothesis of no interaction between cage color and bee type where we now get a p-value of 0.0334 so the interaction is only borderline significant.

We can create confidence intervals for differences between categories as usual. For example, the 95% confidence interval for the relative risk between the "Kragevig" location and "Havreholm" is

```
> exp(-0.4745 + c(-1,1)*1.96*0.2881) # 95% CI Kragevig/Havreholm
[1] 0.353746 1.094367
```

Thus, we are 95% confident that the interval from 0.35 to 1.09 contains the true relative risk between the Kragevig and Havreholm locations.

METHODS FOR ANALYSIS OF REPEATED MEASUREMENTS

3.16 FIT A LINEAR MIXED-EFFECTS MODEL

Problem: You want to fit a linear mixed-effects model where some of the model terms are considered random effect.

Solution: Mixed models are statistical models that include not only traditional "fixed effects" terms as used in linear and generalized linear models to model the mean but also "random effect" terms. Random effects are essentially random variables and they enter the model differently from the fixed effects, and the interpretation of their effects is different from fixed effect parameters. Random effects are appropriate for representing extra variation and clustering — and hence to model dependent observations or introduce extra sources of variation around the population mean.

Interpretation of the random effects is different from the fixed effects. In particular, the random effects are assumed to be a random sample from a population such that any results can be generalized to the full population from which the sample was drawn. For example, if we have multiple measurements for each individual then we might not be interested in making conclusions for just these individuals, but instead consider the individuals to be a random sample of individuals from the population. For a categorical explanatory variable, the number of parameters in the model increases with the number of categories when the variable is included as a fixed effect, whereas it is fixed for random effects.

The `lmer` in the `lme4` package extends the `lm` function to fit linear mixed-effect models. The input syntax for `lmer` is almost identical to `lm` except that `lmer` uses a special notation for the random effect terms in the model formula. `lmer` also adds an `REML` argument which is `TRUE` by default and ensures that `lmer` computes restricted maximum-likelihood estimates. If `REML=FALSE` then maximum-likelihood estimates are produced.

Random effects are specified in the model formula by the vertical bar ("`|`") character. The grouping factor for the random effect term (typically just the name of a variable) is specified at the right of the vertical bar. The expression to the left of the vertical bar defines the model for the random effect that is generated for each level of the grouping factor. A simple random-effects model where there is one random intercept for each category of the variable is specified by setting the random effect term to `(1|variable)`; i.e., the expression to the left of the vertical bar corresponds to the normal formula for an intercept, `1`.

Several random effect terms can be included simply by adding more terms to the model. Nested effects are handled by including a random effects term for all levels of nesting. For example, if you have two variables, `city` and `country`, where city is nested within country then `(1|city) + (1|country)` should be added to the model. Correlated random effects can be specified by changing the expression to the left of the vertical bar. For example, if `x` is a categorical variable then the random effects term `(1+x|country)` allows for a random effect for each country and allows for a correlation between the random effects for the levels of `x`.

In the following we use the `ChickWeight` data frame to model the logarithm of chicken weight over time for different diets. There are up to 12 measurements on each chicken from time of birth to 21 days old, and each chicken has received one of four experimental feeds. We wish to account for a biological variation between chickens, but we are not interested in the particular chickens from the sample and only consider them a random sample from the population of chickens. By including the term (1|Chick) we allow for a positive correlation among measurements taken on the same chicken after the fixed effects have been taken into account. An interaction between time and diet is included to allow for different diets to influence the rate of the weight gain. Note that even though the diets are numbered as 1–4, it is still coded as a factor in the data frame. Thus, we do not have to worry that the categories are mistaken for their numerical values.

```
> library(lme4)
> data(ChickWeight)
> head(ChickWeight)
Grouped Data: weight ~ Time | Chick
  weight Time Chick Diet
1     42    0     1    1
2     51    2     1    1
3     59    4     1    1
4     64    6     1    1
5     76    8     1    1
6     93   10     1    1
> model <- lmer(log(weight) ~ Time + Diet + Time*Diet + (1|Chick),
+              data=ChickWeight)
> summary(model)
Linear mixed model fit by REML ['lmerMod']
Formula: log(weight) ~ Time + Diet + Time * Diet + (1 | Chick)
   Data: ChickWeight

REML criterion at convergence: -296

Scaled residuals:
    Min      1Q  Median      3Q     Max
-4.2030 -0.5479  0.1269  0.6332  2.8654

Random effects:
 Groups   Name        Variance Std.Dev.
 Chick    (Intercept) 0.02580  0.1606
 Residual             0.02585  0.1608
Number of obs: 578, groups:  Chick, 50

Fixed effects:
            Estimate Std. Error t value
(Intercept) 3.768319   0.041194   91.48
Time        0.067537   0.001639   41.21
Diet2       0.048496   0.071074    0.68
Diet3       0.024561   0.071074    0.35
Diet4       0.104324   0.071125    1.47
Time:Diet2  0.008219   0.002716    3.03
Time:Diet3  0.022093   0.002716    8.13
```

```
Time:Diet4  0.014737   0.002751    5.36

Correlation of Fixed Effects:
           (Intr) Time   Diet2  Diet3  Diet4  Tm:Dt2 Tm:Dt3
Time       -0.401
Diet2      -0.580  0.232
Diet3      -0.580  0.232  0.336
Diet4      -0.579  0.232  0.336  0.336
Time:Diet2  0.242 -0.603 -0.405 -0.140 -0.140
Time:Diet3  0.242 -0.603 -0.140 -0.405 -0.140  0.364
Time:Diet4  0.239 -0.596 -0.138 -0.138 -0.406  0.359  0.359
```

The fixed effect estimates are found in the "Fixed effects" section of the output and are interpreted in the same way as output from the `lm` function. For example, the slope of time for diet 4 is 0.01474 larger than the slope for diet 1 (which can be read to be 0.0675), so the mean effect on logarithm of weight for diet 4 is $0.0675 + 0.01474 = 0.08224$ for each increase in time. In the "Random effects" section, we have parameter estimates associated with the random effects. In our model, there are two sources of variation: a biological variation between the chickens, and a within-chicken variability. The `Chick` estimate of 0.02580 is the variation between chickens and `Residual` (with an estimate of 0.02585) is the variation within chickens. Here we have that the size of the biological variation between chickens is roughly the same as the variation within each chicken.

We are interested in testing if there is any difference in weight gain rate for the four diets. That corresponds to testing if the interaction between diet and time is significant. The `anova` function can be used for model reductions, where the models to be compared are included as arguments. When testing fixed effect parameters, we should set the `REML=FALSE` option to obtain maximum likelihood model fits before comparing model likelihoods.

```
> model <- lmer(log(weight) ~ Time + Diet + Time*Diet + (1|Chick),
+               data=ChickWeight, REML=FALSE)
> noint <- lmer(log(weight) ~ Time + Diet + (1|Chick),
+               data=ChickWeight, REML=FALSE)
> anova(model, noint)
Data: ChickWeight
Models:
noint: log(weight) ~ Time + Diet + (1 | Chick)
model: log(weight) ~ Time + Diet + Time * Diet + (1 | Chick)
      Df     AIC     BIC logLik deviance  Chisq Chi Df
noint  7 -272.43 -241.91 143.22  -286.43
```

```
model 10 -335.24 -291.64 177.62  -355.24 68.807       3
      Pr(>Chisq)
noint
model  7.684e-15 ***
---
Signif. codes:  0 '***' 0.001 '**' 0.01 '*' 0.05 '.' 0.1 ' ' 1
```

The interaction between diet and time is clearly significant ($p < 0.0001$) and we reject the hypothesis that the weight gain rate is the same for all four diets. From the estimates shown above, we find that the weight gain rate increases with diet. Chickens on diet 1 have an estimated rate of $\exp(0.067537) = 1.0699$, which corresponds to an increase in weight of 6.99% per day. Chickens on diet 4 increase their weight by a factor $\exp(0.067537 + 0.014737) = 1.085753$, so on average 8.58% per day.

It is also possible to test if any of the random effect terms are equal to zero by comparing two models using the `anova` function. However, in this situation the p-value result from `anova` has twice the size since the hypothesis that the corresponding variance component is equal to zero is on the boundary of the parameter space. Thus, the p-values computed by `anova` should be halved when testing simple random effects.

See also: The books by Demidenko (2013) and Pinheiro and Bates (2000) give a thorough introduction to mixed effects models (and mixed-effects model formulas) in R. See Problem 3.9 for an example of interpreting output from `lm`.

3.17 FIT A LINEAR MIXED-EFFECTS MODEL WITH SERIAL CORRELATION

Problem: You want to fit a linear mixed-effects model which allows for both serial correlation as well as nested random effects.

Solution: Mixed models are statistical models that include not only traditional "fixed effects" as used in linear and generalized linear models but also "random effect" terms. Here, we will consider linear mixed effect models, which accommodate both nested random effects as well as serial correlation. Random effects are appropriate for representing extra variation and clustering — and hence to model dependent observations or introduce extra sources of variation around the population mean.

Random effects induce correlation between observations from a cluster, for example, when there are repeated measurements on each individual, or when pupils are sampled from different schools nested within different districts. In some situations — in particular if measurements are repeated in time or space — it may be more reasonable to assume that measurements that are close to each other in time/space are potentially more correlated than observations that are further away in time/space. Serial correlation models the residual correlation structure for a cluster and can let the correlation depend on the difference between observed variables such as time or distance.

Problem 3.16 used the `lmer` function to fit a mixed effect model since that easily handles multiple crossed random effects. However, `lmer` cannot accommodate serial correlation so here we use the extremely versatile `lme` function from the `nlme` package instead. `lme` only easily handles nested random effects, but on the other hand it can model serial correlation as well as variance heterogeneity.

Random effects terms are specified differently in `lme` than in `lmer`. `lme` expects two model formulas — one for the fixed effects terms, which are specified as usual, and one for the random effect terms through the `random` argument. The random effect terms can be specified in various ways, and here we only consider the simplest situation, where the random effect has the form `~ 1 | group`. This corresponds to a random intercept model, where `group` determines the grouping or clustering structure. The `~ 1 | group` random effect formula in `lme` corresponds to the random effect term `(1|group)` in the syntax of `lmer`.

Serial correlation is specified through the `correlation` argument which accepts a correlation structure function, and R comes with a few predetermined, including for example `corAR1`, `corExp`, `corGaus`, and `corRatio` representing first-order auto-regressive, exponential, Gaussian and rational quadratic correlation structures, respectively. The correlation structures determine the within-group correlation, and the main argument to the correlation structure function is `form`, which expects a one-sided formula that determines both the clustering of the residual correlation structure — observations from different clusters are assumed to be independent — and how the correlation is affected by the "distance" between observations within the cluster. For example, we should include `form = ~ time|subject` as argument to the correlation structure function to specify that the correlation depends on the (differences between) time within each subject. If the option `nugget` is set to `TRUE`, then a nugget parameter is included in the model for the correlation structure.

In the following we use the `ChickWeight` data frame to model the logarithm of chicken weight over time for different diets. There are up to 12 measurements on each chicken from birth to 21 days old, and each chicken has received one of four experimental feeds. An interaction between time and diet is included to allow for different diets to influence the rate of the weight gain. A random effect of chicken is included to account for a biological variation between chickens since we are not interested in the particular chickens in the data but only consider them a random sample from the population of chickens. We include a serial correlation structure (over time for each chicken) to account for the fact that measurements taken close in time on the same animal are likely to be more correlated than measurements taken further apart in time on the same animal. Initially, we assume that the serial correlation structure can be represented by a Gaussian spatial correlation structure over time. Thus, mathematically the model is

$$y_{ij} = X_{ij}\boldsymbol{\beta} + b_i + \kappa_{ij} + \epsilon_{ij}$$

where y_{ij} is the jth measurement on the ith individual, X_{ij} is the design matrix, $\boldsymbol{\beta}$ is the vector of fixed effect parameters, $b_i \sim N(0, \nu^2)$ is the random effect for each chicken, $\kappa \sim N(0, \boldsymbol{\Psi})$ is the serial correlation, and $\epsilon \sim N(0, \sigma^2)$ is the individual measurement error. Two different measurements on the same chicken will have covariance

$$\text{cov}(y_{ij}, y_{ij'}) = \nu^2 + \psi^2 \cdot \exp(-\frac{(t_j - t_{j'})^2}{\phi})$$

where ν^2 is due to the random effect term and the remaining correlation is due to the Gaussian serial correlation. The inclusion of a nugget effect is the reason the ϵ_{ij}s are included in the model so the total variance for a single observation becomes

$$\text{var}(y_{ij}) = \nu^2 + \psi^2 + \sigma^2.$$

The R code is

```
> library(nlme)
> data(ChickWeight)
> head(ChickWeight)
Grouped Data: weight ~ Time | Chick
  weight Time Chick Diet
1     42    0     1    1
2     51    2     1    1
3     59    4     1    1
```

```
4      64    6     1    1
5      76    8     1    1
6      93   10     1    1
> model <- lme(log(weight) ~ Diet + Time + Time*Diet,
+              random= ~ 1|Chick,
+              correlation=corGaus(form= ~ Time|Chick,
+                                  nugget=TRUE),
+              data=ChickWeight)
> model
Linear mixed-effects model fit by REML
  Data: ChickWeight
  Log-restricted-likelihood: 717.3723
  Fixed: log(weight) ~ Diet + Time + Time * Diet
 (Intercept)        Diet2        Diet3        Diet4
 3.777869867 -0.016298711 -0.016190877  0.005659405
        Time   Diet2:Time   Diet3:Time   Diet4:Time
 0.060117705  0.013134919  0.024021147  0.020730733

Random effects:
 Formula: ~1 | Chick
        (Intercept)  Residual
StdDev:  0.03121003 0.2517433

Correlation Structure: Gaussian spatial correlation
 Formula: ~Time | Chick
 Parameter estimate(s):
     range      nugget
11.02202289  0.01393186
Number of Observations: 578
Number of Groups: 50
```

Model reductions and comparisons of nested models are done using the anova function, and maximum likelihood estimation should be used when testing fixed effects instead of the default restricted maximum likelihood estimates. To get maximum likelihood estimates in lme we need to set the optional argument method="ML".

```
> model2 <- lme(log(weight) ~ Diet + Time + Time*Diet,
+              random= ~ 1|Chick,
+              correlation=corGaus(form= ~Time|Chick,
+                                  nugget=TRUE),
+              method="ML", data=ChickWeight)
> model3 <- lme(log(weight) ~ Diet + Time, random= ~ 1|Chick,
+              correlation=corGaus(form= ~Time|Chick,
+                                  nugget=TRUE),
+              method="ML", data=ChickWeight)
> anova(model3, model2)
```

```
       Model df       AIC       BIC   logLik   Test  L.Ratio
model3     1  9 -1448.402 -1409.166 733.2012
model2     2 12 -1463.379 -1411.064 743.6894 1 vs 2 20.97638
       p-value
model3
model2   1e-04
```

We reject the additive model ($p = 0.0001$), so the effect of time on weight depends on the type of diet. The estimates of the final model are obtained by running the `summary` command on the final model using restricted maximum likelihood.

```
> summary(model)
Linear mixed-effects model fit by REML
 Data: ChickWeight
        AIC       BIC   logLik
  -1410.745 -1358.597 717.3723

Random effects:
 Formula: ~1 | Chick
        (Intercept)  Residual
StdDev:  0.03121003 0.2517433

Correlation Structure: Gaussian spatial correlation
 Formula: ~Time | Chick
 Parameter estimate(s):
     range      nugget
11.02202289  0.01393186
Fixed effects: log(weight) ~ Diet + Time + Time * Diet
                Value  Std.Error  DF  t-value p-value
(Intercept)  3.777870 0.05328731 524 70.89624  0.0000
Diet2       -0.016299 0.09210101  46 -0.17697  0.8603
Diet3       -0.016191 0.09210101  46 -0.17579  0.8612
Diet4        0.005659 0.09211534  46  0.06144  0.9513
Time         0.060118 0.00350522 524 17.15091  0.0000
Diet2:Time   0.013135 0.00592672 524  2.21622  0.0271
Diet3:Time   0.024021 0.00592672 524  4.05302  0.0001
Diet4:Time   0.020731 0.00596089 524  3.47779  0.0005
 Correlation:
           (Intr) Diet2  Diet3  Diet4  Time   Dt2:Tm Dt3:Tm
Diet2      -0.579
Diet3      -0.579  0.335
Diet4      -0.578  0.335  0.335
Time       -0.651  0.377  0.377  0.377
Diet2:Time  0.385 -0.662 -0.223 -0.223 -0.591
Diet3:Time  0.385 -0.223 -0.662 -0.223 -0.591  0.350
Diet4:Time  0.383 -0.221 -0.221 -0.660 -0.588  0.348  0.348
```

```
Standardized Within-Group Residuals:
       Min         Q1        Med         Q3        Max
-3.8888672 -0.1629101  0.2985462  0.8735338  2.8966671

Number of Observations: 578
Number of Groups: 50
```

The weight gain per time is largest for diets 3 and 4 ($0.0601 + 0.0240$ and $0.0601 + 0.02073$, respectively) and smallest for diet 1. Recall that this is on the log scale, so we use the exponential function to back-transform the estimates to the original scale; e.g., the weight increases by a factor $\exp(0.0601 + 0.0240) = 1.088$ per day for diet 3.

The "Random Effects" section of the summary output is split into two parts: one related to the random effect terms and one related to the serial correlation structure. The estimates for the standard deviation found for "Intercept" and "Residual" correspond to estimates of the between chicken ($\hat{\nu}$), and the *combined* measurement error and within-chicken standard deviations $\sqrt{\hat{\sigma}^2 + \hat{\psi}^2}$, respectively. The standard deviation between chickens (0.03121) is much smaller than the standard deviation within chickens (0.2517). Thus, the biological variation between chickens seems to have little impact in this situation.

The serial correlation structure is shown to have the "range" of 11.0220 which is an estimate of ϕ and measures how far the autocorrelation stretches on the time scale (presumably, the autocorrelation is essentially zero beyond the range). Due to the parameterization used in lme, the value of 0.0139 listed under "nugget" effect is $\hat{\sigma}^2/(\hat{\sigma}^2 + \hat{\psi}^2)$. The nugget represents extra variability at distances smaller than the typical sample distance.

We can test the hypothesis that the nugget effect is zero. Since the nugget effect is part of the variance structure, we compare the model without the nugget effect to the model with the nugget effect using restricted maximum likelihood.

```
> model4 <- lme(log(weight) ~ Diet + Time + Time*Diet,
+               random= ~ 1|Chick,
+               correlation=corGaus(form= ~ Time|Chick),
+               data=ChickWeight)
> anova(model4, model)
       Model df        AIC        BIC   logLik   Test L.Ratio
model4     1 11  -963.6855  -915.8835 492.8428
model       2 12 -1410.7445 -1358.5969 717.3723 1 vs 2 449.059
       p-value
```

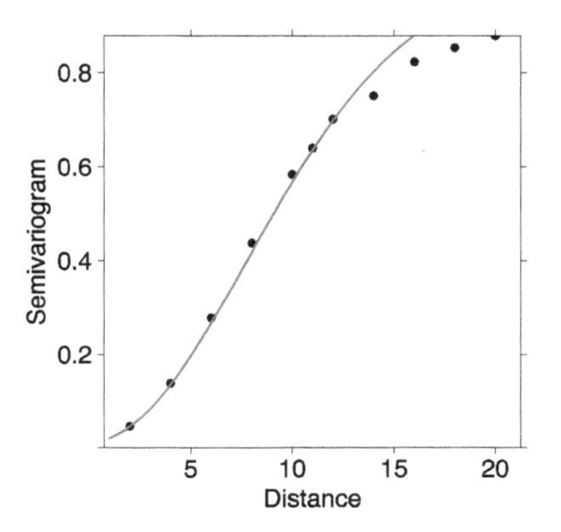

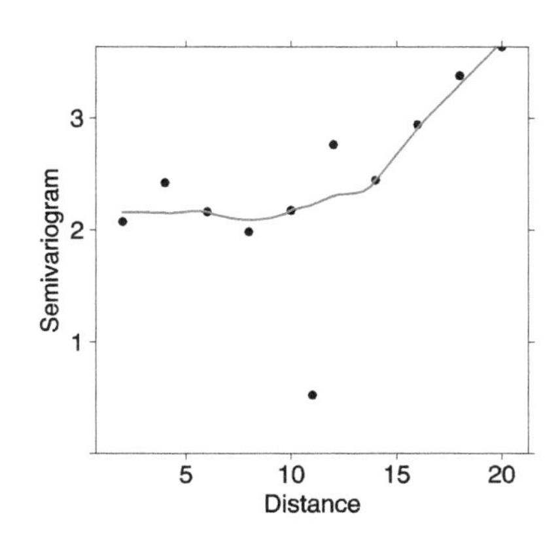

Figure 3.9: The left panel shows the semi-variogram for the empirical serial correlation structure (points) and model-based correlation structure (lines). The right panel shows the semi-variogram with the actual and smoothed empirical correlations after the serial correlation structure of the model has been factored out.

```
model4
model   <.0001
```

We reject the hypothesis that the nugget effect is zero since $p < 0.0001$. Thus, the serial correlation used in the model provides a substantial improvement in the model fit.

Different types of correlation structures can be compared using Akaike's information criterion (AIC), which is extracted from a model fit using the `AIC` function. A smaller AIC indicates a better fit to the data.

```
> model5 <- lme(log(weight) ~ Diet + Time + Time*Diet,
+               random= ~ 1|Chick,
+               correlation=corExp(form= ~ Time|Chick,
+                                  nugget=TRUE),
+               data=ChickWeight)
> AIC(model)
[1] -1410.745
> AIC(model5)
[1] -1177.013
```

The Gaussian correlation structure provides a substantially better fit to the data than the exponential correlation structure.

The adequacy of the serial correlation structure can be assessed graphically using a semi-variogram which uses the within-group residuals from the model to assess the serial correlation structure. The standard semi-variogram plots both the empirical serial correlations from the data as well as the serial correlations from the model. The `Variogram` function computes a semi-variogram which can be plotted directly. By default, Pearson residuals are computed (an example of which are shown in the left plot of Figure 3.9) but "normalized" residuals (where the serial correlation structure from the model has been factored out) can be computed with the `resType="n"` argument (right panel of Figure 3.9).

```
> plot(Variogram(model), lwd=3, pch=20)
> plot(Variogram(model, resType="n"), lwd=3, pch=20)
```

The left plot of Figure 3.9 shows that the empirical (the points) and model-based (the curve) semi-variograms match nicely until a distance around 12 where the observed correlation is slightly *larger* (lower on the graph) than the model predicts. The same trend is shown for the normalized residuals, where the model-based correlation structure has been factored out. Ideally, the right panel of Figure 3.9 should be a flat line to show that the normalized residuals are constant, but there seems to be change in variance for large distances which correspond.

See also: `lme` can handle crossed random effects; they are just complicated to specify. The books by Demidenko (2013) and Pinheiro and Bates (2000) give a thorough introduction to mixed-effects models (and mixed-effects model formulas) in R.

3.18 FIT A GENERALIZED LINEAR MIXED MODEL

Problem: You want to fit a generalized linear mixed-effects model where some of the model terms are considered random effect.

Solution: The class of generalized linear mixed models extends linear mixed-effects models in the same way as the class of generalized linear models extends linear models.

The `glmer` function from the `lme4` package implements generalized linear mixed-effects models in R. `glmer` extends the `lmer` function described in Problem 3.16 in the same way as `glm` extends `lm` to generalized linear models. The `family` option sets the error distribution for `glmer`, and the same families as for `glm` can be applied except for the quasi-binomial and quasi-Poisson families which were used to account for overdispersion.

The summary and anova functions should be used to extract information on parameter estimates and for model reductions, respectively. By default, glmer uses maximum likelihood to find the parameter estimates except if the family is Gaussian in which case restricted maximum likelihood is used. The nACQ argument sets the number of points used for the adaptive Gauss–Hermite approximation of the likelihood. It takes a positive integer, and when increased it increases the number of points and therefore the precision at the cost of speed.

In this example we use the cbpp data frame from the lme4 package to illustrate a generalized linear logistic regression mixed-effects model. cbpp contains information on the incidence of the contagious bovine pleuropneumonia (cbpp) disease from 15 different commercial herds in Africa measured over four periods. Blood samples were collected quarterly to determine cbpp status and we would like to include a random effect of herd to account for any correlation between incidence measurements taken on the same herd over time. Our primary focus is on the changes that occur over time.

```
> library(lme4)
> data(cbpp)
> model <- glmer(cbind(incidence, size-incidence) ~
+                period + (1|herd), family=binomial, data=cbpp)
> summary(model)
Generalized linear mixed model fit by maximum likelihood
  (Laplace Approximation) [glmerMod]
 Family: binomial  ( logit )
Formula:
cbind(incidence, size - incidence) ~ period + (1 | herd)
   Data: cbpp

     AIC      BIC   logLik deviance df.resid
   194.1    204.2    -92.0    184.1       51

Scaled residuals:
    Min      1Q  Median      3Q     Max
-2.3816 -0.7889 -0.2026  0.5142  2.8791

Random effects:
 Groups Name        Variance Std.Dev.
 herd   (Intercept) 0.4123   0.6421
Number of obs: 56, groups:  herd, 15

Fixed effects:
            Estimate Std. Error z value Pr(>|z|)
(Intercept)  -1.3983     0.2312  -6.048 1.47e-09 ***
```

```
period2       -0.9919     0.3032  -3.272 0.001068 **
period3       -1.1282     0.3228  -3.495 0.000474 ***
period4       -1.5797     0.4220  -3.743 0.000182 ***
---
Signif. codes:  0 '***' 0.001 '**' 0.01 '*' 0.05 '.' 0.1 ' ' 1

Correlation of Fixed Effects:
        (Intr) perid2 perid3
period2 -0.363
period3 -0.340  0.280
period4 -0.260  0.213  0.198
```

The contrasts for periods 2–4 relative to period 1 are all negative and significantly different from zero so it appears as if period 1 has the largest incidence of cbpp, after which it decreases. The odds ratio for contagious bovine pleuropneumonia when comparing period 2 to period 1 is

```
> exp(-0.9923)
[1] 0.370723
```

so the average odds of cbpp are a factor $2.697(= 1/0.3707)$ *smaller* for period 2 compared to period 1. The `herd` estimate found in the "Random effects" section is 0.4125 which is the (Gaussian) variance of the herd random effects on the logit scale. Based on this random effects estimate, the odds for cbpp are $\exp(2 \cdot 0.64226) = 3.6129$ times higher for moderately high incidence herds (one standard deviation above the average) compared to moderately low incidence herds (one standard deviation below the average).

We can test the hypothesis that the odds for cbpp are the same in all four periods using the `anova` function by first fitting another model without `period` as an explanatory variable and comparing the model fits.

```
> model2 <- glmer(cbind(incidence, size-incidence) ~ 1 +
+                 (1|herd), family=binomial, data=cbpp)
> anova(model2, model)
Data: cbpp
Models:
model2: cbind(incidence, size - incidence) ~ 1 + (1 | herd)
model: cbind(incidence, size - incidence) ~ period + (1 | herd)
       Df    AIC    BIC   logLik deviance Chisq Chi Df
model2  2 213.66 217.71 -104.832   209.66
model   5 194.05 204.18  -92.027   184.05 25.61      3
```

```
       Pr(>Chisq)
model2
model   1.151e-05 ***
---
Signif. codes:  0 '***' 0.001 '**' 0.01 '*' 0.05 '.' 0.1 ' ' 1
```

We can see from the anova output that the incidence rate at the four periods are significantly different ($p < 0.00001$) which corresponds to the substantial decrease we found in the summary output earlier. Thus, at least one time period has a different level of odds than the other periods.

As mentioned above, glmer cannot use the quasi-binomial and quasi-Poisson families to accommodate overdispersion. Instead we can include a random effect for each observation to introduce extra, individual, additive variation to each observation. We do that in the code below.

```
> cbpp$obs <- 1:nrow(cbpp)   # Construct vector for overdispersion
> model3 <- glmer(cbind(incidence, size - incidence) ~ period +
+                 (1|herd) + (1|obs), family=binomial, data=cbpp)
> summary(model3)
Generalized linear mixed model fit by maximum likelihood
  (Laplace Approximation) [glmerMod]
 Family: binomial  ( logit )
Formula:
cbind(incidence, size - incidence) ~ period + (1 | herd) + (1 |
    obs)
   Data: cbpp

     AIC      BIC   logLik deviance df.resid
   186.6    198.8    -87.3    174.6       50

Scaled residuals:
    Min      1Q  Median      3Q     Max
-1.2866 -0.5989 -0.1181  0.3575  1.6216

Random effects:
 Groups Name        Variance Std.Dev.
 obs    (Intercept) 0.79400  0.8911
 herd   (Intercept) 0.03384  0.1840
Number of obs: 56, groups:  obs, 56; herd, 15

Fixed effects:
            Estimate Std. Error z value Pr(>|z|)
(Intercept)  -1.5003     0.2967  -5.056 4.27e-07 ***
period2      -1.2265     0.4803  -2.554  0.01066 *
period3      -1.3288     0.4939  -2.690  0.00713 **
period4      -1.8662     0.5936  -3.144  0.00167 **
```

```
---
Signif. codes:  0 '***' 0.001 '**' 0.01 '*' 0.05 '.' 0.1 ' ' 1

Correlation of Fixed Effects:
        (Intr) perid2 perid3
period2 -0.559
period3 -0.537  0.373
period4 -0.441  0.327  0.314
```

The fixed effects estimates change slightly when overdispersion is introduced but the largest change is for the random effects. The majority of the variation from the random effects is due to overdispersion and not because of difference among the herds.

See also: Problems 3.11 and 3.15 give examples of generalized linear models and Problem 3.16 gives an introduction to linear mixed-effects models and how to include random effect terms in the model formula. See Elston et al. (2001) for a worked example of using individual-level random effects as a means for accommodating overdispersion.

3.19 FIT A GENERALIZED ESTIMATING EQUATION MODEL

Problem: You want to use generalized estimating equations (GEE) to estimate the parameters in a model for clustered/dependent observations, where a correlation structure may be present.

Solution: Correlated or dependent data are common in many experiments. Sometimes it is possible to correctly model the source of the dependency among observations that give rise to the correlation structure. In other situations, however, it may be assumed that some observations are correlated for example due to clustering but the exact correlation structure may be largely unknown. Thus, additional assumptions may be needed to formulate a full likelihood for the data and even then the likelihood may prove to be intractable.

Generalized estimating equations (GEE) are a convenient and general approach to the analysis of, for example, generalized linear models when correlation is present but where the underlying process that generates the correlation is unspecified. The main advantage of the generalized estimating approach is that it produces unbiased estimates of the fixed-effect parameters even if the correlation structure is misspecified.

Generalized estimating equations can be solved with the `geeglm` function found in the `geepack` package. The `geeglm` function works as the `glm` function and accepts a model formula as its first argument, as well as the arguments `id`, `corstr`, and `family` which are used to define the cluster indicator, the correlation structure within clusters, and the statistical model family, respectively. The `id` argument should be a vector or a factor of the same length as the number of observations that defines the individual clusters.

The simplest form of correlation structure is the independence model which is also the default correlation structure for `geeglm`. It assumes no correlation within clusters and an identity matrix is used to represent the correlation structure within a cluster. Other correlation structures include `corstr="exchangeable"` for the exchangeable correlation structure (i.e., when observations within a cluster are assumed to have a common correlation), a completely unstructured correlation structure (`"unstructured"`), auto-regressive with band 1 (`"ar1"`), and a user defined (`"userdefined"`). The user defined correlation structure, `corstr="userdefined"`, allows the user to specify the correlation structure directly by providing the lower triangular correlation matrix as a vector through the argument `zcor`.

The `family` argument defines the error distribution family just as for the `glm` function. The default distribution is `gaussian`, but `geeglm` also accepts, for example, `binomial`, `poisson`, as well as their `quasi`-versions.

As an example we use the `ohio` dataset from the `geepack` package which contains information from a longitudinal study of the health effects of air pollution on children. We wish to use a logistic regression model to model the odds of wheezing among children and see how that depends on the age of the child and whether the mother was smoking at the start of the study. Wheezing status (`resp`) was registered yearly for each child from age 7 to age 10 (age 9 is scored in the data frame as 0), and we would like to account for possible dependency of observations taken on the same child. Thus, child id should be considered a cluster in the analysis and we use the exchangeable correlation structure to model the correlation among measurements taken on the same child.

Note that the input data frame to `geeglm` must be sorted according to the clustering variable so observations within a cluster are in contiguous rows in the data frame. Otherwise, `geeglm` assumes a new cluster every time it encounters a change in the `id` vector. The `ordered.clusters` function from the `MESS` package can be used to check if the clusters are indeed ordered contiguously.

```
> library(geepack)
> library(MESS)
> data(ohio)
> ordered.clusters(ohio$id)       # Check that clusters are ordered
[1] TRUE
```

The clusters are ordered so we can continue with the analysis. We include an interaction between age and mother's smoking status to allow the age effect to depend on smoking status, and we consider age as a categorical explanatory variable.

```
> fage <- factor(ohio$age+9)      # Add 9 to get the right label
> model <- geeglm(resp ~ fage + smoke + fage:smoke, id=id, data = ohio,
+                  family = binomial, corstr="exchangeable")
> summary(model)

Call:
geeglm(formula = resp ~ fage + smoke + fage:smoke, family = binomial,
    data = ohio, id = id, corstr = "exchangeable")

 Coefficients:
             Estimate  Std.err    Wald Pr(>|W|)
(Intercept)  -1.65823  0.14580 129.347   <2e-16 ***
fage8        -0.08762  0.16967   0.267   0.6056
fage9        -0.13353  0.17803   0.563   0.4532
fage10       -0.47706  0.18963   6.329   0.0119 *
smoke         0.04236  0.24480   0.030   0.8626
fage8:smoke   0.36984  0.27103   1.862   0.1724
fage9:smoke   0.28087  0.28374   0.980   0.3222
fage10:smoke  0.26962  0.29879   0.814   0.3669
---
Signif. codes:  0 '***' 0.001 '**' 0.01 '*' 0.05 '.' 0.1 ' ' 1

Estimated Scale Parameters:
            Estimate Std.err
(Intercept)        1   0.116

Correlation: Structure = exchangeable  Link = identity

Estimated Correlation Parameters:
      Estimate Std.err
alpha   0.3553 0.06403
Number of clusters:   537   Maximum cluster size: 4
```

There are no built-in functions to test parameters, so we could rely on the robust z values listed in the `Wald` column in the output from `summary` which are approximately normally distributed. To test several parameters simultaneously, for example in the case of a categorical explanatory variable with more than two categories, we can use a generalized Wald or score test. These tests are not part of the `geeglm` package but they have been implemented in the `drop1` function from the `MESS` package.

For example to test the hypothesis of no interaction between age and smoking status using the generalized Wald test, we use `drop1` with the fitted model object.

```
> drop1(model)               # Generalized Wald test
Single term deletions

Model:
resp ~ fage + smoke + fage:smoke
           DF Wald Pr(>Chi)
fage:smoke  3 1.97     0.58
```

We fail to reject the hypothesis that the age effect is the same regardless of mother's smoking status ($p = 0.58$) and conclude that the age effect is independent of smoking status. The score test might be slightly more robust and that is obtained by setting the `test="score"` argument.

```
> drop1(model, test="score") # Score test
Single term deletions

Model:
resp ~ fage + smoke + fage:smoke
           DF Score Pr(>Chi)
fage:smoke  3  2.01     0.57
```

The output from the score test yields virtually the same conclusion so we fit a model with no interaction. The estimates and their standard errors are obtained through the `summary` function

```
> model2 <- geeglm(resp ~ fage + smoke, id=id, data = ohio,
+                   family = binomial, corstr="exchangeable")
> summary(model2)

Call:
geeglm(formula = resp ~ fage + smoke, family = binomial, data = ohio,
    id = id, corstr = "exchangeable")

 Coefficients:
            Estimate Std.err   Wald Pr(>|W|)
(Intercept)  -1.7434  0.1374 161.00   <2e-16 ***
fage8         0.0540  0.1323   0.17     0.68
```

```
fage9        -0.0278 0.1388   0.04      0.84
fage10       -0.3755 0.1467   6.55      0.01 *
smoke         0.2712 0.1781   2.32      0.13
---
Signif. codes:  0 '***' 0.001 '**' 0.01 '*' 0.05 '.' 0.1 ' ' 1

Estimated Scale Parameters:
            Estimate Std.err
(Intercept)        1   0.115

Correlation: Structure = exchangeable  Link = identity

Estimated Correlation Parameters:
      Estimate Std.err
alpha    0.354 0.0636
Number of clusters:   537   Maximum cluster size: 4
```

The output shows an increase in odds of wheezing for children of smokers but the effect is not significant. However, the estimated coefficients suggest that the odds of wheezing decrease as the child ages. In particular children aged 10 have substantially lower odds of wheezing. The odds for wheezing for a random child aged 7 relative to a random child aged 10 are $\exp(-0.3755) = 0.687$. Thus, the odds are 0.687 times higher for wheezing at age 7 compared to age 10 with a confidence interval that is $\exp(-0.3755 \pm 1.96 \times 0.1467) = (0.916, 4.21)$.

The output from `summary` has the same format as for other R objects but the standard errors listed are the robust standard errors for the parameter estimates. The naive standard errors can be extracted from the fitted object as the matrix `geese$vbeta.naiv` in the following way

```
> sqrt(diag(model$geese$vbeta.naiv))
[1] 0.146 0.168 0.170 0.183 0.245 0.272 0.276 0.295
```

The naive estimate is the standard error under the assumption that the correlation matrix has been correctly specified and estimated. Generally, the robust standard errors provided by `summary` are preferred but here we see that there is almost no difference in the estimates of the standard errors so in this situation the choice of exchangeable correlation appears not to be problematic.

Even though the generalized estimating equation approach provides robust estimates even for misspecifications of the correlation structure, we can try to use an auto-regressive correlation structure to see how different the estimated correlation becomes when we allow measurements close together in time to be potentially more correlated than measurements that are further apart in time.

```
> model3 <- geeglm(resp ~ fage + smoke, id=id, data = ohio,
+                  family = binomial, corstr="ar1")
> summary(model3)

Call:
geeglm(formula = resp ~ fage + smoke, family = binomial, data = ohio,
    id = id, corstr = "ar1")

 Coefficients:
            Estimate Std.err   Wald Pr(>|W|)
(Intercept)  -1.7322  0.1373 159.28   <2e-16 ***
fage8         0.0540  0.1322   0.17     0.68
fage9        -0.0277  0.1387   0.04     0.84
fage10       -0.3754  0.1467   6.55     0.01 *
smoke         0.2420  0.1799   1.81     0.18
---
Signif. codes:  0 '***' 0.001 '**' 0.01 '*' 0.05 '.' 0.1 ' ' 1

Estimated Scale Parameters:
            Estimate Std.err
(Intercept)    0.999   0.113

Correlation: Structure = ar1  Link = identity

Estimated Correlation Parameters:
      Estimate Std.err
alpha     0.49  0.0662
Number of clusters:   537   Maximum cluster size: 4
```

We see that the auto-regressive correlation structure gives virtually the same estimates. However, the robust standard errors of the parameter estimates are not vastly different.

See also: The `geeM` package provides another implementation of generalized estimating equations. Generalized estimating equations for ordinal regression are implemented by the `ogee` function from the `repolr` package.

3.20 ANALYZE TIME SERIES USING AN ARIMA MODEL

Problem: You have a time series and want to analyze it using an auto-regressive moving average (ARIMA) model.

Solution: Time series analysis is concerned with analysis of an ordered sequence of values of a variable typically measured at equally spaced time intervals. Frequently sampled time series data are common in many fields, for example in stock market quotes, daily precipitation or temperature, annual measurements of water levels, etc.

Problem 3.21 presents a way to decompose a periodic time series into trend, seasonality, and remainder components to describe the nature of the phenomenon represented by the observations. Here we will look at a more formal statistical models for time series.

R uses a special data type for time series analysis and the `ts` function creates time series objects from vectors, arrays, or data frames. For a periodic/seasonal time series the `frequency` argument should be set to an integer greater than 1 when calling `ts`. `frequency` states how frequently observations are measured per natural time unit, and its default value is 1 so there is one measurement per natural time unit. For example, with monthly measurements we set `frequency=12`, and for quarterly measurements we can use `frequency=4`. The optional `start` and `end` arguments set the time of the first and last observation, respectively. Either option should be a single number or a vector of two integers, which specify a natural time unit and then number of samples into that time unit.

We start by looking at the `greenland` dataset from the `MESS` package which contains the average yearly summer air temperatures from Tasiilaq, Greenland, from 1960 to 2010. We only have one measurement per year so there is no seasonality.

```
> library(MESS)
> data(greenland)
> temp <- ts(greenland$airtemp, start=1960)
> temp
Time Series:
Start = 1960
End = 2010
Frequency = 1
 [1] 6.2 6.6 5.8 5.9 5.9 5.9 6.1 4.9 6.0 5.5 4.7  NA  NA 5.1 6.0
[16] 5.8 5.7 6.3 6.5 5.2 5.4 5.3 5.1 4.0 5.1 5.6 5.0 4.8 5.6 4.9
[31] 6.1 6.1 4.6 4.9 6.2 5.8 5.6 5.7 5.6 5.9 5.8 6.1 6.3 7.6 6.4
[46] 6.9 6.1 7.0 6.6 6.3 7.8
```

The `plot` function plots the time series, `acf` will estimate (and by default also plot) the autocorrelation function, while `pacf` estimates and plots the partial autocorrelation function. They all accept a time series object as input. The autocorrelation plot is useful for identifying non-stationary time series (i.e., time series with no trend and seasonality).

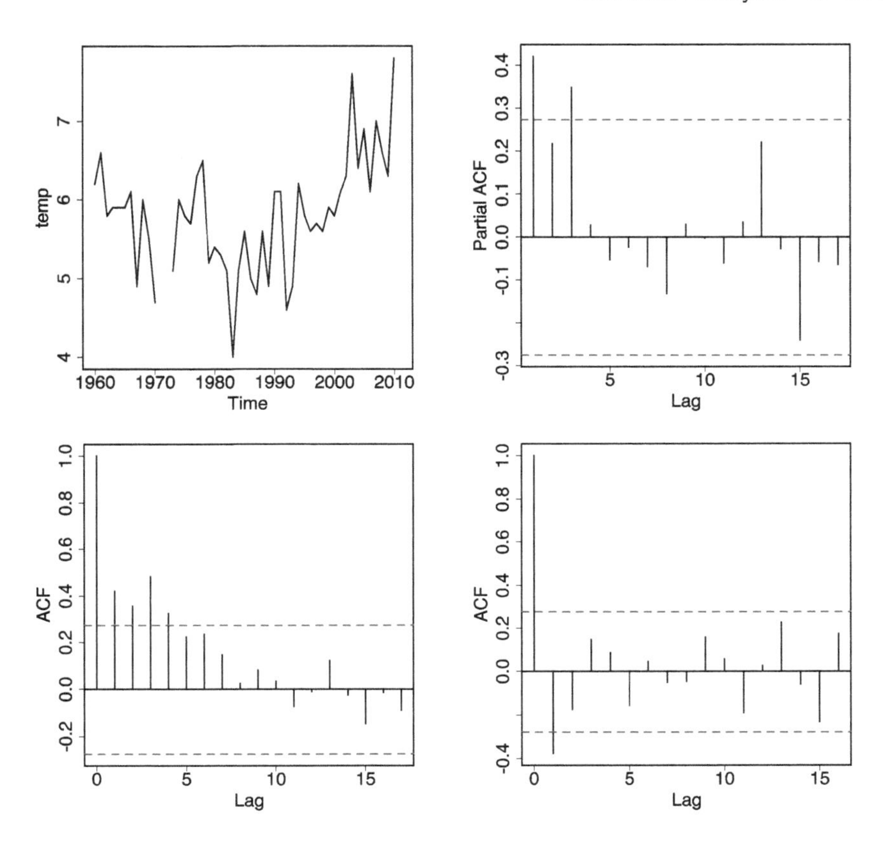

Figure 3.10: Time series with average summer air temperature from Greenland from 1960–2010 (left plot). The upper left plot shows the original time series, the upper right plot shows the partial autocorrelation, while the two lower plots show the autocorrelation for the original and first-order lagged time series.

The autocorrelation function quickly drops to zero for stationary time series while it decreases slowly for non-stationary time series. Below we plot both the autocorrelation for the original time series as well as for the first-order lagged time series.

```
> plot(temp)
> acf(temp, na.action=na.pass)
> acf(diff(temp), na.action=na.pass)
> pacf(temp, na.action=na.pass)
```

The original temperature time series is shown in the upper left plot of Figure 3.10. The autocorrelation plot for the original time series (lower left plot) shows a slow decay towards zero indicating that an autoregressive model may be appropriate or that the data should be lagged to obtain stationarity. The autocorrelation plot for the lagged time series (the lower right plot) suggests that the yearly *change* in temperature (i.e., the lagged time series) behaves like a white noise time series. Thus, a lag 1 time series may be appropriate for modeling these data. Similarly, the partial autocorrelation function quickly becomes small, so apart from the peak at lag 3 a first-order autoregressive model might be relevant to include.

The auto-regressive integrated moving average (ARIMA) model class handles both serial correlation in the response variable (i.e., an autoregressive process), serial correlation in the error term (i.e., a moving average process), and lagged time series. Models can be summarized as ARIMA(p, d, q), where the three parameters are non-negative integers that correspond to the number of autoregressive parameters, the order of differencing needed to stationarize the time series, and the number of moving average parameters, respectively. White noise, random-walk, random-trend models, autoregressive models, and exponential smoothing models (i.e., exponential weighted moving averages) are all special cases of ARIMA models, corresponding to ARIMA$(0, 0, 0)$, ARIMA$(0, 1, 0)$, ARIMA$(0, 1, 0)$ with constant drift, ARIMA$(p, 0, 0)$, and ARIMA$(0, 1, 1)$, respectively.

R has a built-in function, `arima`, for fitting auto-regressive integrated moving average models with Gaussian errors but we will use the improved `Arima` function from the `forecast` package. `Arima` expects a univariate time series as first argument, and the number of non-seasonal parameters is set by the `order` argument, which expects a vector with three integers corresponding to the p, d, and q parameters described above. `Arima` handles individual modeling of the time series periodicity through the `seasonal` option which accepts a list with components `order` and `period` to describe the model within each natural non-seasonal time unit. Other options include `xreg` which should be a vector or matrix containing external explanatory variables with the same number of rows as the number of observations, and `include.mean` which is `TRUE` by default for undifferenced series where it includes a parameter for the overall mean level of the time series and it is `FALSE` for lagged series where a mean value would cancel out. Finally, `Arima` allows for setting the argument `include.drift=TRUE` to include a drift or trend in the time series. Based on Figure 3.10 it seems relevant to use an ARIMA$(1, 1, 1)$ model to fit to the data in order to account for the lagged series (which suggests $d = 1$), large negative first-lag in the autocorrelation plot for the lagged series lower (suggesting $q = 1$), and possibly a first-order auto-regressive process based on the partial autocorrelation plot. We also allow for drift in the model.

```
> library(forecast)
> model <- Arima(temp, order=c(1,1,1), include.drift=TRUE)
> model
Series: temp
ARIMA(1,1,1) with drift

Coefficients:
         ar1     ma1  drift
      -0.033  -0.650  0.018
s.e.   0.217   0.158  0.030

sigma^2 estimated as 0.378:  log likelihood=-44.7
AIC=97.4   AICc=98.3   BIC=105
```

The model gives that the observed lagged difference at time t is estimated as

$$y_t - y_{t-1} = 0.018 - 0.033 \cdot (y_{t-1} - y_{t-2}) + \epsilon_t - 0.650 \cdot \epsilon_{t-1},$$

where ϵ_t and ϵ_{t-1} are the independent Gaussian errors for the measurements at time t and $t-1$, respectively. The drift is clearly not substantial, but the large (close to 1) negative moving average is, so this suggests that there will be a lot of fluctuations from one period to the next which is indeed what we see in the original data.

The `forecast` function can be used to predict future values of the time series. It takes an `Arima` fit as input, and the `h` argument sets the number of periods ahead to predict. Below we predict the air temperature for the next 15 years (and corresponding 80% and 95% prediction intervals), and plot the original time series as well as the prediction (with prediction intervals). The graph is shown in Figure 3.11.

```
> forecast(model, h=15)
     Point Forecast Lo 80 Hi 80 Lo 95 Hi 95
2011           6.97  6.19  7.76  5.77  8.18
2012           7.02  6.19  7.85  5.76  8.28
2013           7.04  6.17  7.91  5.71  8.37
2014           7.06  6.15  7.96  5.67  8.44
2015           7.07  6.13  8.02  5.62  8.52
2016           7.09  6.11  8.07  5.59  8.60
```

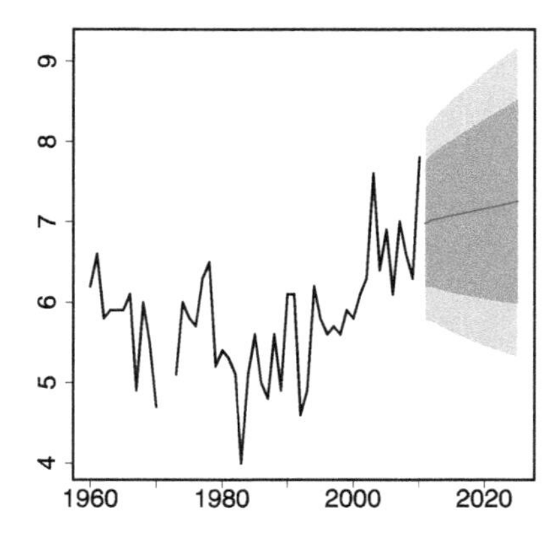

Figure 3.11: Observed temperature time series and 15-year prediction interval with 80% and 95% prediction intervals.

```
2017            7.11  6.09  8.13  5.55  8.67
2018            7.13  6.07  8.18  5.51  8.74
2019            7.14  6.06  8.23  5.48  8.81
2020            7.16  6.04  8.28  5.45  8.87
2021            7.18  6.03  8.33  5.42  8.94
2022            7.20  6.02  8.38  5.39  9.00
2023            7.22  6.00  8.43  5.36  9.07
2024            7.23  5.99  8.47  5.34  9.13
2025            7.25  5.98  8.52  5.31  9.19
> plot(forecast(model, h=15))
```

The `forecast` package also includes the `auto.arima` function which somewhat automates the process of identifying and estimating the type and number of parameters in an ARIMA model. The `auto.arima` function considers a number of information criteria to identify the best model — for more information see the corresponding help page.

```
> auto.arima(temp)
Series: temp
ARIMA(0,1,1)

Coefficients:
         ma1
      -0.656
s.e.   0.108

sigma^2 estimated as 0.366:  log likelihood=-44.9
AIC=93.8   AICc=94.1   BIC=97.6
```

The suggested model is an ARIMA$(0, 1, 1)$ without drift which is in accordance with our previous findings.

See also: The books by Shumway and Stoffer (2010) and Hyndman and Athanasopoulos (2013) give good introductions to time series modeling in R. The package `its` implements irregular time series.

3.21 DECOMPOSE A TIME SERIES INTO TREND, SEASONAL, AND RESIDUAL COMPONENTS

Problem: You have a time series and want to decompose the time series into a trend, seasonal, and residual component.

Solution: Time series analysis is concerned with analysis of an ordered sequence of values of a variable typically measured at equally spaced time intervals. Frequently sampled time series data are common in many fields, for example in stock market quotes, daily precipitation or temperature, quarterly sales revenue, annual measurements of water levels, etc.

One of the common goals of time series analysis is to decompose the series into three components: trend, seasonality, and residual/irregular. The trend component is the long-term changes in the series; seasonality is a systematic, short-term calendar related effect. The residual component is what remains after the seasonal and trend components of a time series are removed, and it results from short-term fluctuations.

R has a special data type for use with time series analysis and the `ts` function creates time series objects from vectors, arrays or data frames. See Problem 3.20 for more information on the `ts` function.

The `stl` function makes a seasonal decomposition of a time series into trend, seasonal, and residual components using local regression. It needs a periodic time series object (i.e., the frequency should be greater than 1) as first argument. The only other required argument to `stl` is `s.window` which should either be the character string `"periodic"` or an odd number giving the span in lags used by the local regression window for seasonal extraction. Additionally, it is possible to change the degree and window size for the seasonality, the trend and each subseries through the `s.degree`, `t.degree`, `l.degree`, and `t.window`, `l.window` options, respectively. The `stl` function assumes an additive seasonal component but we can log-transform the response to use a multiplicative seasonal component for `stl`.

We will illustrate the `stl` decomposition using the monthly totals of

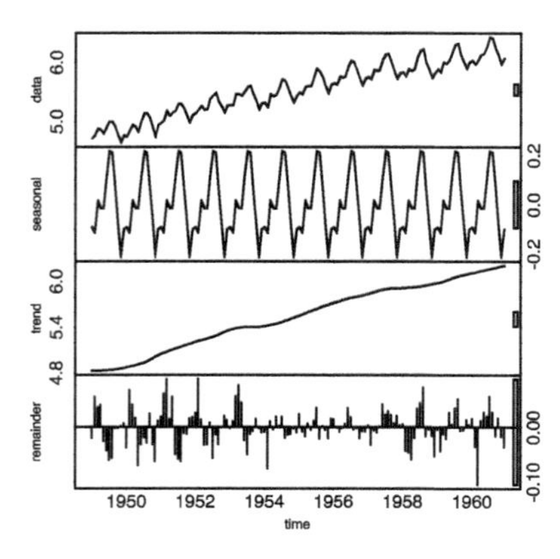

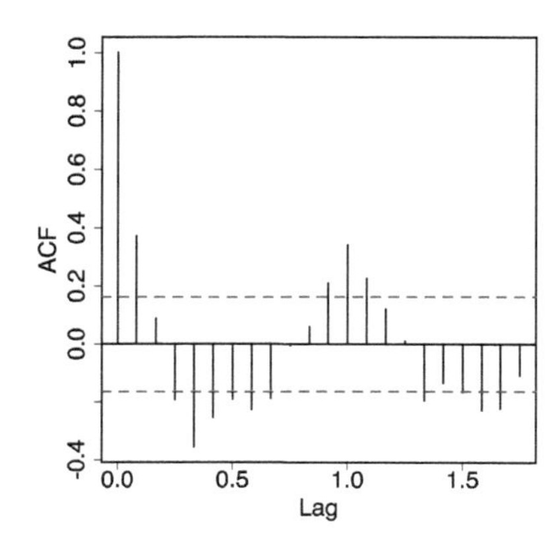

Figure 3.12: Decomposition of a time series using the `stl` function (left panel). The thick gray vertical bars along the right-hand side of the frame give a relative scale of comparison to determine the relative importance of the different components. The effect of components with smaller bars dominates the time series. The right panel shows the autocorrelation function for the residual component.

international airline passengers (in thousands) from 1949 to 1960 found in the `AirPassengers` data frame. A total of 144 observations are available. The data are already stored as a periodic R time series object so we do not have to convert it using `ts`.

```
> data(AirPassengers)
> model <- stl(log(AirPassengers), s.window="periodic")
> plot(model)
```

The four graphs in the left panel of Figure 3.12 show plots of the original time series as well as the seasonal, trend, and residual components obtained from `stl`. The data, seasonal, trend, and residual curves are all plotted on their individual scale so the thick gray vertical bars along the right-hand side of the frame give a relative scale of comparison to determine the relative importance of the different components. To directly compare the contributions from the individual components, the components should be scaled such that the gray bars have the same size, so smaller bars suggest that the component plays a larger role in the time series. Thus, the left panel in Figure 3.12 suggests that the trend dominates the data series and that the effect of the residual component is relatively small.

The same information is conveyed in the `summary` output which shows the inter-quartile range of the individual components relative to the inter-quartile range of the original data. Thus, if we were to model the data we should certainly use a method that includes a trend component.

```
> summary(model)
 Call:
 stl(x = log(AirPassengers), s.window = "periodic")

 Time.series components:
    seasonal              trend            remainder
 Min.    :-0.2135    Min.   :4.83    Min.    :-0.1050
 1st Qu.:-0.0939    1st Qu.:5.20    1st Qu.:-0.0183
 Median :-0.0145    Median :5.55    Median :-0.0008
 Mean   : 0.0000    Mean   :5.54    Mean    :-0.0003
 3rd Qu.: 0.0781    3rd Qu.:5.91    3rd Qu.: 0.0179
 Max.   : 0.2164    Max.   :6.20    Max.    : 0.0865
 IQR:
     STL.seasonal STL.trend STL.remainder data
     0.172        0.710     0.036         0.695
   %  24.8        102.3       5.2         100.0

 Weights: all == 1

 Other components: List of 5
 $ win  : Named num [1:3] 1441 19 13
 $ deg  : Named int [1:3] 0 1 1
 $ jump : Named num [1:3] 145 2 2
 $ inner: int 2
 $ outer: int 0
```

The seasonal, trend, and residual values are stored as three time series in the `time.series` element of the result from `stl`. We can plot the autocorrelation function of the residual term using the `acf` function as shown below. The output is shown in the right plot of Figure 3.12.

```
> acf(model$time.series[,3])  # Plot autocorrelation of residuals
```

Based on the plot of the autocorrelation function there still seems to be some periodic variation left in the residual component of the time series which could possibly be modeled with a moving average ARIMA model (see Problem 3.20).

See also: `decompose` makes a seasonal decomposition of a time series using moving averages. The `zoo` package can handle irregularly spaced time series.

SPECIFIC METHODS

3.22 COMPARE POPULATIONS USING T TEST

Problem: You wish to compare two population means or test a single population mean using a t test.

Solution: The t test tests the hypothesis that two population means are identical or that a single population mean has a hypothesized value. The `t.test` function computes one- and two-sample t tests and it accepts input as either a model formula or as one or two vectors of quantitative observations.

For the two-sample test, we can input the two vectors of observations as the first two arguments. The optional argument `alternative` specifies the alternative hypothesis and can be one of `two.sided` (the default) for a two-sided test, or `less` or `greater` depending on the alternative hypothesis if a one-sided test is desired. The `var.equal` argument is by default `FALSE` which produces Welch's two-sample t test where the variances in the two populations are not assumed to be equal. If the `paired=TRUE` argument is set, then a paired t test is computed.

The `cats` data frame from the `MASS` package contains information on the body weight (in kg) and heart weight (in grams) for 144 adult cats. Here, we wish to examine if the body weight is the same for male and female cats.

```
> library(MASS)
> data(cats)
> head(cats)
  Sex Bwt Hwt
1   F 2.0 7.0
2   F 2.0 7.4
3   F 2.0 9.5
4   F 2.1 7.2
5   F 2.1 7.3
6   F 2.1 7.6
> t.test(Bwt ~ Sex, data=cats)

Welch Two Sample t-test

data:  Bwt by Sex
t = -9, df = 100, p-value = 9e-15
alternative hypothesis: true difference in means is not equal to 0
95 percent confidence interval:
 -0.663 -0.418
sample estimates:
```

```
mean in group F mean in group M
           2.36            2.90
```

The mean body weights for the two groups are found to be 2.36 and 2.90 kg for the female and male cats, respectively. The p-value is virtually zero so we reject the hypothesis that the body weight for males and females is identical, and a 95% confidence interval for the difference in body weight is $[-0.6631; -0.4177]$.

We set the argument `var.equal=TRUE` if we assume equal variances in the two populations.

```
> t.test(Bwt ~ Sex, data=cats, var.equal=TRUE)

Two Sample t-test

data:  Bwt by Sex
t = -7, df = 100, p-value = 2e-11
alternative hypothesis: true difference in means is not equal to 0
95 percent confidence interval:
 -0.686 -0.395
sample estimates:
mean in group F mean in group M
           2.36            2.90
```

The p-value becomes slightly larger in this situation but we still clearly reject the null hypothesis of equal body weights for male and female cats.

To test the mean of a single population, we supply `t.test` with a single numeric vector as input and set the `mu` argument to the hypothesized mean value (default value is zero).

```
> x <- c(1.3, 5.5, -2.1, 0.9, -0.4, 1.1)
> t.test(x, mu=0)

One Sample t-test

data:  x
t = 1, df = 5, p-value = 0.4
alternative hypothesis: true mean is not equal to 0
95 percent confidence interval:
 -1.6  3.7
sample estimates:
mean of x
     1.05
```

We fail to reject the null hypothesis that the population mean is zero ($p = 0.3554$). The 95% confidence interval for the population mean is given by $[-1.601; 3.701]$, and the mean of the sample is 1.05.

See also: The t test with equal variances can also be analyzed with the `lm` function — see Problem 3.6. See Problem 3.50 for a non-parametric version of the one-sample t test and Problem 3.51 for a non-parametric two-sample test.

3.23 FIT A NONLINEAR MODEL

Problem: You wish to fit a nonlinear model to your data and use least squares to estimate the corresponding parameters.

Solution: Nonlinear models are an extension of linear models where the response variable is modeled as a nonlinear function of the model parameters but where the residual errors are still assumed to be Gaussian distributed. Nonlinear models are often derived on the basis of physical and/or biological considerations, e.g., from differential equations, where there is some knowledge of the data-generating process. In particular, the parameters of a nonlinear model usually have direct interpretation in terms of the process under study.

Least-squares estimates of the parameters from a nonlinear model can be obtained with the `nls` function. `nls` allows for more complex model formulas than we usually find for model formulas with linear predictors since `nls` treats the right-hand side of the formula argument as a standard algebraic expression rather than as a linear-model formula.

The first argument to `nls` should be an algebraic formula that expresses the relationship between the response variable and the explanatory variables and parameters. The `start` argument is a named list of parameters and corresponding starting estimates. The names found in the `start` list also determine which terms in the algebraic formula are considered parameters to be estimated.

As an example, consider the following four-parameter extension to the logistic growth model which is sometimes used to fit fluorescence readings from real-time polymerase chain reaction (PCR) experiments. If we let t indicate the cycle number, then the reaction fluorescence at cycle t is defined as

$$F(t) = \frac{F_{\max}}{1 + \exp\left(-\frac{t-c}{b}\right)} + Fb,$$

where $F_{\max}$ is the maximal increase in reaction fluorescence, c is the fractional cycle at which reaction fluorescence reaches half of $F_{\max}$, b is related to the slope of the curve, and Fb is the background reaction fluorescence.

The qpcr data frame from the MESS package contains information on fluorescence levels at different cycles from a real-time PCR experiment. We choose sensible starting values based on prior knowledge or from the experiment. Reasonable values for the maximum, $F_{\max}$, the background, Fb and c are easily determined in this case since we can plot the data and get decent guesses for these parameters from the plot (see Figure 3.13).

```
> library(MESS)
> data(qpcr)
> # Use data from just one of the 14 runs
> run1 <- subset(qpcr, transcript==1 & line=="wt")
> model <- nls(flour ~ fmax/(1+exp(-(cycle-c)/b))+fb,
+              start=list(c=25, b=1, fmax=100, fb=0),
+              data=run1)
> model
Nonlinear regression model
  model: flour ~ fmax/(1 + exp(-(cycle - c)/b)) + fb
   data: run1
     c      b   fmax     fb
29.593  0.841 96.756  3.723
 residual sum-of-squares: 53.8

Number of iterations to convergence: 9
Achieved convergence tolerance: 7.75e-06
```

The least squares estimates for the four parameters are seen in the output with $\hat{c} = 29.59, \hat{b} = 0.8406, \widehat{F_{\max}} = 96.756$ and $\widehat{Fb} = 3.723$.

We can plot the data easily since the logistic growth model is a function of cycle time. Figure 3.13 shows a plot of the observed data and we observe a nice fit of the model to the data when the predicted curve is added.

```
> plot(run1$cycle, run1$flour,
+      xlab="Cycle", ylab="Fluorescence")
> lines(run1$cycle, predict(model)) # Add the fitted line
```

As usual, summary provides information on the individual parameters and their standard error and lists the residual standard error. Note that — as usual — the p values listed in the output from summary test the hypotheses that the individual parameters are equal to zero, which in most cases will probably not be relevant hypotheses. A better idea is to look at confidence intervals, which may be computed for all parameters using the `confint` function.

```
> summary(model)

Formula: flour ~ fmax/(1 + exp(-(cycle - c)/b)) + fb

Parameters:
     Estimate Std. Error t value Pr(>|t|)
c     29.5932     0.0284  1041.3   <2e-16 ***
b      0.8406     0.0247    34.0   <2e-16 ***
fmax  96.7559     0.3965   244.0   <2e-16 ***
fb     3.7226     0.2247    16.6   <2e-16 ***
---
Signif. codes:  0 '***' 0.001 '**' 0.01 '*' 0.05 '.' 0.1 ' ' 1

Residual standard error: 1.15 on 41 degrees of freedom

Number of iterations to convergence: 9
Achieved convergence tolerance: 7.75e-06
> confint(model)

Waiting for profiling to be done...

       2.5%  97.5%
c    29.536 29.651
b     0.791  0.892
fmax 95.958 97.555
fb    3.267  4.178
```

The `nls` function can be quite sensitive to starting values, and even though the function will run without specifying values for the parameters with the `start` argument, it is not uncommon to run into convergence problems. The `control` argument sets control settings for `nls`, and commonly used control settings are `maxiter`, which changes the maximum number of iterations allowed and `tol` which sets the tolerance levels. The control parameters are easily changed using the `nls.control` function since it returns a list with the five components that the `control` argument in `nls` accepts.

The `algorithm` option to `nls` specifies the algorithm to use for the least-squares estimation. The default algorithm is Gauss–Newton, but other options are `port` for the nl2sol algorithm and `plinear` for the Golub–Pereyra algorithm for partially linear least-squares models.

Constraints on the parameter values can be built into the nonlinear

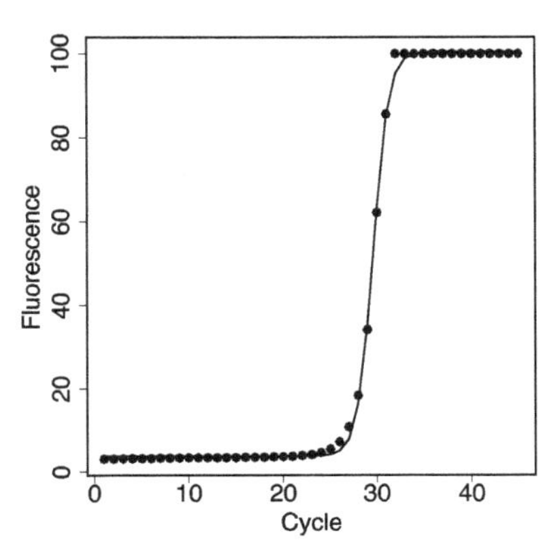

Figure 3.13: Fit of a four-parameter nonlinear regression model to real-time PCR data.

least squares fit through the `lower` and `upper` arguments. If either of them is specified they should be a vector with the same length as the `start` list, and they represent the lower and upper bounds, respectively, of the parameters. Bounds can only be used with the `port` algorithm.

For example, to fit the same nonlinear model but with a lower bound for b at 0.9, with the maximum number of iterations set to 100, and relative tolerance to 10^{-8}, we use the following code.

```
> model2 <- nls(flour ~ fmax/(1+exp(-(cycle-c)/b))+fb,
+              start=list(c=25, b=1, fmax=100, fb=0),
+              lower=c(-Inf, .9, -Inf, -Inf), algorithm="port",
+              control=nls.control(maxiter=100, tol=1e-08),
+              data=run1)
> model2
Nonlinear regression model
  model: flour ~ fmax/(1 + exp(-(cycle - c)/b)) + fb
   data: run1
     c      b   fmax     fb
29.589  0.900 97.051  3.609
 residual sum-of-squares: 60.98

Algorithm "port", convergence message: both X-convergence and
relative convergence (5)
```

The parameter estimates have changed a little since the b parameter reached its lower boundary. The restriction on the b parameters has also caused the residual sum-of-squares to increase from 53.79 to 60.98.

See also: The `nls` package provides additional functions for nonlinear regression. The `nls2` package adds a brute-force algorithm to `nls` and allows for multiple starting values. The `selfStart` function can be used to create a self-starting nonlinear model fit, where the starting values do not have to be specified by the user.

3.24 FIT A CENSORED REGRESSION MODEL

Problem: You want to fit a Tobit regression model to a dataset because some of the observed responses are left censored.

Solution: The censored regression model (also called the Tobit regression model) estimates a linear regression model for a left- or right-censored quantitative response variable. Censored data are common in situations where data can only be observed under certain conditions (e.g., when observations can only be measured when they are above a detection threshold or when negative values are not possible).

Tobit regression can be fitted directly using the `survreg` function from the `survival` package, but the `tobit` function from the `AER` package provides a wrapper function to `survreg` as well as some additional functionality including easy handling of both left and right censoring.

The `tobit` function takes a regular model formula as first argument, and the `left` and `right` arguments set the left and right limit for censoring of the response variable. By default, the left value is set to zero and the right value to infinity to indicate that there is no right censoring and that values are left censored at zero. Censoring of the response variable is done automatically based on the `left` and `right` options so we do not have to modify the data frame directly.

The `nh4` data from the `MESS` package shows the monthly levels of ammonia nitrogen in a river over two periods (one starting January 1978 and the other starting January 1998). For each month the amount of ammonia nitrogen in mg/l is measured but there is a lower limit of detection so the lowest possible actual measurement is 0.1 mg/l. In the data there are a number of zeroes but they are all corresponding to censored observations (i.e., we only know that they are <0.01). We log-transform the nh4 measurements since the data appear to be skewed to the right.

```
> library(AER)
> library(MESS)
> data(nh4)
> model <- tobit(log(nh4) ~ factor(year), left=log(0.00999), data=nh4)
```

```
> summary(model)

Call:
tobit(formula = log(nh4) ~ factor(year), left = log(0.00999),
    data = nh4)

Observations:
         Total  Left-censored     Uncensored Right-censored
           120             36             84              0

Coefficients:
                 Estimate Std. Error z value Pr(>|z|)
(Intercept)       -4.1722     0.1068  -39.06   <2e-16 ***
factor(year)1998  -0.0417     0.1506   -0.28   0.7816
Log(scale)        -0.2347     0.0826   -2.84   0.0045 **
---
Signif. codes:  0 '***' 0.001 '**' 0.01 '*' 0.05 '.' 0.1 ' ' 1

Scale: 0.791

Gaussian distribution
Number of Newton-Raphson Iterations: 3
Log-likelihood: -132 on 3 Df
Wald-statistic: 0.0769 on 1 Df, p-value: 0.8
```

`summary` provides a table of left-, right-, and uncensored observations and we note that 36 of our observations are censored at just under 0.01 since that is our detection limit. The estimates for the explanatory variables are interpreted as usual for a linear model, i.e., the estimate of `year1998` suggests that on average the logarithm of the nh4 level was 0.0417 below the corresponding level in 1978, but the difference in years is not significant (since the p-value is 0.7816). The `Log(scale)` line shows the logarithm of the estimate for the residual variance, which corresponds to the value found for `Scale`. Hence, the residual standard error is $\sqrt{0.791} = 0.889$.

The `drop1` function tests each explanatory variable by removing the terms from the model formula and comparing the fit to the original model. `drop1` can be used for model reduction for tobit regression by setting the option `test="Chisq"` to ensure that `drop1` computes likelihood ratio test statistics and corresponding p-values based on the χ^2 distribution.

```
> drop1(model, test="Chisq")
Single term deletions
```

```
Model:
Surv(ifelse(log(nh4) <= log(0.00999), log(0.00999), log(nh4)),
    log(nh4) > log(0.00999), type = "left") ~ factor(year)
              Df AIC    LRT Pr(>Chi)
<none>           270
factor(year)  1 268 0.0769     0.78
```

Here we get the same conclusion about the difference in years as we did above in the summary output.

See also: The `zelig` function from the `Zelig` package also implements tobit regression models.

3.25 FIT A ZERO-INFLATED REGRESSION MODEL

Problem: You want to fit a zero-inflated regression model to a dataset where there is observed more zeros values than expected from a given distribution.

Solution: Zero-inflated models can be used when we observe more zero values than would be expected for a given model. Zero-inflated data can arise if there are two underlying types of observations: one type which will never give a varied response (we can only observe a zero) and another type which will follow a pre-specified distribution (typically referred to as the count model). Thus, there are two sources of zeros: they may arise both from the zero-inflation process or from "natural zeros" from the count model which typically is a Poisson distribution or some other non-negative discrete distribution. Zero-inflated data are typically modeled as a two-component mixture model that combines two processes: a binomial distribution to add extra point mass at zero and a proper count distribution that is used for the non-zeroes from the binomial distribution.

Zero-inflated regression models for Poisson, negative-binomial, or geometric-distributed count model data can be fitted with the `zeroinfl` function from the `pscl` package. The `zeroinfl` function takes a regular model formula as first argument, and the `dist` and `link` arguments set the distribution of the count model and the link function for the binomial zero-inflation model. The link function for the count model is always the logarithm and the distribution for the zero-inflation model is always a binomial.

The `Affairs` data from the `AER` package shows infidelity statistics from 601 individuals, where the `affairs` response variable contains information on the number of extramarital sexual intercourses during the previous year. The majority of observations are zeroes since 451 out of the 601 observations have had no extramarital intercourse so it makes sense to use a zero-inflated model. We are modeling the "number of affairs" so we use a Poisson distribution for the non-zero distribution. We would like to examine how the variables `gender`, `age`, and `children` affect the number of affairs.

```
> library(pscl)
> library(AER)
> data(Affairs)
> model <- zeroinfl(affairs ~ age + gender + children,
+                   dist="poisson", data=Affairs)
> summary(model)

Call:
zeroinfl(formula = affairs ~ age + gender + children, data = Affairs,
    dist = "poisson")

Pearson residuals:
   Min     1Q Median     3Q    Max
-0.600 -0.583 -0.542 -0.378  5.078

Count model coefficients (poisson with log link):
            Estimate Std. Error z value Pr(>|z|)
(Intercept)  1.36898    0.14249    9.61   <2e-16 ***
age          0.01635    0.00398    4.11    4e-05 ***
gendermale  -0.15684    0.06883   -2.28    0.023 *
childrenyes -0.10377    0.09341   -1.11    0.267

Zero-inflation model coefficients (binomial with logit link):
            Estimate Std. Error z value Pr(>|z|)
(Intercept)  1.70455    0.36226    4.71  2.5e-06 ***
age          0.00209    0.01127    0.19   0.8530
gendermale  -0.20668    0.19377   -1.07   0.2861
childrenyes -0.76354    0.25438   -3.00   0.0027 **
---
Signif. codes:  0 '***' 0.001 '**' 0.01 '*' 0.05 '.' 0.1 ' ' 1

Number of iterations in BFGS optimization: 17
Log-likelihood: -816 on 8 Df
```

`summary` provides a table of the count and zero-inflation model coefficients. The zero-inflation model is a logistic regression model and the coefficients can be interpreted as for a regular logistic regression model. Likewise, the coefficients for the count model can be interpreted as for a Poisson regression model since that is the model class we used here. However, it should be noted that the coefficients for the count model correspond to the effects *conditional* on not observing a zero from the zero-inflation process.

Negative estimates for the coefficients in the zero-inflation model suggests an increase in the probability that the result will be a zero from the zero-inflation part. In the output above we get that the presence of children decreases the risk of an extramarital affair compared to not having children since it both lowers the expected mean for the count model and increases the risk of zero-inflation.

Note that by default, the same model formula is used for both the zero-inflation as well as the count model. Individual models can be specified for the two components by separating the model formulas by a vertical bar. For example, to have every single observation have the same risk of being a zero-inflation we could write the following code.

```
> m2 <- zeroinfl(affairs ~ age + gender + children | 1,
+                dist="poisson", data=Affairs)
> summary(m2)

Call:
zeroinfl(formula = affairs ~ age + gender + children |
    1, data = Affairs, dist = "poisson")

Pearson residuals:
   Min     1Q Median     3Q    Max
-0.543 -0.523 -0.519 -0.501  4.671

Count model coefficients (poisson with log link):
            Estimate Std. Error z value Pr(>|z|)
(Intercept)  1.36596    0.14304    9.55   <2e-16 ***
age          0.01633    0.00398    4.11    4e-05 ***
gendermale  -0.15568    0.06880   -2.26    0.024 *
childrenyes -0.09991    0.09386   -1.06    0.287

Zero-inflation model coefficients (binomial with logit link):
            Estimate Std. Error z value Pr(>|z|)
(Intercept)   1.0955     0.0944    11.6   <2e-16 ***
---
Signif. codes:  0 '***' 0.001 '**' 0.01 '*' 0.05 '.' 0.1 ' ' 1

Number of iterations in BFGS optimization: 19
Log-likelihood: -823 on 5 Df
```

The model formula above can be read with the vertical bar as "conditional on"; i.e., here we model the expected number of affairs as a Poisson regression with age, gender, and children as explanatory variables conditional on that we do not see a zero-inflated zero, and that all observations have the same risk of being a zero-infalted zero.

The drop1 function tests each explanatory variable by removing the term from the model formula and comparing the fit to the original model. If a term is present in both formulas then drop1 removes the term from both models simultaneously. The argument test="Chisq" should be set to ensure that drop1 computes likelihood ratio test statistics and corresponding p-values based on the χ^2 distribution.

```
> drop1(model, test="Chisq")
Single term deletions

Model:
affairs ~ age + gender + children
         Df  AIC   LRT Pr(>Chi)
<none>       1649
age       2 1661 16.39  0.00028 ***
gender    2 1651  6.29  0.04316 *
children  2 1655 10.77  0.00457 **
---
Signif. codes:  0 '***' 0.001 '**' 0.01 '*' 0.05 '.' 0.1 ' ' 1
```

All three explanatory variables are statically significant although there is barely any difference between the two sexes. If anything, men are slightly less inclined to have extramarital affairs than women since men's coefficients are both negative.

See also: The glmmADMB package can be used to specify zero-inflated generalized mixed effect models.

3.26 FIT A SMOOTH CURVE

Problem: You want to fit a smooth curve to describe the relationship between two quantitative variables.

Solution: Sometimes it is desirable to describe the overall relationship between two quantitative variables with a smooth curve without specifying a statistical model for the relationship. Curve fitting is useful when you want the data (instead of the model) to show the relationship between the two variables or when you just want to improve the appearance of the plot by drawing a smooth curve through the data.

R provides several functions for fitting smooth curves to describe the

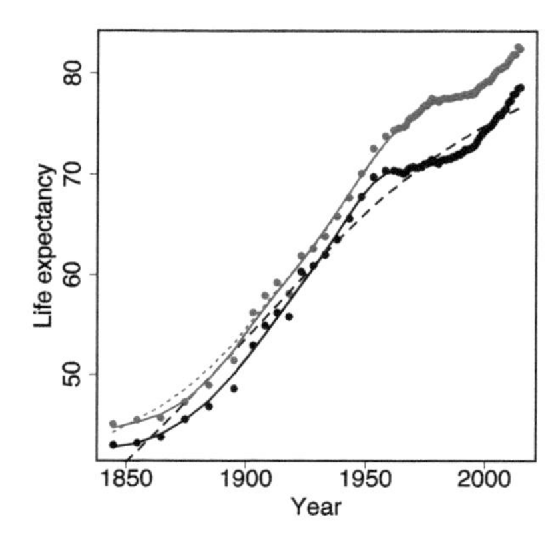

Figure 3.14: Smoothed curves showing the life expectancy for newborn Danish boys and girls. Boys are black and the curves correspond to loess estimates with span 0.25 (solid black line), and span 1 (dashed black line). Girls are blue with the solid line estimated using smoothing spline while the dashed blue line is Friedman's super smoother.

relationship between two quantitative variables. The `loess` function uses local polynomial regression fitting, `supsmu` employs Friedman's "super smoother," and `smooth.spline` fits a cubic smoothing spline to the data.

We will fit smooth curves to the expected life expectancy for newborn male and female Danes found in the `lifeexpect` dataset in the `MESS` package. We first plot the estimated life expectancies and then fit a local polynomial regression model using the `loess` function. The `span` argument accepts a positive number and controls the flexibility of the smoothed curve with larger values corresponding to a more smoothed curve. The predicted values for the smoothed curve is extracted from the loess fit by the `predict` function and the results are shown in Figure 3.14.

```
> library(MESS)
> data(lifeexpect)
>
> matplot(lifeexpect$myear, cbind(lifeexpect$male, lifeexpect$female),
+         xlab="Year", ylab="Life expectancy",
+         pch=20, col=c("black", "blue"))
>
> lo <- loess(male ~ myear, span=.25, data=lifeexpect)
> lo2 <- loess(male ~ myear, span=1, data=lifeexpect)
> lines(lifeexpect$myear, predict(lo))
> lines(lifeexpect$myear, predict(lo2), lty=2)
```

The two loess curves follow the overall trend of the male life expectancy closely but the one with the lowest span (the solid line) has a much closer fit to the data while still being smooth. When the span increases the line is not quite as flexible and the solid line does not capture the changes as nicely.

A spline is a function that is piecewise-defined by polynomials with certain restrictions to make them smooth. Smoothing splines are fit with the `smooth.spline` function which takes two numeric vectors for the c and y values as input. The flexibility of the curve is set by supplying a number to the `df` argument where larger numbers correspond to a more flexible curve. The number of piecewise polynomials is set by the `nknots` argument.

```
> smo <- smooth.spline(lifeexpect$myear, lifeexpect$female, df=8)
> lines(predict(smo), col="blue")
```

The spline function estimates the major changes in female life expectancy closely without being too fluctuating. Finally, Friedman's super smoother — called by the `supsmu` function — provides a close fit towards the end of the curve where there are yearly data but does not provide as good a fit at the yearly years. The flexibility of the super smoother can be modified with the `span` argument, which should be given the fraction of observations used for smoothing (smaller numbers allow for greater flexibility).

```
> supersmooth <- supsmu(lifeexpect$myear, lifeexpect$female)
> lines(supersmooth, col="blue", lty=3, lwd=3)
```

See also: See Problem 3.27 for an introduction to generalized additive models, which can also be used for smoothing data.

3.27 FIT A GENERALIZED ADDITIVE MODEL

Problem: You wish to use a generalized additive model to model nonlinear, smooth, data-driven relationships between predictors and the outcome.

Solution: Generalized additive models can be seen as generalizations of generalized linear models (see for example Problems 3.11 and 3.15) where the relationships between the individual predictors and the outcome follow smooth patterns that can be linear or nonlinear and where the shape of the predictor function is determined by the data. Thus, the linear predictor is given as a linear combination of smoothed and non-smoothed functions of predictor variables, e.g.,

$$g(\mathbb{E}Y_i) = \underbrace{s_1(x_{1i}; \lambda_1) + s_2(x_{2i}; \lambda_2)}_{\text{smoothed predictors}} + \cdots + \underbrace{\beta_k x_{ki} + \beta_l x_{li}}_{\text{regular predictors}},$$

where the functions $s_1, \ldots, s_2$ denote smooth, non-parametric functions which means that the shape of the predictor functions will be determined by the data for the relevant variable. The λ parameters control the individual smoothness with higher values of λ corresponding to smoother curves. $g(\mathbb{E}Y_i)$ is the link function of the linear predictor which allows the generalized additive model framework to handle both Gaussian errors, as well as logistic regression, Poisson regression, etc., just like typical generalized linear models.

The `gam` function from the `mgcv` package fits generalized additive models with both parametric coefficients and smoothed terms. `gam` expects a model formula (of the same type as for `glm`) as first argument. Smooth terms are included in the model formula with the `s` (spline smoothers) or `te` (tensor smoothers) functions in the form of expressions like `s(x1,x2,...)` where `x1,x2,...` are the covariates that are smoothed over. The `s` and `te` functions both accept the `k` and `sp` arguments which set the dimension of the basis used to identify the function and the degree of smoothing (λ in the formula above). Positive numbers fix the dimension and smoothing parameter — otherwise they will be estimated from the data (the default). Both influence the effective degrees of freedom where smaller values of `k` and larger values of `sp` both result in the curve becoming smoother. Finally, just as for `glm`, the `family` argument is used to set the model family and by default `gam` assumes that the family is Gaussian.

We use the `fev` data from the `isdals` package to illustrate how forced expiratory volume — a surrogate for lung capacity — is influenced by gender (0=female, 1=male), exposure to smoking (0=no, 1=yes), age, and height for 654 children aged 3–19 years. We do not know the shape of the relationships between age, height, and forced expiratory volume, so we use smoothed, non-parametric functions of those two variables. A scatter plot of the data is shown in Figure 3.5.

```
> library(mgcv)
> library(isdals)
> data(fev)
```

```
> model <- gam(FEV ~ Gender + Smoke + s(Age) + s(Ht), data=fev)
> summary(model)

Family: gaussian
Link function: identity

Formula:
FEV ~ Gender + Smoke + s(Age) + s(Ht)

Parametric coefficients:
            Estimate Std. Error t value Pr(>|t|)
(Intercept)   2.6031     0.0242  107.70   <2e-16 ***
Gender        0.0926     0.0335    2.76   0.0059 **
Smoke        -0.1399     0.0576   -2.43   0.0155 *
---
Signif. codes:  0 '***' 0.001 '**' 0.01 '*' 0.05 '.' 0.1 ' ' 1

Approximate significance of smooth terms:
        edf Ref.df    F p-value
s(Age) 2.43   3.14 18.7 6.5e-12 ***
s(Ht)  6.52   7.64 73.0 < 2e-16 ***
---
Signif. codes:  0 '***' 0.001 '**' 0.01 '*' 0.05 '.' 0.1 ' ' 1

R-sq.(adj) =  0.796   Deviance explained =   80%
GCV =  0.156  Scale est. = 0.15315    n = 654
```

The results for the non-smoothed parameters should be interpreted as for the corresponding generalized linear model. Thus, here we find that there is a negative effect of smoking with an average difference in lung capacity of -0.1399 liters and that this difference in lung capacity between smokers and non-smokers appears to be just about significant ($p = 0.0155$). There is also a difference in boys and girls with boys having larger lung capacity.

For the smoothed functions we see information on whether they are significant (essentially testing the hypotheses that a smoothed function could be replaced with a constant function). It also lists the effective degrees of freedom (edf), which is a combination of the dimension of the bases and the smoothing parameters and larger values of the effective degrees of freedom correspond to more flexibility, i.e., more wiggly curves.

If we plot the fitted model we are shown the smoothed relationship between the smoothed predictors and the outcome.

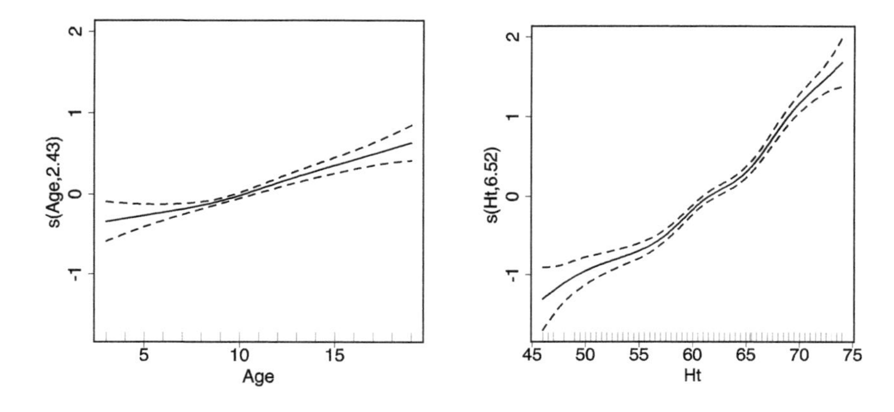

Figure 3.15: Smoothed functions for age and height estimated from the generalized additive model applied on the forced expiratory volume dataset.

```
> plot(model)
```

The result is shown in Figure 3.15 where we see that the estimated relationship between age and lung capacity is almost linear while it is slightly more wiggly and monotonously increasing for height.

We can test the linearity of, say, age by fitting a model with a linear effect of age and comparing the fit to the model with a smoothed effect of age. This is done with the anova function, and we set the test="Chisq" argument to obtain a p-value based on the difference in model fits.

```
> linmodel <- gam(FEV ~ Gender + Smoke + Age + s(Ht), data=fev)
> anova(linmodel, model, test="Chisq")
Analysis of Deviance Table

Model 1: FEV ~ Gender + Smoke + Age + s(Ht)
Model 2: FEV ~ Gender + Smoke + s(Age) + s(Ht)
  Resid. Df Resid. Dev   Df Deviance Pr(>Chi)
1       642       98.8
2       640       98.3 2.05    0.495     0.21
```

With a p-value of 0.21 it is clearly okay to let the effect of age be linear when modeling its impact on lung capacity.

The predict function can be used to predict the expected outcome from a generalized additive model. Besides the fitted model object we can set the newdata argument with a data frame that contains the variables necessary for making a prediction. For example, to predict the lung capacity for a 14-year-old female who is exposed to smoking and with a height of 60 inches we call

```
> predict(model, newdata=data.frame(Age=14, Ht=60, Smoke=1, Gender=0))
   1
2.58
```

Thus, the expected lung capacity for the 14-year-old girl is 2.58 liters.

Thin-plate splines generalize the one-variable spline smoothers presented above. They can be defined in gam by including two variables within the s function. For example, to create a smooth function of both age and height we include both in the call to s.

```
> model2 <- gam(FEV ~ Gender + Smoke + s(Age, Ht), data=fev)
> summary(model2)

Family: gaussian
Link function: identity

Formula:
FEV ~ Gender + Smoke + s(Age, Ht)

Parametric coefficients:
            Estimate Std. Error t value Pr(>|t|)
(Intercept)   2.6021     0.0241  108.08   <2e-16 ***
Gender        0.0945     0.0334    2.83   0.0049 **
Smoke        -0.1395     0.0578   -2.42   0.0160 *
---
Signif. codes:  0 '***' 0.001 '**' 0.01 '*' 0.05 '.' 0.1 ' ' 1

Approximate significance of smooth terms:
           edf Ref.df   F p-value
s(Age,Ht) 12.2   16.5 135  <2e-16 ***
---
Signif. codes:  0 '***' 0.001 '**' 0.01 '*' 0.05 '.' 0.1 ' ' 1

R-sq.(adj) =  0.798   Deviance explained = 80.2%
GCV = 0.15552  Scale est. = 0.15191   n = 654
```

There is very little difference in the output from the summary function and the output also shows that including the more flexible mode (where the smoothed function depends on two variables) only results in a very small increase in deviance explained.

If the pers=TRUE argument is added when we plot the fitted model then we see the perspective plot of the two-dimensional smoothed function shown in Figure 3.16.

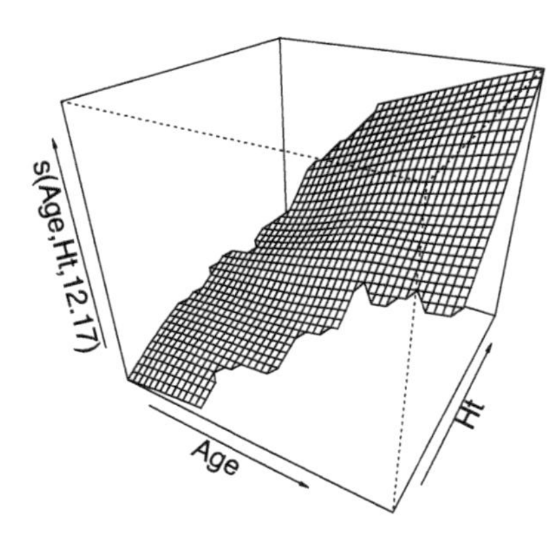

Figure 3.16: Smoothed two-dimensional function for age and height estimated from the generalized additive model applied on the forced expiratory volume dataset.

```
> plot(model2, pers=TRUE)
```

The plot shows the same result as the `summary` function: the effects of the two variables age and height on lung capacity appears to work independently of each other since the smoothed two-dimensional surface appears to be very close to a plane.

See also: The `gam` package also implements generalized additive models in R. The book by Wood (2006) presents a thorough introduction to generalized additive models in R.

3.28 USE META ANALYSIS TO COMBINE AND SUMMARIZE THE RESULTS FROM SEVERAL STUDIES

Problem: You want to combine and synthesize the effect from several different studies.

Solution: Meta analysis is used to combine and summarize compatible effects from several individual studies. It is relevant when there exist multiple studies with conflicting results and we wish to obtain a consensus result, or when the effects from several smaller studies should be combined to provide greater power. An effect can be almost any statistic of interest and the meta analysis assigns greater weight to studies that are more precise (i.e., have smaller variance).

Two types of models are typically used for meta analysis: fixed effect and random effects. The fixed effect model is generally applicable when it is assumed that all the studies that are part of the analysis share a common true effect and that the differences between the study results are therefore due to sampling variation. Under the random effects model the true effects in the studies are assumed to be sampled from an underlying distribution that represents the variation in true effects among the studies caused by studies being slightly different, for example due to differences in populations or design implementations. Essentially the random effects setup is identical to the mixed model situation from Problem 3.16.

The `metafor` package is one of many R packages that can be used for meta analysis. The main function in the `metafor` package is `rma` which performs both fixed and random effect meta analysis. `rma` expects the first argument to be a vector of the observed outcomes or effects for each study, and the individual study standard errors as argument `sei` (alternatively the individual study *variances* could be given as the `vi` argument). Two additional arguments are of importance: the `method` argument determines if a fixed or random effect analysis is performed, while the `measure` argument accepts a character string that specifies the *type* of effect measure (e.g., `RR` for log relative risk, `OR` for log odds ratio, `RD` for risk difference, or `GEN` for a general setup (the default), etc.).

Our first analysis is from a simple meta analysis situation with six studies where we know the effect size (the mean difference between treated and control groups with positive values suggesting that the treatment effect is better) and the corresponding standard errors for each study (data are adapted from Borenstein et al. (2011, Table 14.1)). We do not need any more information such as the individual standard deviation and sample sizes as they are already part of the standard error.

```
> library(metafor)
> yi  <- c(0.095, 0.277, 0.367, 0.664, 0.462, 0.185)
> sei <- c(0.033, 0.031, 0.050, 0.011, 0.043, 0.023)
> res <- rma(yi, sei=sei)
> res

Random-Effects Model (k = 6; tau^2 estimator: REML)

tau^2 (estimated amount of total heterogeneity): 0.0417 (SE = 0.0271)
tau (square root of estimated tau^2 value):      0.2042
I^2 (total heterogeneity / total variability):   98.40%
H^2 (total variability / sampling variability):  62.44
```

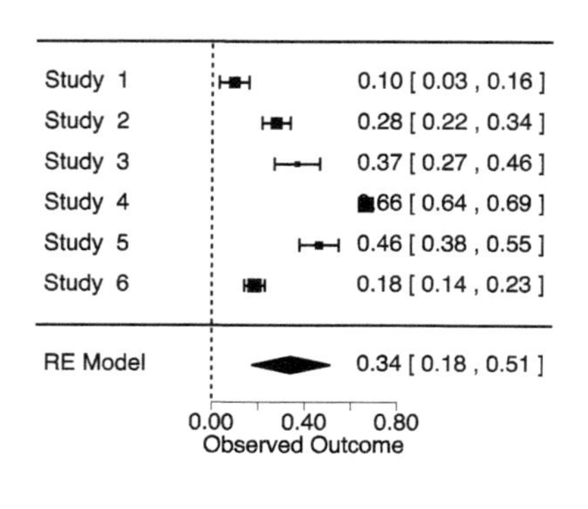

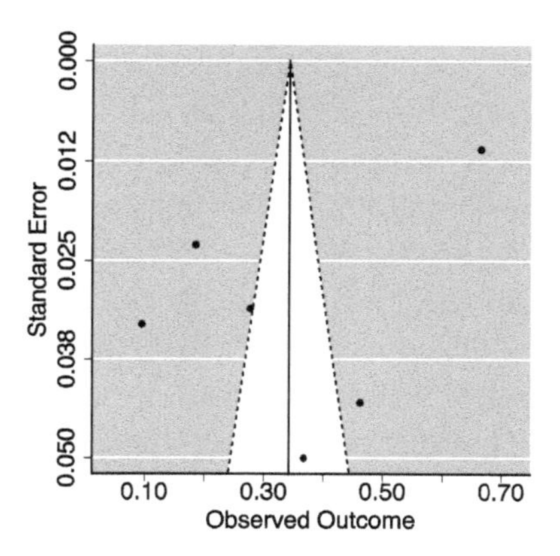

Figure 3.17: Forest and funnel plot.

```
Test for Heterogeneity:
Q(df = 5) = 619.5577, p-val < .0001

Model Results:

estimate       se      zval      pval     ci.lb     ci.ub
  0.3420   0.0845    4.0468    <.0001    0.1764    0.5077      ***

---
Signif. codes:  0 '***' 0.001 '**' 0.01 '*' 0.05 '.' 0.1 ' ' 1
```

The first part of the output is almost self-explanatory and shows information about the heterogeneity (since we used a random-effect model) and tests the hypothesis of no heterogeneity. In our case the variation in true effects among studies is quite substantial, and the hypothesis of no heterogeneity is clearly rejected. The last part of the output shows the overall estimate of the effect and its confidence interval. In this case the overall effect size is 0.342, with a 95% confidence interval for the difference between treated and controls ranging from 0.1764 to 0.5077. Thus, the overall effect is different from zero.

We can view the results graphically in a forest plot, where the effect size and confidence error for each study is shown. A forest plot is created from the meta analysis fit using the `forest` function. An example is given in the left plot of Figure 3.17 where we also see the overall estimate and confidence interval as the diamond towards the bottom.

```
> forest(res)
```

It is also customary to evaluate the possible publication bias by creating a funnel plot, which the `funnel` function creates. In a funnel plot the standard error is plotted against the effect for each study and we should see a fairly symmetric distribution of points around the overall effect estimate with larger variation when the standard error is large. The right-hand plot in Figure 3.17 shows a funnel plot where the approximate 95% confidence interval is shown in gray. There does not appear to be any publication bias (although it is almost impossible to evaluate with only six studies) but most of the points are in the gray area which corresponds to the heterogeneity we saw above.

```
> funnel(res)
```

A fixed-effects model can be fitted by setting `method="FE"`.

```
> res2 <- rma(yi, sei=sei, method="FE")
> res2

Fixed-Effects Model (k = 6)

Test for Heterogeneity:
Q(df = 5) = 619.5577, p-val < .0001

Model Results:

estimate       se     zval     pval    ci.lb    ci.ub
  0.5063   0.0088  57.8477   <.0001   0.4891   0.5235      ***

---
Signif. codes:  0 '***' 0.001 '**' 0.01 '*' 0.05 '.' 0.1 ' ' 1
```

The output resembles the output we got from the random effects model except that we get no information about the among study variance component (since there are not any random effects in the model). For the fixed-effect model the estimate of the overall effect becomes slightly larger and — as we would expect due to the different weights of the studies — the width of the confidence interval becomes narrower.

The major part of a meta analysis is to collect and summarize the results from the individual studies. The `escalc` can help with calculating the effect size and study weights (or standard errors) for common situations, but we can also include the tables directly in `rma`. Many meta analyses are based on results that can be summarized in 2×2 tables such as the one below.

	Outcome 1	Outcome 2
Group 1	ai	bi
Group 2	ci	di

The variable names given in the table correspond to the arguments for the `rma` function if `sei` or `vi` are not given. The function is very flexible and can also accept row and column totals instead, but we will just enter the table elements as in the table above.

The `metafor` package provides the `dat.bcg` which contains results from 13 studies on the effectiveness of the BCG vaccine against tuberculosis. The data can be seen below.

```
> data("dat.bcg")
> dat.bcg
   trial               author year tpos  tneg cpos  cneg ablat
1      1              Aronson 1948    4   119   11   128    44
2      2     Ferguson & Simes 1949    6   300   29   274    55
3      3      Rosenthal et al 1960    3   228   11   209    42
4      4    Hart & Sutherland 1977   62 13536  248 12619    52
5      5 Frimodt-Moller et al 1973   33  5036   47  5761    13
6      6      Stein & Aronson 1953  180  1361  372  1079    44
7      7     Vandiviere et al 1973    8  2537   10   619    19
8      8           TPT Madras 1980  505 87886  499 87892    13
9      9     Coetzee & Berjak 1968   29  7470   45  7232    27
10    10      Rosenthal et al 1961   17  1699   65  1600    42
11    11       Comstock et al 1974  186 50448  141 27197    18
12    12   Comstock & Webster 1969    5  2493    3  2338    33
13    13       Comstock et al 1976   27 16886   29 17825    33
        alloc
1      random
2      random
3      random
4      random
5   alternate
6   alternate
7      random
8      random
9      random
10 systematic
11 systematic
12 systematic
13 systematic
```

The four columns `tpos`, `tneg`, `cpos`, and `cneg` represent the values `ai`, `bi`, `ci`, and `di`, respectively. Since the data are from prospective studies it is relevant to use log relative risk as the effect size, which we specify with the `measure="RR"`

```
> rma(measure="RR", ai=tpos, bi=tneg, ci=cpos, di=cneg, data=dat.bcg)

Random-Effects Model (k = 13; tau^2 estimator: REML)

tau^2 (estimated amount of total heterogeneity): 0.3132 (SE = 0.1664)
tau (square root of estimated tau^2 value):      0.5597
I^2 (total heterogeneity / total variability):   92.22%
H^2 (total variability / sampling variability):  12.86

Test for Heterogeneity:
Q(df = 12) = 152.2330, p-val < .0001

Model Results:

estimate      se     zval    pval    ci.lb    ci.ub
 -0.7145  0.1798  -3.9744  <.0001  -1.0669  -0.3622      ***

---
Signif. codes:  0 '***' 0.001 '**' 0.01 '*' 0.05 '.' 0.1 ' ' 1
```

Our conclusion here is that estimated average log relative risk is -0.7145 which translates back to a relative risk estimate of $\exp(-0.7145) = 0.49$. Thus, the overall risk of infection is halved for the individuals that are vaccinated compared to non-vaccinated, and this effect is clearly significant since the confidence interval is far from zero.

See also: `rma` also accepts moderators, i.e., external variables that might influence the results. The `metaplus` package fits random effects models where the assumption that the random effects follow a normal distribution is relaxed.

3.29 USE RANDOM FOREST FOR CLASSIFICATION AND REGRESSION

Problem: You want to predict an outcome by combining information from multiple classification or regression trees into a random forest ensemble learner.

Solution: Classification and regression trees are decision trees used to predict an outcome and identify important explanatory variables. In random forest, a large number of decision trees generated from bootstrapped data is combined to improve predictive accuracy by aggregating the results across all of the trees.

R has several packages that can be used for random forest analysis. Here we will focus on the `caret` package (the name is short for classification and regression trees) which provides a unified interface to most of the existing methods, and it contains all of the functionality that is needed for tuning random forest analysis.

For illustration we use the Vinho verde red wine data provided by Cortez et al. (1998) which contains information on 11 different variables as well as the wine quality score ranging from 0 (very bad) to 10 (very excellent) as graded by a number of expert raters. We convert the quality to a categorical variable using the `cut` function such that quality values 5 and 6 are called "normal", below is "bad", and above is "good".

```
> url <- "http://www.biostatistics.dk/data/winequality-red.csv"
> indata <- read.csv(url, sep=";", header=TRUE)
> indata$taste <- cut(indata$quality, breaks=c(0, 4, 6, 10),
+                     labels=c("Bad", "Normal", "Good"))
> head(indata)
  fixed.acidity volatile.acidity citric.acid residual.sugar
1           7.4             0.70        0.00            1.9
2           7.8             0.88        0.00            2.6
3           7.8             0.76        0.04            2.3
4          11.2             0.28        0.56            1.9
5           7.4             0.70        0.00            1.9
6           7.4             0.66        0.00            1.8
  chlorides free.sulfur.dioxide total.sulfur.dioxide density
1     0.076                  11                   34   0.998
2     0.098                  25                   67   0.997
3     0.092                  15                   54   0.997
4     0.075                  17                   60   0.998
5     0.076                  11                   34   0.998
6     0.075                  13                   40   0.998
    pH sulphates alcohol quality  taste
1 3.51      0.56     9.4       5 Normal
2 3.20      0.68     9.8       5 Normal
3 3.26      0.65     9.8       5 Normal
4 3.16      0.58     9.8       6 Normal
5 3.51      0.56     9.4       5 Normal
6 3.51      0.56     9.4       5 Normal
```

In order to evaluate the model fitted by the random forest we randomly split the data into two parts — a training and a test dataset —

and fit the model on the training dataset, while we evaluate its performance on the test dataset. Below, we include 63% of the observations in the training dataset and the remaining observations in the test dataset. Also, we should remove the `quality` variable from the training and test datasets since that variable perfectly predicts the outcome since we constructed `taste` that way.

```
> pick <- sample(nrow(indata), 0.63 * nrow(indata))
> traindata <- indata[pick,]  # Select a random sample for training
> testdata <- indata[-pick,]  # and put the rest in test data
> traindata$quality <- NULL   # Remove the quality variable
> testdata$quality <- NULL    # from training and test data
> table(traindata$taste)

   Bad Normal   Good
    49    815    143
```

A random forest is grown and trained with the `train` function from the `caret` package. It accepts a model formula as input to identify the outcome and the predictors that the random forest algorithm can choose from. A period, ., can be used on the right-hand side of the model formula to automatically include all variables except the outcome.

`train` accepts several arguments that influence the random forest generation. `ntree` sets the number of trees to grow (default 500) and `method` sets the method used for computing the tree. At the time of writing there are more than 200 possible methods, all of which can be seen by typing `names(getModelInfo())`. Here we will use the default method `rf` which corresponds to random forest.

The default values of `train` are set to optimize and tune the number of variables that are randomly sampled as candidates at each node split.The `metric` argument sets the summary statistics used to select the optimal model. By default, the root-mean-squared-error `metric="RMSE"` is used for regression trees while `metric="Accuracy"` is used for classification trees. Finally, the `trControl` argument accepts a list of values that determine the parameters used for training and tuning the data. The `trainControl` function returns a list of all possible control parameters. By default 25 bootstrap samples are used for tuning but setting the `method="cv"` in the call to `trainControl` changes this to cross-validation.

We fit a random forest to the taste classification based on all the variables in the data frame.

```
> library(caret)
>
> fit <- train(taste ~ . , data=traindata, method="rf",
+              trControl=trainControl(method="cv",number=5))
> print(fit)
Random Forest

1007 samples
  11 predictor
   3 classes: 'Bad', 'Normal', 'Good'

No pre-processing
Resampling: Cross-Validated (5 fold)
Summary of sample sizes: 805, 805, 807, 805, 806
Resampling results across tuning parameters:

  mtry  Accuracy  Kappa
   2    0.839     0.346
   6    0.834     0.380
  11    0.843     0.438

Accuracy was used to select the optimal model using
 the largest value.
The final value used for the model was mtry = 11.
```

Information about the random forest generation process is provided in the output as well as information on the tuning parameters. In our case the cross-validation shows that using all 11 predictors provides slightly better accuracy than using a smaller number of predictors. However, there does not seem to be a huge difference if the number of sampled predictors is lessened. We can force `train` to use a specific value (or set of values) by supplying the `tuneGrid` argument in the call to `train`. The input to `tuneGrid` should be a data frame with possible tuning parameters and we can autogenerate this by giving a vector to the `expand.grid` function. The variable names in the data frame should match the name of the tuning parameter of interest, but here we only worry about the `mtry` parameter.

```
> mtry <- expand.grid(mtry = 5)
> train(taste ~ . , data=traindata, method="rf", tuneGrid=mtry,
+               trControl=trainControl(method="cv",number=5))
Random Forest

1007 samples
  11 predictor
   3 classes: 'Bad', 'Normal', 'Good'
```

```
No pre-processing
Resampling: Cross-Validated (5 fold)
Summary of sample sizes: 805, 806, 806, 806, 805
Resampling results:

  Accuracy  Kappa
  0.84      0.401

Tuning parameter 'mtry' was held constant at a value of 5
```

In this case the out-of-bag prediction error is 16% which is not substantially different from the tuned result we got with all 11 predictors. The `confusionMatrix` extracts the confusion matrix of the training data which is part of the fitted object, so we can see how and where our trained forest goes wrong.

```
> confusionMatrix(fit)
Cross-Validated (5 fold) Confusion Matrix

(entries are percentual average cell counts across resamples)

          Reference
Prediction  Bad Normal Good
    Bad     0.5    1.3  0.0
    Normal  4.3   76.1  6.5
    Good    0.1    3.6  7.7

 Accuracy (average) : 0.8431
```

Here we see that we have very few incidences where the trained model classifies a bad wine as good or vice versa.

The relative variable importance is extracted by the `varImp` function. It automatically scales the importance scores to be between 0 and 100 so we set `scale=FALSE` to avoid the normalization step.

```
> varImp(fit, scale=FALSE)
rf variable importance

                     Overall
alcohol                 58.9
volatile.acidity        44.5
sulphates               35.0
total.sulfur.dioxide    27.8
citric.acid             27.0
```

```
fixed.acidity            24.8
residual.sugar           24.8
pH                       22.7
chlorides                20.2
free.sulfur.dioxide      20.0
density                  19.9
```

alcohol and volatile.acidity are clearly more important than the rest of the predictors, which appear to have roughly the same importance. However, there are really no predictors that we can completely disregard. Note that it is generally the relative size of the importance that is of interest when looking at these numbers.

Finally, we can use our random forest created from the training dataset to evaluate the prediction on the test dataset using predict function on the test data.

```
> pred <- predict(fit, newdata=testdata)
> confusionMatrix(pred, testdata$taste) # Compare
Confusion Matrix and Statistics

          Reference
Prediction Bad Normal Good
    Bad      1      0    0
    Normal  13    492   36
    Good     0     12   38

Overall Statistics

               Accuracy : 0.897
                 95% CI : (0.87, 0.92)
    No Information Rate : 0.851
    P-Value [Acc > NIR] : 7e-04

                  Kappa : 0.513
 Mcnemar's Test P-Value : NA

Statistics by Class:

                     Class: Bad Class: Normal Class: Good
Sensitivity             0.07143         0.976      0.5135
Specificity             1.00000         0.443      0.9768
Pos Pred Value          1.00000         0.909      0.7600
Neg Pred Value          0.97800         0.765      0.9336
Prevalence              0.02365         0.851      0.1250
Detection Rate          0.00169         0.831      0.0642
Detection Prevalence    0.00169         0.914      0.0845
Balanced Accuracy       0.53571         0.710      0.7452
```

In this case the `confusionMatrix` function provides a ton of information. We can see that we have some problems correctly identifying the bad wines that have an abysmal sensitivity. We also see that our random forest model provides statistically significant better accuracy than the no information rate which is simply the relative frequency of the most common outcome. For the test data our best bet on the wine quality without knowing anything about it would be that it was a normal wine since $(492+12)/592 = 0.8514$ so roughly 85% of the wines are in the normal group. Thus, our random forest model only improves the classification probability by 4.6 percentage points.

See also: The `ranger` package provides a faster implementation of random forests than the standard random forest algorithm. It can be used with `caret` by setting `method="ranger"` after loading the package. The `randomForestSRC` package gives a slightly different implementation that is also fast and both packages also allow for more complicated outcomes such as survival regression trees.

3.30 FIT A LINEAR QUANTILE REGRESSION MODEL

Problem: You want to fit a quantile regression model to describe the linear relationship between a quantile of a quantitative response variable and one or more explanatory quantitative variables.

Solution: Quantile regression models are useful for modeling a conditional quantile (e.g., the conditional median) of a response variable given one or more explanatory variables. Traditional regression models such as those used in linear and multiple regression (see Problems 3.3 and 3.4) estimate the *average* effect of predictors on the outcome: the regression coefficient represents the average increase in the outcome for a one-unit increase in the corresponding predictor variable. In quantile regression we estimate the change for a specific *quantile* of the outcome when the corresponding predictor variable is changed by one unit. This is particularly interesting when some quantiles are influenced more by a predictor than other quantiles. Thus, quantile regression enables us to study and compare the impact of predictors on different quantiles of the response distribution, and therefore provides a more complete picture of the relationship between the predictors and the outcome.

The `quantreg` package implements functions for fitting and testing hypotheses in quantile regression models. The `rq` function fits a quantile regression model and accepts a model formula as its first argument. By default, only the median (50th quantile) is fitted but specific quantiles can be set with the `tau` argument. In the example below, we use the `fev` data from the `isdals` package to examine if and how age in years, gender (0=female, 1=male), and exposure to smoking (0=no, 1=yes) influence the forced expiratory volume (FEV) in children. Initially we only fit the median quantile and as usual, we can get more information from the model fit with the `summary` function.

```
> library(quantreg)
> library(isdals)
> data(fev)
> res <- rq(FEV ~ Age + Gender + Smoke, data=fev)
> summary(res)

Call: rq(formula = FEV ~ Age + Gender + Smoke, data = fev)

tau: [1] 0.5

Coefficients:
            coefficients lower bd upper bd
(Intercept)  0.2050       0.0511   0.4505
Age          0.2300       0.2016   0.2457
Gender       0.2410       0.1386   0.3147
Smoke       -0.0970      -0.2452   0.1714
```

The output shows the estimated coefficients for the conditional effect of the predictors on the median. The estimates should be interpreted as the marginal change in the conditional median due to a marginal change in the corresponding predictor. There is no guarantee that a person will remain in the same quantile when the predictor changes — it is the overall distribution that we model. The quantile regression results suggest that smoking has a negative effect on the median forced expiratory volume. The conditional median for smokers is 0.097 liters smaller than for non-smokers. The lower and upper boundary columns show 95% confidence interval limits for the respective coefficients so it is clear that the median lung capacity is not substantially different for smokers versus non-smokers.

Note that the `rq` function throws an error that the solution may be non-unique. This is not an error *per se* but is a consequence of the quantile regression approach: multiple values of the coefficients might give rise to the same quantile. The error is quite common with discrete covariates and interpolation between values can remedy this. The default method used to fit the model, `br`, does not use interpolation but setting the argument `method="fn"` ensures that the model fitting approach does. These warnings have been removed in the output below for brevity.

There is no standard procedure to compute standard errors in quantile regression, but the `se` argument to `summary` allows the user to choose between different methods. The default is `se="rank"` which produces boundaries based on inverting ranks. Other possibilities for `se` include `iid` for asymptotic standard error based on an independence assumption, `nid` which computes a local sandwich estimate, `ker` for kernel-based sandwich estimates of the standard error, and `boot` for a bootstrap approach for standard error estimation. If the bootstrap approach is chosen then it is possible to use different versions of the bootstrap and to choose the number of bootstrap samples through the `bsmethod` and `R` arguments, respectively.

As we see below the output changes slightly when we request the local sandwich estimates of the standard errors, but the overall conclusions do not change compared to the rank-based boundaries computed above.

```
> summary(res, se="nid")

Call: rq(formula = FEV ~ Age + Gender + Smoke, data = fev)

tau: [1] 0.5

Coefficients:
            Value     Std. Error t value  Pr(>|t|)
(Intercept)  0.20500  0.09211     2.22570  0.02638
Age          0.23000  0.00990    23.23264  0.00000
Gender       0.24100  0.05262     4.57997  0.00001
Smoke       -0.09700  0.12881    -0.75305  0.45169
```

We can compare our quantile regression estimates to the corresponding estimates from a multiple regression model (see Problem 3.4).

```
> summary(lm(FEV ~ Age + Gender + Smoke, data=fev))

Call:
lm(formula = FEV ~ Age + Gender + Smoke, data = fev)

Residuals:
    Min      1Q  Median      3Q     Max
-1.4671 -0.3543 -0.0381  0.3220  1.9494

Coefficients:
            Estimate Std. Error t value Pr(>|t|)
(Intercept)  0.23777    0.08023    2.96   0.0032 **
```

```
Age          0.22679    0.00788   28.76  < 2e-16 ***
Gender       0.31527    0.04271    7.38  4.8e-13 ***
Smoke       -0.15397    0.07798   -1.97   0.0487 *
---
Signif. codes:  0 '***' 0.001 '**' 0.01 '*' 0.05 '.' 0.1 ' ' 1

Residual standard error: 0.543 on 650 degrees of freedom
Multiple R-squared:  0.609,Adjusted R-squared:  0.608
F-statistic:  338 on 3 and 650 DF,  p-value: <2e-16
```

We can see that the average effect of gender from the multiple linear model (0.315) is substantially different from the median effect found from the quantile regression model (0.241). The difference is due to the fact that the gender effect varies a lot among the different quantiles and that the effect from the multiple regression model has to capture all of those different effects in a single number to cover all the quantiles.

If we fit a quantile regression model for multiple quantiles and plot the summary then we can visually inspect how the quantile regression coefficients differ. Figure 3.18 shows the quantile regression estimates for the 10%, 25%, 50%, 75%, and 90% quantiles for each predictor together with the corresponding estimate from the multiple regression model and the corresponding 95% confidence intervals (set by the `level` argument). The age and gender effects clearly differ for different parts of the distribution such that the effect of age is larger for the larger quantiles than it is for smaller quantiles.

```
> res <- rq(FEV ~ Age + Gender + Smoke, data=fev,
+           tau=c(0.1, 0.25, 0.5, 0.75, 0.9))
> plot(summary(res), lcol="blue", level=0.95)
```

Two types of hypotheses are generally relevant for quantile regression models: We might want to test that the effect of a predictor is constant across a range of quantiles (corresponding to testing the hypothesis that the black “curves” in Figure 3.18 are constant) or it may be of interest to examine if a predictor is significant for a given quantile.

The `anova` function uses a Wald test to examine the hypothesis that the parameters are the same for each of the quantiles examined. Below we test the combined hypothesis that each of the effects of age, gender, and smoking status are constant across the quantiles considered. The argument `joint=FALSE` is included in the call to `anova` to ensure that we get a separate test for each of the coefficients instead of a combined test.

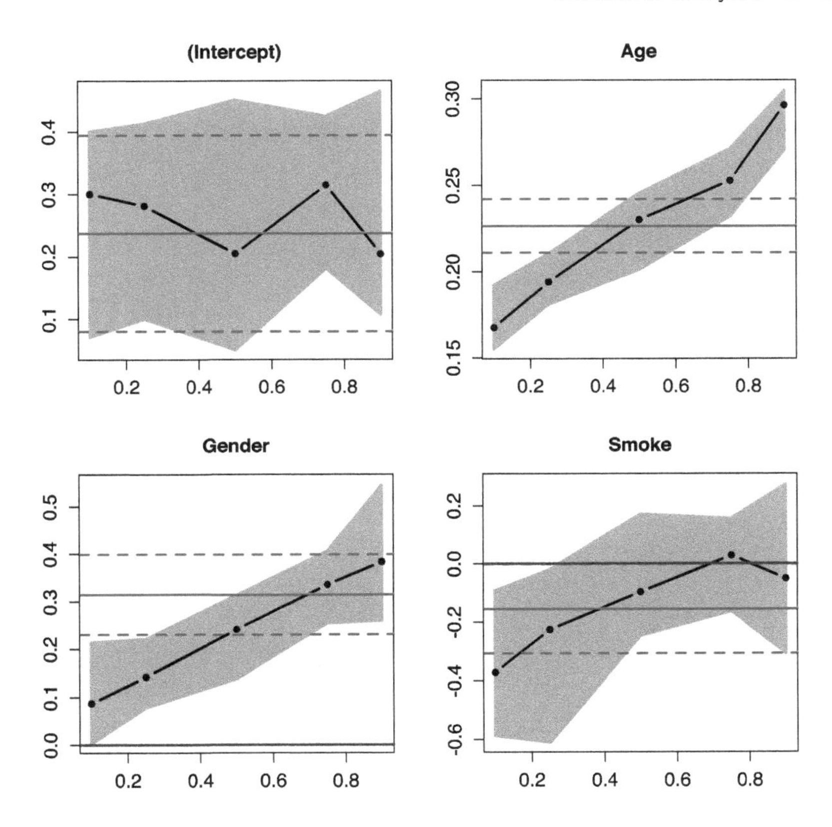

Figure 3.18: The black lines show the estimated coefficients for different quantiles ranging from 0.1 to 0.9 for each of the variables in the model with corresponding 95% confidence interval (gray areas). Blue lines are the estimates and corresponding 95% confidence interval obtained from an ordinary least square estimate.

```
> anova(res, joint=FALSE)
Quantile Regression Analysis of Deviance Table

Model: FEV ~ Age + Gender + Smoke
Tests of Equality of Distinct Slopes: tau in {  0.1 0.25 0.5 0.75 0.9  }

       Df Resid Df F value  Pr(>F)
Age     4     3266   16.05 5.1e-13 ***
Gender  4     3266    4.58  0.0011 **
Smoke   4     3266    1.23  0.2960
---
Signif. codes:  0 '***' 0.001 '**' 0.01 '*' 0.05 '.' 0.1 ' ' 1
```

The results from the call to anova is identical to the results seen in Figure 3.18. There is a clear change in both the age and gender coefficients across the quantiles which is also reflected in the corresponding p values. Smoking status is not significant which is also in accordance with Figure 3.18 where the effects are roughly constant.

Hypotheses about the predictors for a single, fixed quantile are tested by fitting the two nested models — the one with and the one without the predictor of interest — and using the anova function to compare them. For example, to formally test the effect of age on the median we fit the two models and compare them.

```
> start <- rq(FEV ~ Age + Gender + Smoke, data=fev, tau=0.5)
> nested <- rq(FEV ~ Gender + Smoke, data=fev, tau=0.5)
> anova(start, nested)
Quantile Regression Analysis of Deviance Table

Model 1: FEV ~ Age + Gender + Smoke
Model 2: FEV ~ Gender + Smoke
  Df Resid Df F value Pr(>F)
1  1      650     540 <2e-16 ***
---
Signif. codes:  0 '***' 0.001 '**' 0.01 '*' 0.05 '.' 0.1 ' ' 1
```

Age is clearly significant for the median which is also what we could see from the confidence interval for the age effect on the median in Figure 3.18.

See also: Nonlinear quantile regression models can be fitted with the nlrq function also found in the quantreg package. The modQR can accommodate quantile regression models with multivariate responses, while the rqpd package handles quantile regression methods for longitudinal data.

MODEL VALIDATION

3.31 TEST FOR NORMALITY OF A SINGLE SAMPLE

Problem: You wish to test if a sample of residuals or observations follows a normal distribution.

Solution: Many statistical models require the residuals to follow a normal distribution for the resulting p-values and confidence intervals to be correct.

Normality of a sample can be assessed using graphical methods but normality can also be tested statistically. Here, we will cover four omnibus tests for normality of a single sample: Shapiro–Wilk's test, Lilliefors–Kolmogorov–Smirnov's test, Cramér–von Mises test, and the Anderson–Darling test. The four tests consider different departures from normality and collectively they provide a good overall examination of different aspects of normality.

Shapiro–Wilk's test is computed by the `shapiro.test` function while the Anderson–Darling, Cramér–von Mises, and Lilliefors–Kolmogorov–Smirnov tests are found in the `nortest` package as functions `ad.test`, `cvm.test`, and `lillie.test`, respectively. All four functions accept a single vector of observations as input.

Below we compute the tests for two different datasets — one from a normal distribution and one from a right-skewed distribution.

```
> library(nortest)
> x <- rnorm(100, mean=5, sd=2)          # Normal distribution
> y <- (rnorm(100, mean=5, sd=1.5))**2  # Right-skewed distribution
> shapiro.test(x)

Shapiro-Wilk normality test

data:  x
W = 1, p-value = 0.3
> ad.test(x)

Anderson-Darling normality test

data:  x
A = 0.5, p-value = 0.2
> cvm.test(x)

Cramer-von Mises normality test

data:  x
W = 0.09, p-value = 0.2
> lillie.test(x)

Lilliefors (Kolmogorov-Smirnov) normality test

data:  x
D = 0.07, p-value = 0.2
```

All four tests show no compelling evidence for non-normality for the normally distributed data.

For the non-normal data, all four tests reject the hypothesis of normality as seen below. In both of the examples here the four tests provide the same conclusion but that is not always the case.

```
> shapiro.test(y)

Shapiro-Wilk normality test

data:  y
W = 0.9, p-value = 1e-04
> ad.test(y)

Anderson-Darling normality test

data:  y
A = 2, p-value = 2e-04
> cvm.test(y)

Cramer-von Mises normality test

data:  y
W = 0.3, p-value = 5e-04
> lillie.test(y)

Lilliefors (Kolmogorov-Smirnov) normality test

data:  y
D = 0.1, p-value = 8e-04
```

Note that the `ks.test` function from base R cannot directly be used to compute the Kolmogorov–Smirnov test since it does not account for the fact that the parameters used with the theoretical distribution are estimated from the data.

See also: Problems 4.6, 4.20, and 4.21 show examples of ways to examine the distribution of a sample graphically.

3.32 TEST VARIANCE HOMOGENEITY ACROSS GROUPS

Problem: You wish to test the assumption of variance homogeneity for an analysis of variance model.

Solution: One of the assumptions of analysis of variance is that the variances of the observations in the individual groups are equal. Homogeneity of variances can be checked graphically as described in Problem 4.21, but it can also be tested statistically using for example Bartlett's test, Levene's test, or Fligner–Killeen's test. These tests are also meaningful in their own right if the interest is in knowing whether population group variances are different.

Bartlett's test tests the null hypothesis that the variances of the response variable are the same for the different groups by comparing the pooled variance estimate with the sum of the variances of individual groups. The `bartlett.test` function computes Bartlett's test in R. The input to `bartlett.test` should be either a numeric vector of response values and a numeric or factor vector that defines the groups or a regular model formula where the right-hand side of the model formula defines the groups.

We illustrate Bartlett's test using the `ToothGrowth` dataset where the tooth lengths are modeled as a function of the dose of vitamin C.

```
> data(ToothGrowth)
> bartlett.test(len ~ factor(dose), data=ToothGrowth)

Bartlett test of homogeneity of variances

data:  len by factor(dose)
Bartlett's K-squared = 0.7, df = 2, p-value = 0.7
```

Based on Bartlett's test, we fail to reject the null hypothesis of variance homogeneity for the three doses since the p-value is large (0.717).

Bartlett's test is sensitive to both departures from normality as well as heteroscedasticity and Levene's test is an alternative test for equality of variances which is more robust against departures from the normality assumption. Levene's test computes the absolute differences of the observations in each group from the group median and then tests if these deviations are equal for all groups.

The `leveneTest` from the `car` package computes Levene's test for homogeneity of variance across groups. `leveneTest` accepts the same input as `bartlett.test`.

```
> library(car)
> leveneTest(len ~ factor(dose), data=ToothGrowth)
Levene's Test for Homogeneity of Variance (center = median)
      Df F value Pr(>F)
group  2    0.65   0.53
      57
```

Levene's test for homogeneity has a p-value of 0.5281 which is slightly smaller than the p-value from Bartlett's test but the conclusion is still the same: we fail to reject the hypothesis of variance homogeneity for the three doses.

Instead of looking at absolute deviations from the group median we can look at absolute deviations from the group mean. The `center` option to `leveneTest` sets the function that is used to compute the absolute deviations. For example, if we set `center=mean` then the mean will be used instead of the median. Generally, the median provides good robustness against many types of non-normal data while retaining good statistical power.

If the data are not believed to be normally distributed, then the non-parametric Fligner–Killeen test can be used to test the hypothesis that the variances in each group are the same. `fligner.test` computes Fligner–Killeen's test statistic and the function accepts the same type of input as `bartlett.test` or `leveneTest`.

```
> fligner.test(len ~ factor(dose), data=ToothGrowth)

Fligner-Killeen test of homogeneity of variances

data:  len by factor(dose)
Fligner-Killeen:med chi-squared = 1, df = 2, p-value =
0.5
```

We reach the same conclusion and fail to reject the hypothesis that the variances are equal.

If there is more than one grouping factor, then all factors must be completely crossed before using `bartlett.test`, `leveneTest`, or `fligner.test`. The `interaction` function can be used to create a complete cross of all factors. If the `ToothGrowth` data is to be analyzed using a two-way analysis of variance, we should check for variance homogeneity for the cross of dose and supplement type.

```
> bartlett.test(len ~ interaction(supp, dose), data=ToothGrowth)

Bartlett test of homogeneity of variances

data:  len by interaction(supp, dose)
Bartlett's K-squared = 7, df = 5, p-value = 0.2
```

There is no indication of variance heterogeneity when the groups are defined by both supplement type and dose.

See also: The `oneway.test` function can test equality of means for normally distributed data without assuming equal variances in each group.

3.33 VALIDATE A LINEAR OR GENERALIZED LINEAR MODEL

Problem: You want to validate a linear or generalized linear model to check if the underlying assumptions appear to be fulfilled.

Solution: Residuals commonly form the basis for graphical model validation for linear and generalized linear models (see Problem 4.21). However, the cumulative sum of residuals or other aggregates of residuals (e.g., smoothed residuals or moving sums) can be used to provide objective criteria for model validation.

The cumres function from the gof package implements model-checking techniques based on cumulative residuals and calculates Kolmogorov–Smirnov and Cramér–von Mises test statistics based on linear and generalized linear models (currently cumres supports linear, logistic, and Poisson regression models with canonical links).

cumres takes a model object fitted by lm or glm as first argument and calculates the test statistic for the cumulative residuals against the predicted values as well as against each of the quantitative explanatory variables. The R argument defaults to 500 and sets the number of samples used for the simulations, and the b option determines the bandwidth used for moving sums of cumulative residuals. The default value of b=0 corresponds to an infinitely wide bandwidth, i.e., standard cumulated residuals.

In the cherry tree dataset trees we seek to model the volume of the trees as a linear function of the height and diameter. We use the cumres function to check the validity of the multiple regression model.

```
> library(gof)
> data(trees)
> attach(trees)
> model <- lm(Volume ~ Girth + Height)  # Multiple regression
> cumres(model)

Kolmogorov-Smirnov-test: p-value=0.067
Cramer von Mises-test: p-value=0.007
Based on 1000 realizations.
Cumulated residuals ordered by predicted-variable.
---
Kolmogorov-Smirnov-test: p-value=0.069
Cramer von Mises-test: p-value=0.008
Based on 1000 realizations.
Cumulated residuals ordered by Girth-variable.
---
Kolmogorov-Smirnov-test: p-value=0.244
```

```
Cramer von Mises-test: p-value=0.271
Based on 1000 realizations.
Cumulated residuals ordered by Height-variable.
---
```

The output shows quite low p-values for both the predicted and girth variables which suggests that a better model should be found. An improved model might be based on the formula for the volume of a geometric cone shape, which suggests we log transform the volume, girth, and height.

```
> lVol <- log(Volume)        # Log transform variables
> lGir <- log(Girth)
> lHei <- log(Height)
> model2 <- lm(lVol ~ lGir + lHei)     # New model
> cumres(model2)

Kolmogorov-Smirnov-test: p-value=0.601
Cramer von Mises-test: p-value=0.602
Based on 1000 realizations.
Cumulated residuals ordered by predicted-variable.
---
Kolmogorov-Smirnov-test: p-value=0.574
Cramer von Mises-test: p-value=0.494
Based on 1000 realizations.
Cumulated residuals ordered by lGir-variable.
---
Kolmogorov-Smirnov-test: p-value=0.464
Cramer von Mises-test: p-value=0.61
Based on 1000 realizations.
Cumulated residuals ordered by lHei-variable.
---
```

We see that the p-values for the cumulative residual goodness-of-fit statistics have improved substantially with this transformed multiple regression model.

The `gof` package also extends the generic `plot` function to produce plots of the cumulative residuals. The plots show the cumulated residuals, a sample of realizations from the asymptotic distribution (the number of samples is determined by the `plots` argument to `cumres`), and the corresponding 95% confidence interval. The `level` argument sets the confidence level when plotting `cumres` objects and the default value is 0.95. Note that the `plot` function creates one plot for the predicted values as well as one for each of the quantitative explanatory variables. The

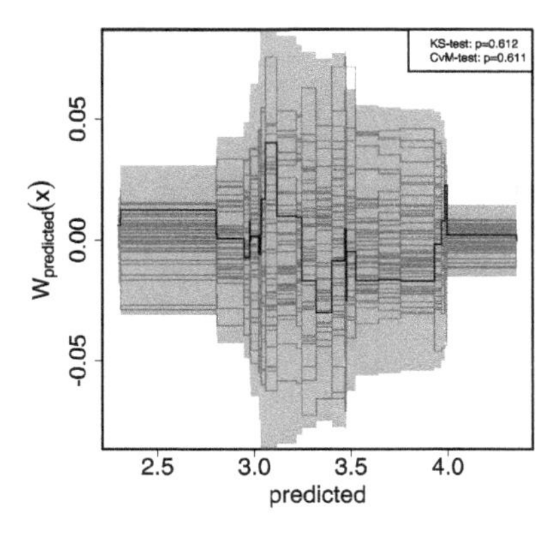

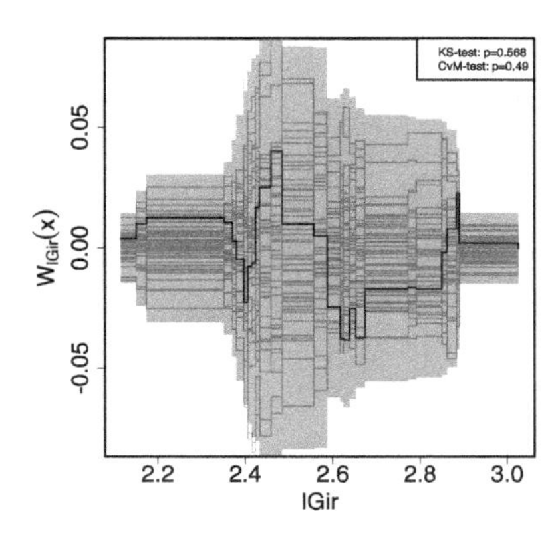

Figure 3.19: The left panel shows the cumulated residuals (black line), 50 sample realizations (dark gray lines), and the corresponding 95% confidence interval (light gray area) against the predicted values. The right panel shows the cumulated residuals against logarithm of tree girth.

following code is used to plot the cumulative residuals and the result for two of the three plots are shown in Figure 3.19.

```
> cr2 <- cumres(model2)  # Save the cumulated residuals
> plot(cr2)              # and make the plots
```

The observed cumulated residuals are well within the confidence band in both plots, so there is nothing here that suggests the transformed model is inadequate.

See also: The `cox.aalen` function in the `timereg` package implements a similar method for survival data.

CONTINGENCY TABLES

3.34 ANALYZE TWO-DIMENSIONAL CONTINGENCY TABLES

Problem: You wish to test for homogeneity or independence in a two-dimensional contingency table.

Solution: A two-dimensional contingency table is a table of counts that is formed by classifying observations by two categorical variables, where one variable determines the row categories and the other variable defines the column categories. The observations are assumed to be independent. R uses the `chisq.test` function for chi-squared tests and the `fisher.test` function for exact tests in two-dimensional contingency tables. Both functions accept a contingency table as input.

As an example, we use the following data from the county of Århus in Denmark where the concentration of the pesticide dichlorobenzamide (also called BAM) in drinking water was examined from two different municipalities within the county. The allowable limit of BAM is $0.10\mu g/l$, and the data are summarized below.

	Concentration (in μg/l)			
Municipality	< 0.01	0.01–0.10	> 0.10	Total
Hadsten	23	12	6	41
Hammel	20	5	9	34

We can test for independence between the two municipalities using both the chi-square and exact test with the following code.

```
> m <- matrix(c(23, 20, 12, 5, 6, 9), ncol=3)   # Input data
> m
     [,1] [,2] [,3]
[1,]   23   12    6
[2,]   20    5    9
> chisq.test(m)                                 # Chi-square test

Pearson's Chi-squared test

data:  m
X-squared = 3, df = 2, p-value = 0.2
> fisher.test(m)                                # Fisher's exact test

Fisher's Exact Test for Count Data

data:  m
p-value = 0.2
alternative hypothesis: two.sided
```

The chi-square and exact analyses show virtually the same results and the conclusion is that there is no difference in the distribution of BAM between the Hadsten and Hammel areas.

Additional options are relevant for the special case of 2×2 contingency tables. By default, the `chisq.test` function applies a continuity correction for an improved test statistic unless `correct=FALSE` is specified. For 2×2 contingency tables, we also have the possibility to specify the alternative hypothesis for the `fisher.test` function with the `alternative` option. The default is `two.sided` and other possibilities are that the true odds ratio is either `greater` or `less` than one for the one-sided alternatives.

If we collapse the table above such that concentrations below 0.10μg/l are grouped together we can analyze the resulting 2×2 contingency table as follows:

```
> m2 <- matrix(c(35, 25, 6, 9), ncol=2)  # Input data
> m2
     [,1] [,2]
[1,]   35    6
[2,]   25    9
> chisq.test(m2)

Pearson's Chi-squared test with Yates' continuity
correction

data:  m2
X-squared = 1, df = 1, p-value = 0.3
> chisq.test(m2, correct=FALSE)          # No continuity correction

Pearson's Chi-squared test

data:  m2
X-squared = 2, df = 1, p-value = 0.2
> fisher.test(m2, alternative="greater") # One-sided alternative

Fisher's Exact Test for Count Data

data:  m2
p-value = 0.2
alternative hypothesis: true odds ratio is greater than 1
95 percent confidence interval:
 0.688   Inf
sample estimates:
odds ratio
      2.08
```

The first example shows how Yates' continuity correction automatically is used with 2×2 tables, and the conclusions from both analyses indicate that the data gives no reason to doubt the independence assumption between the two regions. The one-sided exact test gives the same result when we only consider the one-sided alternative hypothesis.

The `chisq.test` function prints a warning message if the χ^2-approximation of the test statistic may be incorrect. In those situations it may be necessary to collapse categories or just generally use the exact test (see Problem 3.35 for an example of collapsing categories).

3.35 ANALYZE TWO-DIMENSIONAL CONTINGENCY TABLES WITH ORDINAL CATEGORIES

Problem: You wish to test for association in a two-dimensional contingency table with ordered classes.

Solution: A two-dimensional contingency table is a table of counts formed by classifying observations by two categorical variables. For nominal categories we can use chi-square or Fisher's exact test (see Problem 3.34) to analyze the data. Goodman–Kruskal's γ is a measure of association for two-dimensional contingency tables when the categories are ordered, and Goodman–Kruskal's γ is more powerful than ordinary nominal tests when the classes are ordered.

As an example, we use the `smokehealth` dataset from the `MESS` package. The data consists of a table from a larger study from Glostrup, and the idea is to examine the effect of smoking at 45 years of age on self-reported health five years later for males. Smoking and self-reported health are both ordered categorical variables.

```
> library(MESS)
> data(smokehealth)
> smokehealth
        VeryGood Good Fair Bad
Never         16   73    6   1
No more       15   75    6   0
1-14          13   59    7   1
15-24         10   81   17   3
25+            1   29    3   1
```

The data table shows the smoking categories as rows and the self-reported health status as columns. There appears to be a clear relationship between increased smoking and a decline in self-reported health status. The number of observations is also rather sparse for the "Bad" and "25+" categories, so it may be necessary to collapse some of the categories and we start by doing that before running the analysis.

```
> m <- smokehealth
> m[,3] <- m[,3]+ m[,4]    # Collapse last two columns
> m[4,] <- m[4,] + m[5,]   # Collapse last two rows
> m <- m[1:4,1:3]          # Extract result
> m
        VeryGood Good Fair
Never         16   73    7
No more       15   75    6
1-14          13   59    8
```

```
15-24          11  110   24
> gkgamma(m)

Goodman-Kruskal's gamma for ordinal categorical data

data:  m
Z = 3, p-value = 0.001
95 percent confidence interval:
 0.106 0.405
sample estimates:
Goodman-Kruskal's gamma
                  0.256
```

The gamma coeffiecient of 0.256 measures the strength of the monotonous relationship between the two variables, and the value is positive so there is positive association as we also saw from the table.

While the gamma coefficient is not very large it still has a 95% confidence interval that is substantially larger than zero, and the corresponding p-value is also statistically significant. Thus, we reject the hypothesis that the gamma coefficient is equal to zero.

For comparison we can see the results from the `chisq.test` function where the ordering of the classes is not taken into account.

```
> chisq.test(m)

Pearson's Chi-squared test

data:  m
X-squared = 10, df = 6, p-value = 0.05
```

Here we get that the association is not strong enough to be statistically significant, so using this test results in a drop in power.

3.36 ANALYZE TWO-DIMENSIONAL CONTINGENCY TABLES WITH PAIRED MEASUREMENTS

Problem: You have paired categorical data in a contingency table and want to test for marginal homogeneity.

Solution: The standard chi-square test (see Problem 3.34) requires that the numbers in the two-dimensional contingency table represent independent individuals, and thus it cannot be used for paired data. However, it is not uncommon to encounter data, where the same categorical measurement has been scored twice for observations that are paired — for example because an individual is scored twice, or because the two observations are related (e.g., twins, spouses, or if the individuals in the pair were matched in some way).

McNemar's test compares the marginal distributions of the two-dimensional contingency table. In R, the `mcnemar.test` function computes McNemar's test, and it takes a two-dimensional matrix or table as input. By default, `mcnemar.test` uses a continuity correction, but that can be prevented by setting the argument `correct=FALSE`.

In the example below we consider 86 study participants assessed twice for plaque index (PI) on their teeth, at baseline and 4 weeks later. We wish to assess whether the proportion of patients with high PI changes from the beginning of the study to four weeks later.

```
> m <- matrix(c(29, 17, 5, 35), 2,
+              dimnames = list("Baseline"=c("Low PI", "High PI"),
+                              "4 weeks"=c("Low PI", "High PI")))
> m
         4 weeks
Baseline  Low PI High PI
  Low PI      29       5
  High PI     17      35
> mcnemar.test(m)

McNemar's Chi-squared test with continuity correction

data:  m
McNemar's chi-squared = 6, df = 1, p-value = 0.02
```

The p-value is 0.02 which suggests that there might be a difference between the baseline distribution of PI and the 4-week distribution of PI. Only by looking at the numbers in the input table we can see that the plaque index must be generally lower at 4 weeks compared to baseline.

`mcnemar.test` uses asymptotic statistics to evaluate the hypothesis and that approximation might not be realistic if the numbers in the table are low. When we analyze data from a 2×2 table (i.e., we only have two categories in the contingency table) we can use the `binom.test` function instead to make an exact test for the same hypothesis of identical marginal distributions. To make an exact test we extract one of the off-diagonal numbers in the table and compare that to the sum of the off-diagonal numbers (and test the hypothesis that that probability should be equal to 0.5).

```
> binom.test(17, (5+17))

Exact binomial test

data:  17 and (5 + 17)
number of successes = 20, number of trials = 20, p-value
= 0.02
alternative hypothesis: true probability of success is not equal to 0.5
95 percent confidence interval:
 0.546 0.922
sample estimates:
probability of success
                 0.773
```

In this case there is no difference between the asymptotic test and the exact test, and the conclusion is still that we reject the hypothesis that the two distributions are the same.

The `mcnemar.test` function is not restricted to a 2×2 table and can easily handle larger contingency tables.

See also: Problem 3.12 for more information on the `clogit` function and conditional logistic regression, which is a generalization of McNemar's test to multiple explanatory variables (but restricted to binary outcomes).

3.37 ANALYZE CONTINGENCY TABLES USING LOG-LINEAR MODELS

Problem: You wish to test for independence in a multi-dimensional contingency table.

Solution: Contingency tables can be analyzed as a log-linear model where the dependency between categorical variables is modeled. All k categorical variables used to define the contingency table are treated as responses and the table shows their joint distribution. A multinomial (or product multinomial) model is appropriate for describing the joint distribution of the categorical variables when the total number of observations (or the total number of observations for specific subsets) is considered fixed. However, the relevant multinomial model happens to be equivalent to a Poisson generalized linear model with the appropriate terms included as explanatory variables so we can use Poisson regression with a log link (see Problem 3.15) to analyze the counts in the table as realizations from a Poisson random variable.

We use the `survey` data frame from the `MASS` package to investigate three of the categorical traits reported by 237 students. The `xtabs` and `as.data.frame` functions are used to transform the responses to a data frame that contains the frequency counts. In this example we consider three categorical variables: the gender of the student, which hand is on top when hands are clapped (`Clap` with levels "Right", "Left", and "Neither"), and how frequently the student exercises (`Exer` with levels "Freq", "Some", and "None"). Note that because of missing observations we end up with a total of 235 observations after using `xtabs`.

```
> library(MASS)
> data(survey)
> mydata <- as.data.frame(xtabs(~ Sex + Clap + Exer,
+                                data=survey))
> head(mydata)
     Sex    Clap Exer Freq
1 Female    Left Freq   11
2   Male    Left Freq    8
3 Female Neither Freq   17
4   Male Neither Freq   16
5 Female   Right Freq   21
6   Male   Right Freq   41
> full <- glm(Freq ~ Sex*Clap*Exer, family=poisson, data=mydata)
```

The coefficients for *the interactions* in the model can be interpreted as log odds ratios describing the size of the association between the variables (output not shown but see Problem 3.11 for examples of interpretation).

Testing the hypothesis of complete independence in the multinomial model is equivalent to comparing the additive Poisson model to the fully saturated model (i.e., the model where the interaction between all variables is included) since the additive Poisson model corresponds to independence among all categorical variables.

The `anova` function can test specific hypotheses for the log-linear Poisson model, and the option `test="Chisq"` should be set to obtain p-values from the likelihood-ratio test statistic.

```
> indep <- glm(Freq ~ Sex+Clap+Exer, family=poisson, data=mydata)
> anova(indep, full, test="Chisq")
Analysis of Deviance Table

Model 1: Freq ~ Sex + Clap + Exer
Model 2: Freq ~ Sex * Clap * Exer
  Resid. Df Resid. Dev Df Deviance Pr(>Chi)
```

```
1         12         26.7
2          0          0.0 12       26.7    0.0086 **
---
Signif. codes:  0 '***' 0.001 '**' 0.01 '*' 0.05 '.' 0.1 ' ' 1
```

We reject the hypothesis of complete independence in the contingency table since likelihood-ratio test gives a p-value of 0.0086.

Formulating the contingency table as log-linear model enables us to test for block and conditional independence among the categorical variables. Block independence refers to the situation where the distribution of the categorical variables separates into disjoint blocks such that variables in one block are independent of the variables in other blocks. There are several possible hypotheses corresponding to different sets of variables, and here we will just look at one of them. The hypothesis of block independence is tested by comparing the fit of the blocked model to the full, saturated model. Here we want to test the hypothesis, that gender is independent of hand clapping and exercise.

```
> block1 <- glm(Freq ~ Sex + Clap*Exer, family=poisson,
+               data=mydata)
> anova(block1, full, test="Chisq")
Analysis of Deviance Table

Model 1: Freq ~ Sex + Clap * Exer
Model 2: Freq ~ Sex * Clap * Exer
  Resid. Df Resid. Dev Df Deviance Pr(>Chi)
1         8       13.7
2         0        0.0  8     13.7      0.09 .
---
Signif. codes:  0 '***' 0.001 '**' 0.01 '*' 0.05 '.' 0.1 ' ' 1
```

We fail to reject the hypothesis that the gender of the student is independent of hand clapping and exercise ($p = 0.08954$).

Conditional independence refers to the situation where two (or more) categorical variables are independent of each other conditional on a third variable. For example, if gender and exercise are associated and hand clapping and exercise are associated (but gender and hand clapping are not associated) then that corresponds to the log-linear model where there are interactions between gender and exercise, as well as clapping and exercise as shown in the following code. Gender and clapping are both dependent on exercise but there is no interaction between gender and clapping so there is no direct effect between those two variables.

```
> cond1 <- glm(Freq ~ Sex*Exer+Clap*Exer, family=poisson,
+              data=mydata)
> anova(cond1, full, test="Chisq")
Analysis of Deviance Table

Model 1: Freq ~ Sex * Exer + Clap * Exer
Model 2: Freq ~ Sex * Clap * Exer
  Resid. Df Resid. Dev Df Deviance Pr(>Chi)
1         6       7.55
2         0       0.00  6     7.55     0.27
```

We test the conditional independence model against the fully saturated model and fail to reject the hypothesis that gender and hand clapping are conditionally independent given exercise. Thus, we could use the conditional independence model for subsequent analyses.

See also: Problem 2.18 shows how to convert a contingency table to a data frame. Alternatively, contingency tables can often be analyzed as logistic regression (see Problem 3.11), or multinomial logistic regression (see Problem 3.14) depending on the statistical design.

AGREEMENT

3.38 CREATE A BLAND–ALTMAN PLOT OF AGREEMENT TO COMPARE TWO QUANTITATIVE METHODS

Problem: You want to create a Bland–Altman plot to evaluate the agreement between quantitative measurements from two different methods or between two raters.

Solution: The Bland–Altman plot (also called Tukey's mean difference plot) can be used to compare how well two different quantitative measurement techniques agree, or to compare a new measurement method with a gold standard or reference method.

In the Bland–Altman plot, the differences between the two methods are plotted against their average value and from the resulting scatter plot we can determine the bias (i.e., the average difference), the limits of agreement, as well as if the underlying assumption of constant variability is fulfilled (if not, we may see an increase in the scatter of the differences, as the magnitude of the measurements increases).

The Bland–Altman plot is easily created with the following few lines, where `hplc` are measurements of muconic acid in human urine obtained from high performance liquid chromatography (HPLC) from 11 samples

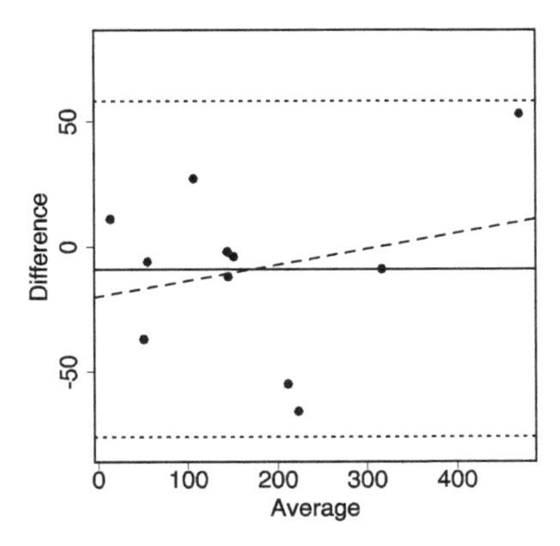

Figure 3.20: Bland–Altman plot of agreement for two methods. The solid line is average bias, dotted lines are 95% limits of agreement, and the dashed line shows bias as function of magnitude.

and `gcms` are the corresponding measurements on the same 11 samples obtained by gas chromatography-mass spectrometry (GC-MS). We wish to compute 95% limits of agreement for the two methods.

```
> hplc <- c(139, 120, 143, 496, 149, 52, 184, 190, 32, 312, 19)
> gcms <- c(151, 93, 145, 443, 153, 58, 239, 256, 69, 321, 8)
> average <- (hplc + gcms)/2                 # Compute average
> dif <- (hplc - gcms)                       # and difference
> plot(average, dif, ylim=c(-80,80),         # Plot diff vs average
+      xlab="Average", ylab="Difference")
> # Calculate limit for 95% agreement
> limit <- qnorm(.975)
> # Add average bias and 95% limits of agreement
> abline(h=mean(dif)+c(-limit,0,limit)*sd(dif),lty=c(3,1,3))
> # Add line showing bias as function of magnitude
> abline(lm(dif~average), lty=2)
```

Figure 3.20 shows the resulting Bland–Altman plot. The limits of agreement tell us how far apart measurements by the two methods are likely to be for most items and we can use these limits to determine if the difference between methods is acceptable. The regression line of difference on average (shown in Figure 3.20 as the dashed line) can be used to check the assumption that the bias is constant with magnitude. In the present case there is a slightly positive slope which indicates that the bias may increase with magnitude, but it is not statistically significant.

If we wish to look at limits of agreement other than 95%, then we should change the corresponding quantile in the call to `qnorm` above.

See also: Problem 3.39 considers agreement of a quantitative measurement among more than two raters and Problem 3.40 computes an index that measures interrater agreement for categorical items. The `BlandAltman` function from the `MethComp` package can also create Bland–Altman plots.

3.39 DETERMINE AGREEMENT AMONG SEVERAL METHODS OF A QUANTITATIVE MEASUREMENT

Problem: You want to evaluate the agreement among several different methods/raters for a quantitative measurement even in the presence of repeated measurements on each item.

Solution: The standard Bland–Altman plot only considers two different measurement methods. If we wish to determine the agreement among several methods/raters or we have repeated measurements of each item (for each method/rater), then we can use a linear mixed model to estimate both the repeatability and the agreement among methods.

The `MethComp` package provides methods for comparison of quantitative measurement methods/raters for the situation where specific methods/raters are of interest. The functions in the `MethComp` package are wrappers that set up the correct mixed-effects models, fits them and presents the results in a nice compact manner. The `MethComp` package fits a linear mixed model with item and method/rater as fixed effects and interactions between method and item (and possibly also between replicate and item if replicates are linked across methods) as random effects. Furthermore, the model allows the different methods to have individual variances corresponding to different precision of the methods.

The `Meth` function sets up a data frame in the format required by the functions in the `MethComp` package. Specifically, the data frame should contain columns entitled `y`, `meth`, `item`, and `repl`, corresponding to a vector of numeric measurements, a factor of methods/raters, a factor of items, and a factor to identify replicates as input. Alternatively, if a data frame is added as the first argument to `Meth` then we can just specify the column numbers that constitute the relevant variables in the data frame.

The `rainman` data frame from the `MESS` package contains guesses from 5 raters of the number of points in 30 pictures (10 items which were — unbeknownst to the raters — each replicated three times). We wish to determine the agreement among raters and the repeatability since we have replicates of each item. The SAND variable contains the true number of points in the image and its value identifies the item. The remaining columns each correspond to a rater and contain the scores given to each image.

```
> library(MethComp)
> library(MESS)
> data(rainman)
> head(rainman)
  SAND  ME  TM  AJ  BM  LO
1  120 175 120 105 100 100
2   48  50  50  45  50  70
3   88 150  75  75  60  80
4   32  45  22  28  30  30
5   24  25  22  25  20  20
6  100 125  80  91  80  70
```

The call to Meth accepts arguments meth, item, repl, and y and they each expect a vector or a number indicating which column in a data frame corresponds to the methods, items, replicates, and values, respectively. For brevity we will just compare three of the raters. In the code below we use column 1 to define items, and input values from columns 3 to 5. Meth automatically uses the original column names for the methods and creates the replicate vector.

```
> mydata <- Meth(item=1, y=3:5, data=rainman)
The following variables from the dataframe
"rainman" are used as the Meth variables:
item: SAND
   y: TM AJ BM
       #Replicates
Method            3 #Items #Obs: 90 Values:  min med max
    AJ           10      10       30           18  57 120
    BM           10      10       30           15  62 120
    TM           10      10       30           20  75 120
```

If the plot function is used on the data frame created by the Meth function, then pairwise Bland–Altman plots (on the original scale) between all methods are created (see the left-hand plot in Figure 3.21). Both simple scatter plots (below the diagonal) and Bland–Altman plots (above the diagonal) are created with the plot function.

The BA.est function estimates and extracts the relevant variance components and computes the limits of agreement between all methods for the situation where the bias between methods is assumed to be constant. BA.est expects a data frame of the form produced by the Meth function as input and the result from BA.est can be plotted to obtain all pairwise Bland–Altman plots. The Transform argument can be set to a function used to transform the response prior to analysis. If the desired transformation is given in the call to BA.est then the plotting functions in the Meth package automatically back-transform the results to the original scale.

The option linked for BA.est defaults to TRUE, which means that the replicates within each item are measured in parallel across methods (i.e., the random interaction between item and replicate is included in the model). If linked=FALSE then the replicates are assumed to be exchangeable within items. The alpha option sets the desired significance level and defaults to 0.05 so 95% limits of agreement are computed. The logarithm of the estimated number of points is used in the analysis since it is believed that the variance increases with increasing number of points in the picture.

```
> result <- BA.est(mydata, Transform="log")  # Use log transform
> result

Note: Response transformed by:  .Primitive("log")


 Conversion between methods:
           alpha   beta sd.pred LoA-lo LoA-up
To: From:
AJ  AJ      0.000  1.000   0.188 -0.376  0.376
    BM      0.005  1.000   0.231 -0.456  0.466
    TM     -0.090  1.000   0.210 -0.511  0.330
BM  AJ     -0.005  1.000   0.231 -0.466  0.456
    BM      0.000  1.000   0.241 -0.481  0.481
    TM     -0.095  1.000   0.221 -0.537  0.347
TM  AJ      0.090  1.000   0.210 -0.330  0.511
    BM      0.095  1.000   0.221 -0.347  0.537
    TM      0.000  1.000   0.199 -0.399  0.399

 Variance components (sd):
    IxR   MxI   res
AJ 0.05 0.081 0.133
BM 0.05 0.000 0.170
TM 0.05 0.000 0.141
```

The output gives the limits of agreement between all pairs of methods. The agreement between methods is considered acceptable if the variability between observations made with different methods on the same subject is not much larger than the variability between observations with the same method on this subject.

The bias between methods/raters is listed in the alpha column of the output and limits of agreements on the transformed scale are shown in the last two columns. The output lines where "To" and "From" correspond to the same method/rater are the estimated repeatability prediction intervals. Thus, rater "BM" has an average bias of 0.095 relative to rater "TM" (i.e., BM scores $\exp(-0.095) = 0.91$ times "TM" so about 9% lower), and the limits of agreement for the ratio between the two raters are $[\exp(-0.347); \exp(0.537)] = [0.707; 1.711]$. Rater "TM" has a repeatability prediction interval for the ratio that goes from $\exp(-0.399) = 0.67$ to $\exp(0.399) = 1.49$.

The estimated variance components used for the calculations are shown at the end of the output. Note they are on the transformed scale.

To create a plot of the conversion between methods based on the `BA.est` result we plot the `BA.est` output object with argument `pl.type="BA"`. The argument `points=TRUE` should be set to plot the individual points, and the `wh.comp` argument should be a vector of length 2 that determines which two methods are compared. The output from `plot` is seen in the right-hand graph of Figure 3.21.

```
> plot(mydata)
> plot(result, wh.comp=c(1,2), points=TRUE, pl.type="BA")
```

In the right graph of Figure 3.21 the limits of agreement are automatically transformed back to the original scale, and the non-constant limits of agreement are due to the logarithmic transformation.

`BA.est` assumes that the methods are fixed in the analysis (i.e., we can only encounter these specific methods). However, when our "methods" are judges or raters it may be more reasonable to assume that the raters are a random sample of raters from a larger group and it is not of interest to describe the bias between particular raters in the sample. Instead we want to consider raters in general. The model for random raters includes an interaction between rater and item (`MxI` in the output below corresponding to a term that allows some raters to be better at scoring some items) and an interaction between replicate and item (if the replicates are not believed to be exchangeable — the `IxR` in the code below). The limits of agreement for the difference between two random raters is then calculated from twice the sum of the variance components

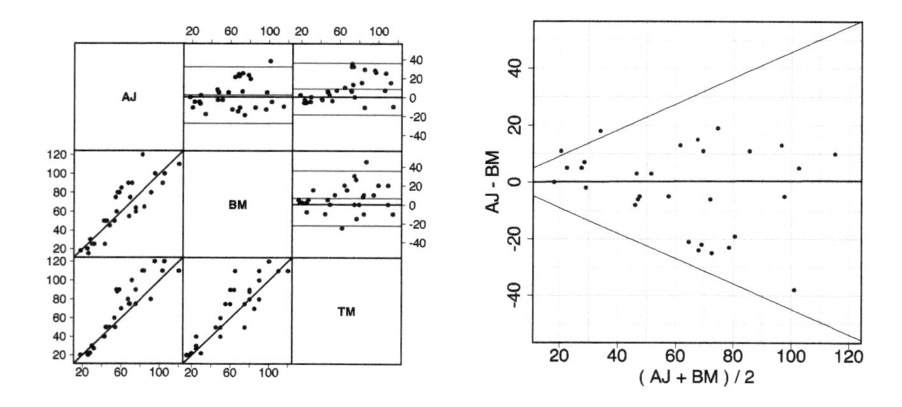

Figure 3.21: The left graph shows Bland–Altman agreement plots for three methods/raters. The upper panels show Bland–Altman plots with limits of agreement. The lower panels show pairwise scatter plots (with identity line included). The right graph shows the limits of agreements between two of the raters.

for the method, residual, and interaction between method and item random effects. The random rater setup is handled by the `BA.est` function by setting the argument `random.raters=TRUE`. We can access the estimated limits of agreement by referencing the element `LoA`.

```
> result2 <- BA.est(mydata, random.raters=TRUE, Transform="log")
> result2

Note: Response transformed by:  .Primitive("log")


 Variance components (sd):
    IxR   MxI     M   res
AJ 0.05 0.079 0.047 0.133
BM 0.05 0.000 0.047 0.170
TM 0.05 0.000 0.047 0.142
> result2$LoA
                             Mean Lower Upper   SD
Rand. rater - rand. rater    0 -0.46  0.46 0.23
```

The output shows the estimated variance components as well as the computed limits of agreement on the transformed — here logarithmic — scale. By definition there is only *one* set of limits since the raters are a random sample and hence their differences must be centered around zero. Also notice that the default arguments assume that there is variance heterogeneity: the `res` variance component is the individual (residual) variance component and it is allowed to differ between raters to accommodate different skill precisions.

Thus the average limits of agreement for two random raters (assuming that all raters have the same variance and that replicates are not exchangeable) are $[-0.460; 0.460]$. This is on the log scale so the limits of agreement for the ratio between two random raters are $[0.631; 1.584]$.

See also: Problem 3.17 introduces linear mixed effect models.

3.40 CALCULATE COHEN'S KAPPA FOR AGREEMENT

Problem: You wish to calculate Cohen's kappa as a measure of inter-rater agreement between two raters/methods on categorical (or ordinal) data.

Solution: Cohen's kappa, κ, is a measure of inter-rater agreement or concordance between two raters/methods who classify each of n observations into k categories. $\kappa = 1$ when there is perfect agreement and $\kappa = 0$ when there is no agreement.

The `irr` package provides the `kappa2` function, which calculates Cohen's kappa for two raters. The `anxiety` dataset from the `irr` package contains anxiety ratings from 20 subjects, rated by 3 raters. Values range from 1 (not anxious at all) to 6 (extremely anxious). Initially, we just consider raters one and two, and we can use `ftable` to convert the data frame to a concordance matrix.

```
> library(irr)
> data(anxiety)
> head(anxiety)
  rater1 rater2 rater3
1      3      3      2
2      3      6      1
3      3      4      4
4      4      6      4
5      5      2      3
6      5      4      2
> ftable(anxiety[,1:2]) # Print the concordance table
       rater2 1 2 3 4 6
rater1
1             1 0 1 0 0
2             0 4 2 0 0
3             0 0 1 3 1
4             0 0 1 0 1
```

```
5              0 1 2 1 0
6              0 0 1 0 0
```

The input to kappa2 should be a matrix or data frame with n rows — one for each item — and two columns representing the ratings or categories given by each of the two raters. The two columns of ratings are internally converted to factors when the kappa measure is calculated and the categories are therefore ordered according to factor levels.

```
> kappa2(anxiety[,1:2]) # Compute kappa for raters 1 and 2
 Cohen's Kappa for 2 Raters (Weights: unweighted)

 Subjects = 20
   Raters = 2
    Kappa = 0.119

        z = 1.16
  p-value = 0.245
```

The estimated kappa value is 0.119 which suggests poor agreement between raters one and two and the hypothesis of setting kappa equal to zero has a test statistic of 1.16 with a corresponding p-value of 0.245.

If data are available in the form of a contingency table representing the concordance matrix (and not a data frame) then we can create the relevant data frame using the expand_table function from the MESS package.

```
> library(MESS)
> m <- matrix(c(10, 4, 1, 5, 9, 4, 0, 3, 13), 3)
> m
     [,1] [,2] [,3]
[1,]   10    5    0
[2,]    4    9    3
[3,]    1    4   13
> kappa2(expand_table(m))
 Cohen's Kappa for 2 Raters (Weights: unweighted)

 Subjects = 49
   Raters = 2
    Kappa = 0.479

        z = 4.76
  p-value = 1.98e-06
```

Cohen's kappa treats classifications as nominal but when the categories are ordered the seriousness of a disagreement depends on the difference between ratings. The kappa2 function uses an unweighted coefficient (i.e., all disagreements between raters are scored the same way regardless of the categories selected) by default which is meaningful for nominal categories. For ordered categories, a weighted kappa measure should be used and kappa2 provides two possibilities through the weight option: weight="equal" (when weights depend linearly on their distance between categories) and weight="squared" (when distance between categories is squared to compute the weights). In the anxiety dataset the categories are ordered, and we could use, for example, squared weights:

```
> kappa2(anxiety[,1:2], weight="squared")
 Cohen's Kappa for 2 Raters (Weights: squared)

 Subjects = 20
   Raters = 2
    Kappa = 0.297

        z = 1.34
  p-value = 0.18
```

The squared weights suggest fair agreement between the two raters although it is not significantly different from no agreement ($p = 0.18$).

Cohen's kappa only handles two raters. When there are more than two raters, we can use Fleiss' kappa to estimate the agreement among raters. It is assumed for Fleiss' kappa that each item has been rated by a fixed number of raters but it is not necessary for all raters to rate all items (e.g., item 1 could be rated by raters 1, 2, and 3 while item 2 could be rated by raters 2, 4, and 9, etc.). Fleiss' kappa is a generalization of the unweighted kappa, and does not take any weighting of disagreements into account. Perfect agreement results in a value of 1 while no agreement yields a Fleiss' kappa value ≤ 0. The kappam.fleiss function from the irr package calculates Fleiss' kappa and takes the same type of input as kappa2: a matrix or data frame with n rows — one for each item — and m columns representing the available ratings for each item. Setting the option exact=TRUE calculates a version of Fleiss' kappa that is identical to Cohen's kappa in the case of two raters.

```
> kappam.fleiss(anxiety)
 Fleiss' Kappa for m Raters

 Subjects = 20
   Raters = 3
    Kappa = -0.0411
```

```
        z = -0.634
  p-value = 0.526
```

Based on Fleiss' kappa we get a value of −0.0411 which is less than zero so we conclude that there is no agreement (other than what would be expected by chance) among the three raters in the anxiety dataset.

See also: The `kappam.light` function from the `irr` package computes Light's kappa as a measure of inter-rater agreement.

MULTIVARIATE METHODS

3.41 FIT A MULTIVARIATE REGRESSION MODEL

Problem: You wish to fit a multivariate linear regression model to two or more dependent quantitative response variables.

Solution: Multivariate regression is an extension of linear normal models to the situation where there are two or more dependent response variables and where each response variable is modeled by the same linear normal model. Multivariate analysis of variance refers to the special case where the explanatory variable(s) are categorical. Multivariate regression allows us to test hypotheses regarding the effect of the explanatory variable(s) on two or more dependent variables simultaneously while maintaining the desired magnitude of type I error, and taking the covariance among the outcome variable into account.

In R, the `lm` function can be used for multivariate regression modeling simply by letting the response variable be a matrix of responses, where each column corresponds to a response variable.

In the following example we wish to model how the ozone, temperature, and the solar radiation are influenced by the day number for the `airquality` dataset. We model the set of response variables as a quadratic function of the day number.

```
> library(lubridate)
> data(airquality)
> dayno <- yday(ymd(paste("1973",
+                         airquality$Month,
+                         airquality$Day, sep="-")))
> head(dayno)
[1] 121 122 123 124 125 126
> model <- lm(cbind(Ozone, Temp, Solar.R) ~ dayno + I(dayno**2),
+             data=airquality)
```

```
> model

Call:
lm(formula = cbind(Ozone, Temp, Solar.R) ~ dayno + I(dayno^2),
    data = airquality)

Coefficients:
             Ozone      Temp       Solar.R
(Intercept)  -2.69e+02  -5.59e+01   1.22e+00
dayno         3.22e+00   1.34e+00   2.17e+00
I(dayno^2)   -7.90e-03  -3.20e-03  -5.91e-03
```

Despite having different measurements we get estimates of roughly the same scale except for the intercept. If we use `summary` on the result from the `lm` call we will see estimates, standard errors, and corresponding p-values for the individual analyses from each of the response variables. That output is not different from making separate calls to `lm` for ozone, temperature, and solar radiation except that we only consider complete cases for *all* response variables.

The test of the overall significance of each variable is computed when the `anova` function is used on the multivariate `lm` object. There exist several test statistics for multivariate regression model and by default, R computes Pillai's trace (also called Pillai–Bartlett's trace). Other test statistics like Wilks' lambda, Hotelling–Lawleys trace, and Roy's maximum root are chosen if the `test` argument for `anova` is set to `"Wilks"`, `"Hotelling-Lawley"`, or `"Roy"`, respectively.

```
> anova(model)
Analysis of Variance Table

             Df Pillai approx F num Df den Df  Pr(>F)
(Intercept)   1  0.995     6829      3    106 < 2e-16 ***
dayno         1  0.293       15      3    106 4.7e-08 ***
I(dayno^2)    1  0.443       28      3    106 1.8e-13 ***
Residuals   108
---
Signif. codes:  0 '***' 0.001 '**' 0.01 '*' 0.05 '.' 0.1 ' ' 1
```

If we look at the individual analyses, it turns out that day number is not significant when we look at the model with solar radiation as response. However, Pillai's simultaneous test for all three response variables shows that both day number and the quadratic polynomial of day number are highly significant with p-values very close to zero.

See also: See Problems 2.19 and 2.20 for more information on the `lubridate` package.

3.42 CLUSTER OBSERVATIONS

Problem: You have a dataset and want to identify clusters or groups of observations that are similar in some sense.

Solution: Cluster analysis is concerned with assignment of a set of observations into groups or clusters such that observations in the same cluster resemble each other in some sense.

There exist numerous packages and functions for cluster analysis in R. Here we will just focus on two functions that are part of base R and can be used for clustering based on numeric variables: `kmeans` for unsupervised k-means clustering and `hclust` for hierarchical clustering.

The first argument to `kmeans` should be a numeric matrix of data or an object that can be coerced to a numeric matrix. The rows represent the observations to be clustered and the columns should contain the variables that are used to determine if the observations are similar. In k-means clustering the investigator specifies the desired number of clusters and that is done by the `centers` argument in R. `centers` should either be an integer that determines the number of clusters or a matrix that sets the initial cluster centers. If only a number is given, then the initial centers are sampled at random from the dataset, and slightly different results may be obtained depending on the initial centers (in particular, the cluster labels may be different from run to run even though the cluster members are the same). The `algorithm` argument sets the algorithm used to partition points into the clusters (default is `Hartigan-Wong` and alternative methods comprise `Lloyd`, `Forgy`, and `MacQueen`).

The `agriculture` data frame from the `cluster` package contains information on the gross national product (GNP) per capita (`x`) and percentage of the population working in agriculture (`y`) for each country belonging to the European Union in 1993.

```
> library(cluster)
> data(agriculture)
> fit2 <- kmeans(agriculture, centers=2)   # 2 cluster solution
> fit2
K-means clustering with 2 clusters of sizes 8, 4

Cluster means:
     x     y
```

```
1 17.8  4.56
2  9.0 16.12

Clustering vector:
  B  DK   D  GR   E   F IRL   I   L  NL   P  UK
  1   1   1   2   2   1   2   1   1   1   2   1

Within cluster sum of squares by cluster:
[1] 71.9 90.8
 (between_SS / total_SS =  77.6 %)

Available components:

[1] "cluster"      "centers"      "totss"        "withinss"
[5] "tot.withinss" "betweenss"    "size"         "iter"
[9] "ifault"
```

The output from kmeans gives information on the cluster sizes, their respective means, the resulting allocation of each row, and the within cluster sum of squares. In particular, the clustering vector shows the allocation of each country to the two clusters.

We can calculate the total within cluster sum of squares for a different number of clusters with the following command if we want to use that to decide on how many clusters to use.

```
> # Calculate the total within sum of squares for different
> # number of clusters (here 2-6)
> sapply(2:6, function(i) {
+        sum(kmeans(agriculture, centers=i)$withinss)})
[1] 162.7 131.4  58.9  36.2  28.9
```

Here we see a decrease in within sum-of-squares until we have three clusters after which it levels off. Thus, it would probably not make sense to have more than three clusters for these data (not to mention that the number of observations is really not that large).

If the variables are measured on very different scales, then the variables in the data matrix may need to be standardized before clustering is undertaken. The scale function standardizes the columns in a data matrix by subtracting the mean and dividing by the standard deviation.

```
> scale.agriculture <- scale(agriculture)
> fit2s <- kmeans(scale.agriculture, centers=2)
> fit2s
K-means clustering with 2 clusters of sizes 4, 8
```

```
Cluster means:
       x      y
1 -1.187  1.196
2  0.593 -0.598

Clustering vector:
  B  DK   D  GR   E   F IRL   I   L  NL   P  UK
  2   2   2   1   1   2   1   2   2   2   1   2

Within cluster sum of squares by cluster:
[1] 2.53 2.43
 (between_SS / total_SS =  77.5 %)

Available components:

[1] "cluster"      "centers"      "totss"        "withinss"
[5] "tot.withinss" "betweenss"    "size"         "iter"
[9] "ifault"
> sapply(2:6, function(i) {
+        sum(kmeans(scale.agriculture, centers=i)$withinss)})
[1] 4.958 2.904 1.662 1.311 0.709
```

Even after scaling we get roughly the same information about the number of clusters, and we also see that the scaling does not change the allocation of countries to clusters.

Hierarchical clustering requires a distance matrix to determine the similarity between observations. The `dist` function computes the pairwise distances between the rows of a data matrix. By default, `dist` uses Euclidean distance but other metrics can be set with the `method` argument (for example `method="maximum"` or `method="manhattan"`).

When the pairwise distances are computed, the resulting matrix can be used as argument to the `hclust` function to undertake hierarchical clustering. The `hclust` also has a `method` argument that determines how the distance between two sub-clusters is measured. Some of the possible grouping methods are `complete` (the default), `average`, and `single`. The help page provides some additional information on these and other clustering methods.

```
> distances <- dist(agriculture)
> hierclust <- hclust(distances)
> hierclust

Call:
hclust(d = distances)
```

```
Cluster method   : complete
Distance         : euclidean
Number of objects: 12
> hierclust.single <- hclust(distances, method="single")
```

The output from `hclust` is not particularly exciting, and it makes more sense to plot the results. Figure 3.22 shows the output dendrograms from `hclust` obtained by using `plot` on the hierarchical cluster analysis object. We can extract the classification from a hierarchical cluster analysis by cutting the tree at a certain level using the function `cutree`. `cutree` requires a tree as first argument and either the option `k` which specifies the desired number of clusters or `h` which determines the height at which to cut the tree. The resulting clusters are printed by `cutree`.

```
> plot(hierclust)
> plot(hierclust.single)
> cutree(hierclust, k=2)           # Cut tree to get 2 groups
  B  DK   D  GR   E   F IRL   I   L  NL   P  UK
  1   1   1   2   2   1   2   1   1   1   2   1
> cutree(hierclust.single, h=5)  # Cut at height 5
  B  DK   D  GR   E   F IRL   I   L  NL   P  UK
  1   1   1   2   3   1   3   1   1   1   3   1
```

```
  B  DK   D  GR   E   F IRL   I   L  NL   P  UK
  1   1   1   2   2   1   2   1   1   1   2   1
  B  DK   D  GR   E   F IRL   I   L  NL   P  UK
  1   1   1   2   3   1   3   1   1   1   3   1
```

There is not much difference in the two clusterings obtained from different methods and when the three is cut we obtain virtually the same clustering as we saw for the k-means clustering method.

See also: The `cluster` package includes several clustering methods that are more advanced than k-means and hierarchical clustering as well as additional methods for plotting cluster results. The `mclust` package provides functions for model-based clustering. See Rule 4.18 on how to draw heat maps.

3.43 USE PRINCIPAL COMPONENT ANALYSIS TO REDUCE DATA DIMENSIONALITY

Problem: You want to use principal component analysis to reduce data dimensionality.

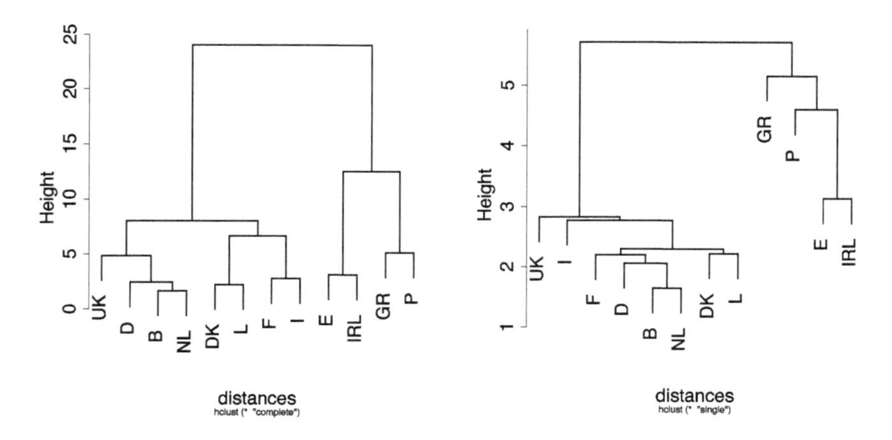

Figure 3.22: The left panel shows the hierarchical clustering dendrogram given by the `hclust` function. The right graph shows the same dendrogram but computed using the option `method=single`.

Solution: Principal component analysis is a useful technique when you have obtained measurements on a number of (possibly correlated) observed variables and you wish to reduce the number of observed variables to a smaller number of artificial variables (called principal components) that account for most of the variance in the observed variables. The principal components may then be used as predictors in subsequent analyses.

The `prcomp` function performs principal component analysis on a given data matrix. By default, R centers the variables in the data matrix, but does not scale them to have unit variance. In order to scale the variables we should set the logical argument `scale.=TRUE`.

In the following code we will make a principal component analysis on the nine explanatory variables found in the `biopsy` breast cancer dataset from the `MASS` package. There are missing data in the data frame so we restrict our attention to the complete cases.

```
> library(MASS)
> data(biopsy)
> names(biopsy)
 [1] "ID"    "V1"    "V2"    "V3"    "V4"    "V5"    "V6"
 [8] "V7"    "V8"    "V9"    "class"
> predictors <- biopsy[complete.cases(biopsy),2:10]
> fit <- prcomp(predictors, scale.=TRUE)
> summary(fit)
Importance of components:
                         PC1    PC2    PC3    PC4    PC5    PC6
Standard deviation     2.429 0.8809 0.7343 0.6780 0.6167 0.5494
```

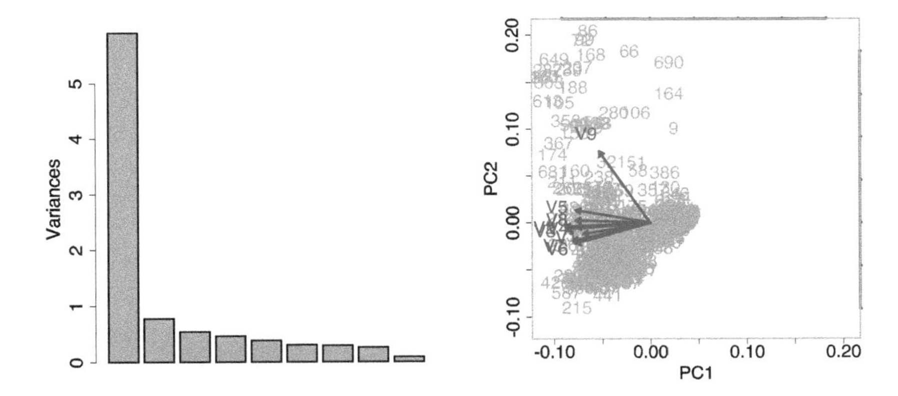

Figure 3.23: Example of `prcomp` output. The scree plot in the left panel is produced by `plot` on the result from the principal component analysis. The right-hand graph is a `biplot` and shows the variables (as arrows) and observations (numbers) plotted against the first two principal components.

```
Proportion of Variance 0.655 0.0862 0.0599 0.0511 0.0423 0.0335
Cumulative Proportion  0.655 0.7417 0.8016 0.8527 0.8950 0.9285
                         PC7   PC8     PC9
Standard deviation     0.5426 0.511 0.29729
Proportion of Variance 0.0327 0.029 0.00982
Cumulative Proportion  0.9612 0.990 1.00000
```

The first principal component (PC1) explains 65.5% of the total variation, the second explains 8.62%, etc. The output also indicates that just a few principal components will explain most of the variation in the explanatory variables.

The `plot` and `biplot` functions take the output from `prcomp` and produce a scree plot and biplot, respectively, and their output can be seen in Figure 3.23. The scree plot clearly suggests that only that first principal component has a major contribution to the variance. From the biplot we see that the variables (indicated by the nine arrows) are quite correlated since most of them have virtually the same direction, and that they contribute more or less the same to the first principal component since they have more or less the same length. Only variable 9, `V9`, has any real contribution to the second principal component.

```
> plot(fit)
> biplot(fit, col=c("gray", "blue"))
```

The loadings factors can be found as the `rotation` component of the model fit object and the pca scores (the original data in the rotated coordinate system) are obtained either from the `x` component of the model fit or by the `predict` function.

```
> fit
Standard deviations:
[1] 2.429 0.881 0.734 0.678 0.617 0.549 0.543 0.511 0.297

Rotation:
      PC1     PC2       PC3    PC4     PC5     PC6      PC7
V1 -0.302 -0.1408  0.866372 -0.108  0.0803 -0.2425 -0.00852
V2 -0.381 -0.0466 -0.019938  0.204 -0.1457 -0.1390 -0.20543
V3 -0.378 -0.0824  0.033511  0.176 -0.1084 -0.0745 -0.12721
V4 -0.333 -0.0521 -0.412647 -0.493 -0.0196 -0.6546  0.12383
V5 -0.336  0.1644 -0.087743  0.427 -0.6367  0.0693  0.21102
V6 -0.335 -0.2613  0.000691 -0.499 -0.1248  0.6092  0.40279
V7 -0.346 -0.2281 -0.213072 -0.013  0.2277  0.2989 -0.70042
V8 -0.336  0.0340 -0.134248  0.417  0.6902  0.0215  0.45978
V9 -0.230  0.9056  0.080492 -0.259  0.1050  0.1483 -0.13212
       PC8      PC9
V1  0.2477 -0.00275
V2 -0.4363 -0.73321
V3 -0.5827  0.66748
V4  0.1634  0.04602
V5  0.4587  0.06689
V6 -0.1267 -0.07651
V7  0.3837  0.06224
V8  0.0740 -0.02208
V9 -0.0535  0.00750
> head(predict(fit))
    PC1     PC2     PC3     PC4     PC5    PC6    PC7    PC8
1  1.47 -0.1042  0.5653 -0.0319  0.1509 -0.060 -0.349  0.420
2 -1.44 -0.5697 -0.2364 -0.4778 -1.6419  0.483  1.115  0.379
3  1.59 -0.0761 -0.0488 -0.0923  0.0597  0.279 -0.233  0.210
4 -1.48 -0.5281  0.6026  1.4098  0.5603 -0.063  0.211 -1.606
5  1.34 -0.0907 -0.0300 -0.3380  0.1087 -0.431 -0.260  0.446
6 -5.01 -1.5338 -0.4607  0.2952 -0.3916 -0.115 -0.384 -0.149
       PC9
1  0.00569
2 -0.02341
3 -0.01336
4 -0.18264
5  0.03879
6  0.04295
```

The first column for the factor loadings show that all nine variables contribute roughly equally to the first principal component since all variables have more or less the same weight. The second column shows that variable 9 V9 dominates principal component 2 which is the same result we saw from the biplot in Figure 3.23.

Input to prcomp can also be specified as a formula with no response variable. In this case, missing variables are handled directly by the na.action argument and we do not have to worry about extracting the complete cases as we did above (output not shown as it is identical to the output shown above).

```
> fit <- prcomp(~ V1 + V2 + V3 + V4 + V5 + V6 + V7 + V8 + V9,
+               data=biopsy, scale.=TRUE)
```

See also: The princomp function can also be used for principal component analysis but the calculations done by prcomp have better numerical accuracy. See Problem 3.44 for an example of principal component regression.

3.44 FIT A PRINCIPAL COMPONENT REGRESSION MODEL

Problem: You want to use principal component regression analysis to investigate the linear relationship between a response variable and a set of explanatory principal components variables.

Solution: Principal component analysis is often used to reduce data dimensionality and to overcome problems with collinearity among observed variables.

Principal component regression uses the principal components identified through principal component analysis as explanatory variables in a regression model instead of the original variables, and principal component regression is often seen as a natural "next step" to principal component analysis. Typically, only a subset of the principal components is used in the regression.

In the following code we start with a principal component analysis of the nine explanatory variables found in the biopsy dataset from the MASS package. We then use the predicted principal components as input to a logistic regression model where the response is the breast tumor type: benign or malignant.

```
> library(MASS)
> data(biopsy)
> names(biopsy)
```

```
 [1] "ID"     "V1"     "V2"     "V3"     "V4"     "V5"     "V6"
 [8] "V7"     "V8"     "V9"     "class"
> fit <- prcomp(~ V1 + V2 + V3 + V4 + V5 + V6 + V7 + V8 + V9,
+               data=biopsy, scale.=TRUE)
> summary(fit)
Importance of components:
                         PC1    PC2    PC3    PC4    PC5    PC6
Standard deviation     2.429 0.8809 0.7343 0.6780 0.6167 0.5494
Proportion of Variance 0.655 0.0862 0.0599 0.0511 0.0423 0.0335
Cumulative Proportion  0.655 0.7417 0.8016 0.8527 0.8950 0.9285
                          PC7   PC8     PC9
Standard deviation     0.5426 0.511 0.29729
Proportion of Variance 0.0327 0.029 0.00982
Cumulative Proportion  0.9612 0.990 1.00000
> pc <- predict(fit)[,1:4]  # Select first 4 principal comp.
> head(pc)
    PC1     PC2     PC3     PC4
1  1.47 -0.1042  0.5653 -0.0319
2 -1.44 -0.5697 -0.2364 -0.4778
3  1.59 -0.0761 -0.0488 -0.0923
4 -1.48 -0.5281  0.6026  1.4098
5  1.34 -0.0907 -0.0300 -0.3380
6 -5.01 -1.5338 -0.4607  0.2952
> y <- biopsy$class[complete.cases(biopsy)] # Complete outcome cases
> model <- glm(y ~ pc, family=binomial)     # Fit the pcr model
> summary(model)

Call:
glm(formula = y ~ pc, family = binomial)

Deviance Residuals:
   Min      1Q  Median      3Q     Max
-3.179  -0.130  -0.062   0.023   2.480

Coefficients:
            Estimate Std. Error z value Pr(>|z|)
(Intercept)   -1.074      0.303   -3.54   0.0004 ***
pcPC1         -2.414      0.256   -9.44   <2e-16 ***
pcPC2         -0.159      0.505   -0.32   0.7525
pcPC3          0.719      0.327    2.20   0.0280 *
pcPC4         -0.915      0.369   -2.48   0.0132 *
---
Signif. codes:  0 '***' 0.001 '**' 0.01 '*' 0.05 '.' 0.1 ' ' 1

(Dispersion parameter for binomial family taken to be 1)

    Null deviance: 884.35  on 682  degrees of freedom
Residual deviance: 106.12  on 678  degrees of freedom
```

```
AIC: 116.1

Number of Fisher Scoring iterations: 8
```

We see from the logistic regression analysis that the first principal component is highly significant while principal components 3 and 4 are barely significant. Principal component 2 is not significant when the three other principal components are part of the model, so even though principal component 2 explains the second-most of the variation (among the variables) it has no statistical significance on the responses.

See also: See Problem 3.43 for principal component analysis and Problem 3.11 for logistic regression modeling. The `pcr` function from the `pls` package can also be used for principal component regression.

3.45 CLASSIFY OBSERVATIONS USING LINEAR DISCRIMINANT ANALYSIS

Problem: You want to find a linear combination of features which can be used for classification into two or more classes.

Solution: Discriminant analysis is a statistical technique for classification of data into mutually exclusive groups. In linear discriminant analysis we assume that the k groups can be separated by a linear combination of features that describe the objects and with k groups we need $k - 1$ discriminators to separate the classes.

The function `lda` from the `MASS` package can be used for linear discrimination analysis. Input for the `lda` function is a model formula of the form `group ~ x1 + x2 +` $\cdots$ where the response `group` is a grouping factor and `x1`, `x2`, ... are quantitative discriminators. The `prior` argument can be set to give the prior probabilities of class membership. If it is unspecified, the probabilities of class membership are estimated from the dataset.

In the following code, we will make a model to classify breast cancer type ("benign" or "malignant") for 699 patients based on tumor clump thickness (`V1`), uniformity of cell size (`V2`), and uniformity of cell shape (`V3`). The variables are found in the `biopsy` data frame found in the `MASS` package.

```
> library(MASS)
> data(biopsy)
> fit <- lda(class ~ V1 + V2 + V3, data=biopsy)
```

```
> fit
Call:
lda(class ~ V1 + V2 + V3, data = biopsy)

Prior probabilities of groups:
   benign malignant
    0.655     0.345

Group means:
             V1   V2   V3
benign     2.96 1.33 1.44
malignant 7.20 6.57 6.56

Coefficients of linear discriminants:
     LD1
V1 0.232
V2 0.257
V3 0.250
```

The prior probability of the groups and the resulting linear discriminator are both seen in the output. There are 458 benign tumors in the dataset so the prior probability corresponds to that ratio, $458/699 = 0.655$. The linear discriminator gives the weights for the linear combination of `V1`, `V2`, and `V3` used to separate the observations into classes.

Plotting the fitted model can be seen in Figure 3.24 and the type of plot produced depends on the number of discriminators. If there is only one discriminator or if the argument `dimen=1` is set, then a histogram is plotted; if there are two or more discriminators, then a `pairs` plot is shown.

```
> plot(fit, col="blue")
```

The `predict` function returns a list of predicted classes (the `class` component from the `predict` function), posterior probabilities (the `posterior` component), and computed linear discriminator(s) (the `x` component) when the result from `lda` is supplied as input.

```
> lapply(predict(fit), head, 4)  # List the first 4
$class
[1] benign    benign    benign    malignant
Levels: benign malignant

$posterior
    benign malignant
```

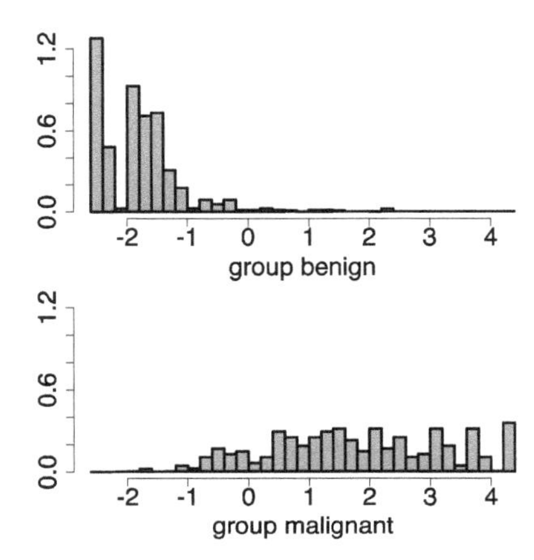

Figure 3.24: Example of `lda` output. Histograms for values of the linear discriminator are shown for observations from both the "benign" and "malignant" group.

```
1 0.997900  0.002100
2 0.659175  0.340825
3 0.999607  0.000393
4 0.000543  0.999457

$x
     LD1
1 -0.966
2  0.556
3 -1.431
4  2.819
```

The predicted classes can be used to create a confusion matrix by comparing them to the true classes in order to evaluate the sensitivity and specificity of the classification.

```
> result <- table(biopsy$class, predict(fit)$class)
> result

            benign malignant
  benign       448        10
  malignant     33       208
> sum(diag(result)) / sum(result)
[1] 0.938
```

We can see here that the linear discriminant analysis correctly classifies 93.84% of the observations. Note, however, that this result is based on a linear discriminator that has been estimated from the same data as the classification is evaluated. In practice, we would like to evaluate the classification on a sample that has not been used to estimate the discriminators. In the following example, we use a third of the original dataset as a training dataset and the remaining 2/3 as a test dataset.

```
> # Split the data up into random training and test samples
> train <- sample(1:nrow(biopsy), ceiling(nrow(biopsy)/3))
> testdata <- biopsy[-train,]
> trainfit <- lda(class ~ V1 + V2 + V3, data=biopsy, subset=train)
> trainfit
Call:
lda(class ~ V1 + V2 + V3, data = biopsy, subset = train)

Prior probabilities of groups:
   benign malignant
    0.639     0.361

Group means:
            V1   V2   V3
benign    2.97 1.21 1.42
malignant 7.01 6.40 6.40

Coefficients of linear discriminants:
     LD1
V1 0.222
V2 0.348
V3 0.194
> result2 <- table(testdata$class,
+                  predict(trainfit, testdata)$class)
> result2

            benign malignant
  benign       300         9
  malignant     19       138
> sum(diag(result2)) / sum(result2)
[1] 0.94
```

Using cross-validation to evaluate the proportion of correctly classified observations did not result in any substantial changes to the overall classification rate. Alternatively, if the `CV` argument in the call to `lda` is set to `TRUE` then results for classes and posterior probabilities are estimated using leave-one-out cross-validation.

Instead of using a model formula, `lda` can also use a data frame or matrix, `x`, that contains the explanatory variables and a factor vector `grouping` that specify the response class for each observation as input.

See also: The `qda` function from the `MASS` package can perform quadratic discrimination analysis. For more advanced discriminant methods check out the mda package.

3.46 USE PARTIAL LEAST SQUARES REGRESSION FOR PREDICTION

Problem: You want to use partial least squares regression to predict a response based on a large set of explanatory variables using partial least squares.

Solution: Partial least squares (pls) regression combines features from principal component analysis and multiple regression and is particularly useful when we want to predict a multivariate response from a (potentially very) large set of explanatory variables.

Pls is commonly applied in situations where the number of explanatory variables is so large compared to the number of observations that traditional regression approaches are no longer feasible or if a large number of the explanatory variables are collinear. The partial least squares technique has seen widespread use in chemometrics where it is used, for example, to study the combination of characteristics of a substance and the output of wavelengths after a large number of near infrared (NIR) wavelengths are passed through the substance.

The pls method extends principal component analysis in the sense that pls extracts orthogonal linear combinations of explanatory variables (also known as "factors" or principal components) that explain the variance in *both* the explanatory variables and the response variable(s) simultaneously.

The pls package implements partial least squares regression as well as principal component regression. The plsr function computes the principal pls components and subsequently makes a regression of the response on the principal components. plsr accepts a model formula as its first argument.

There are four arguments that should be set for the plsr function. The first is ncomp, which sets the maximum number of principal components that are included in the subsequent regression model. The scale option can be set to TRUE or to a numeric vector. In the first case, each of the explanatory variables is divided by its standard deviation. When set to a numeric vector, each explanatory variable is divided by the corresponding element of scale. The method and validation arguments control the algorithm used to compute the principal components and the type of model validation. Their default values are kernel algorithm (`"kernelpls"`) and `"none"` (for no cross-validation), respectively. Alternative computational algorithms that can be chosen with the `method` argument are `"widekernelpls"` (for the wide kernel algorithm), `"simpls"` (for the SIMPLS algorithm), and `"oscorespls"` for the classical orthogonal scores algorithm. Cross-validation may help decide the number of components in the model and cross-validation and leave-one-out cross-validation are performed if `validation="CV"` or if `validation="LOO"`, respectively.

The `gasoline` data frame from the `pls` package contains information on the octane number for 60 gasoline samples as well as corresponding NIR spectra with 401 wavelengths. We use the first 50 samples to create a pls regression model and will evaluate the prediction error on the remaining 10 samples.

```
> library(pls)
> data(gasoline)
> gasTrain <- gasoline[1:50,]  # Training dataset with 50 samples
> gasTest <- gasoline[51:60,]  # Test dataset with 10 samples
```

Below we fit the model to the training dataset and use 5 components in the regression. We use leave-one-out validation with the training data to get estimates of the mean squared error for a different number of components so we can determine how many components are needed, and we will focus on the test dataset later.

```
> gas1 <- plsr(octane ~ NIR, ncomp=5,
+              data=gasTrain, validation="LOO")
> summary(gas1)
Data:   X dimension: 50 401
Y dimension: 50 1
Fit method: kernelpls
Number of components considered: 5

VALIDATION: RMSEP
Cross-validated using 50 leave-one-out segments.
       (Intercept)  1 comps  2 comps  3 comps  4 comps  5 comps
CV           1.545    1.357   0.2966   0.2524   0.2476   0.2398
adjCV        1.545    1.356   0.2947   0.2521   0.2478   0.2388

TRAINING: % variance explained
        1 comps  2 comps  3 comps  4 comps  5 comps
X         78.17    85.58    93.41    96.06    96.94
octane    29.39    96.85    97.89    98.26    98.86
```

The summary output shows the percentage of the variance explained in both the octane response variable and the explanatory variables based on the training data. More than 95% of the variation is explained by the first 4 principal components for both the predictors and the outcome. The summary output also includes a "Validation" section since we used leave-one-out cross-validation. This section shows the estimated root mean squared error from the bias-corrected cross-validation (the adjCV) and the root mean squared error appears to be fairly stable around 0.25 after 3 principal components. Thus we decide on using three components in the subsequent analyses.

summary only provides a summary of the pls result. To get the actual scores and loading matrices from the fit we can use the scores and loadings function, respectively. They both require a fitted plsr object as input. The predict function returns the predictions from the training set and it returns a three-dimensional array of predicted response values. The dimensions correspond to the observations, the response variables, and the model sizes, respectively (none of the output is shown here).

We can also use plot to make a prediction plot directly if we use the fitted object as first argument and set the number of components to include as the ncomp option. The result is seen in the left panel of Figure 3.25 which shows a high level of prediction. plot has been modified for objects fitted with plsr, so if we add the option plottype="loadings" then the different factor loadings are plotted (see the right-hand plot of Figure 3.25, where we have used comps=1:3 to get curves for the first three components).

```
> plot(gas1, ncomp=3, line=TRUE)
> plot(gas1, "loadings", comps=1:3, col=c("black", "blue", "black"))
```

We check for overfitting by predicting the values from the test dataset based on the principal components computed from the training data. The root mean squared error for the test data can be directly computed with the RMSEP function when the fitted object is included as first argument, and the new test data are included as argument newdata.

```
> RMSEP(gas1, newdata=gasTest)
(Intercept)      1 comps      2 comps      3 comps
     1.5369       1.1696       0.2445       0.2341
    4 comps      5 comps
     0.3287       0.2780
```

The root mean squared errors from the test data show that we have

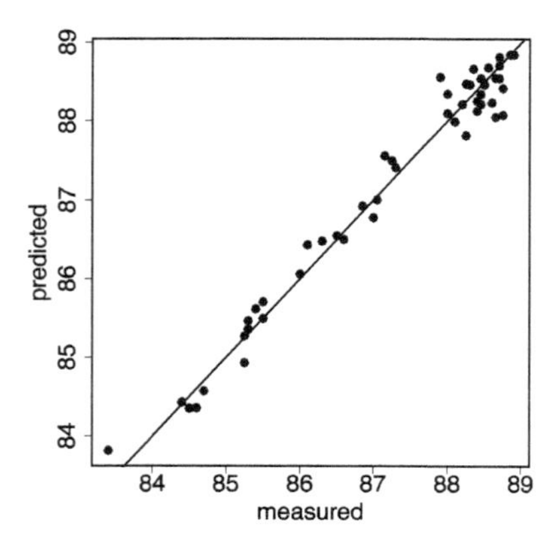

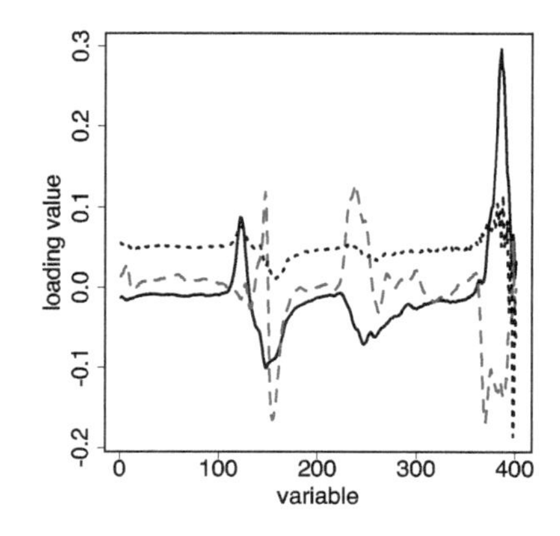

Figure 3.25: Partial least squares prediction of octane from NIR spectra (left plot), and factor loadings (right plot).

the same order of prediction error for three principal components as we saw from the cross-validation of the training data. Thus, there is nothing that suggests that we are overfitting the model when we use three principal components.

See also: An example of principal component regression is shown in Problem 3.44, and cross-validation is also discussed in Problem 3.48.

RESAMPLING STATISTICS AND BOOTSTRAPPING

3.47 NON-PARAMETRIC BOOTSTRAP ANALYSIS

Problem: You want to use non-parametric bootstrap to evaluate the properties and accuracy of a statistic.

Solution: Bootstrap is a simple but computationally intensive method to assess statistical accuracy or to estimate the distribution of a statistic. The idea behind the bootstrap procedure is to repeatedly draw random samples from the original data with replacement and then use these "artificial" samples to compute the statistic of interest a large number of times thereby estimating the distribution of the statistic. Typically the observations should be independent for the standard bootstrap procedure to produce useful results.

The `boot` package provides extensive facilities for bootstrapping and related resampling methods. The `boot` function generates bootstrap samples and calculates the statistic and requires three arguments to be specified: `data` for the data frame, `statistic` for the function that calculates the statistic of interest, and the number of bootstrap replicates, `R`. The data argument can be either a vector, matrix, or data frame and should comprise the original data. If the data is a matrix or data frame, then each row is considered one observation; i.e., we resample from the individual rows. The `statistic` function should be a function which calculates and returns the statistic(s) of interest. The function specified by the `statistic` option should take at least two arguments. The first object passed to the function is the original dataset given by the `data` option, and the second argument is a vector of indices which define the indices that constitute the non-parametric bootstrap sample. Finally, `R` is an integer that sets the number of bootstrap samples. Additional parameters to be passed to the function that produces the statistic of interest can be added to the `boot` call.

We illustrate the non-parametric bootstrap procedure with an example from genetics. In a balanced half-sib animal experiment with n sires and r progeny per sire, we can calculate an estimate of the heritability for a quantitative trait as

$$h^2 = \frac{4(\mathrm{MS}_{\mathrm{sire}} - \mathrm{MS}_{\mathrm{within}})}{(r+1)\mathrm{MS}_{\mathrm{sire}} - \mathrm{MS}_{\mathrm{within}}},$$

where $\mathrm{MS}_{\mathrm{sire}}$ and $\mathrm{MS}_{\mathrm{within}}$ are the mean square values for the among sires and within sires variance sources obtained from an analysis of variance table. We can use the bootstrap analysis to provide confidence limits for the heritability estimate.

We simulate artificial data by first generating the mean level for each sib-ship from a normal distribution and then subsequently generating measurements for the individual half-sibs. Each row in the `mydata` matrix below corresponds to measurements from five half-sibs and there are 30 rows — one for each sire. Here we sample sires and not single observations since the sires are independent and all have exactly 5 measurements, so we are essentially doing a block bootstrap.

```
> library(boot)
> set.seed(1)                # Set seed to reproduce simulations
> mydata <- rnorm(150, mean=rep(rnorm(30), 5))
> mydata <- matrix(mydata,ncol=5)
> head(mydata)
        [,1]   [,2]   [,3]   [,4]   [,5]
[1,]  0.7322  1.775 -1.169 -1.132 -0.176
[2,]  0.0809  0.144  1.392  1.527  0.165
```

```
[3,] -0.4480 -0.146  0.325 -1.050 -1.154
[4,]  1.5415  1.623  2.295  1.416  0.666
[5,] -1.0476 -0.414  1.916  0.229 -1.158
[6,] -1.2355 -0.632 -0.262 -0.108 -1.896
```

The following function calculates the heritability based on the balanced half-sib design. The function should take two arguments: the first should be the data frame or matrix, and the second should be a vector of indices or weights that define the current bootstrap sample. We set a default value for the second argument such that the full original dataset is used if the desired subsample is not specified. The `stack` function is used in combination with the matrix transpose function, `t`, to transform the input data matrix to the form needed for `lm`.

```
> heritability <- function(data, indices=1:nrow(data)) {
+   r <- ncol(data)
+   df <- stack(as.data.frame(t(data[indices,])))
+   result <- anova(lm(values ~ ind, data=df))
+   mssire <- result[1,3]
+   mswithin <- result[2,3]
+   heritab <- 4*(mssire - mswithin) / ((r+1)*mssire - mswithin)
+   return(heritab)
+ }
>
> heritability(mydata)
[1] 0.553
```

The estimated heritability of the original data frame is 0.5533.

Our newly defined function can be used with the `boot` function for computing non-parametric bootstrap estimates of the standard error. If `plot` is applied to the output from the bootstrap procedure, then a histogram and a q-q plot of the estimated statistics are plotted (see Figure 3.26).

```
> result <- boot(mydata, heritability, R=10000)
> result

ORDINARY NONPARAMETRIC BOOTSTRAP


Call:
boot(data = mydata, statistic = heritability, R = 10000)


Bootstrap Statistics :
```

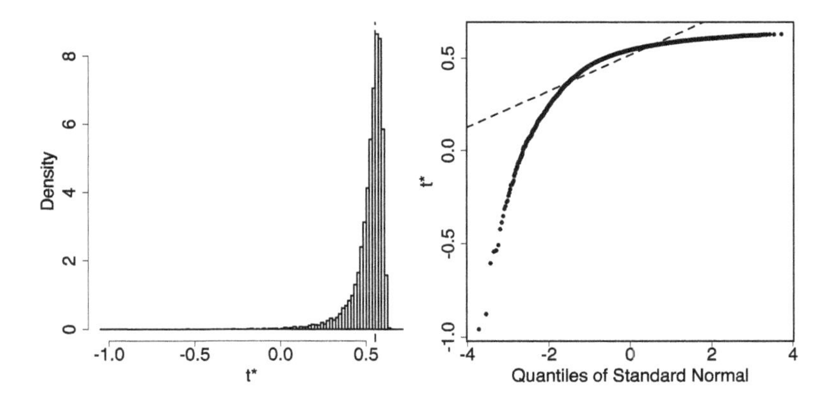

Figure 3.26: Bootstrap estimates of heritability for half-sub design. The left plot shows a histogram of the bootstrap estimates of the heritability, while the right plot shows a q-q plot of the bootstrap estimates.

```
     original      bias    std. error
t1* 0.5533472 -0.03503606  0.09787623
> plot(result)
```

The estimated heritability for the original data is shown in the bootstrap output as well as the bias and the estimated standard error. The bias is -0.03503 which shows that the difference between the original heritability estimate and the mean of the bootstrap sample statistics are very close to zero so there is hardly any bias. The standard error is 0.09788 so in principle we could construct a 95% confidence interval for the heritability by using this value. However, the right-hand plot in Figure 3.26 suggests that the bootstrap heritability estimates are far from symmetric and Gaussian, so the 95% confidence interval will not be well approximated by twice the standard error.

Instead, the `boot.ci` function (also found in the `boot` package) can be used to compute a bootstrap percentile confidence interval and a bias-adjusted bootstrap percentile confidence interval. `boot.ci` needs the output from a `boot` call as input.

```
> boot.ci(result)
BOOTSTRAP CONFIDENCE INTERVAL CALCULATIONS
Based on 10000 bootstrap replicates

CALL :
boot.ci(boot.out = result)
```

```
Intervals :
Level      Normal              Basic
95%   ( 0.397,  0.780 )   ( 0.499,  0.851 )

Level     Percentile            BCa
95%   ( 0.256,  0.608 )   ( 0.384,  0.619 )
Calculations and Intervals on Original Scale
```

The `Normal` interval corresponds to the interval based on twice the standard error (after correction for bias) and the `Percentile` confidence interval is based on the empirical "middle" 95% of the bootstrap heritabilities and it is clearly asymmetric. The `BCa` interval is the bias-adjusted percentile interval. Here we conclude that the 95% bootstrap confidence interval for the heritability would be $(0.3840; 0.6189)$. A fifth type of confidence interval, the studentized interval, requires variance from each bootstrap sample to compute.

See also: The above example describes non-parametric bootstrap. Parametric bootstrap or permutation analysis is handled by `boot` with the `sim="parametric"` or `sim="permutation"` options, respectively.

3.48 USE CROSS-VALIDATION TO ESTIMATE THE PERFORMANCE OF A MODEL OR ALGORITHM

Problem: You want to estimate the performance of a model or algorithm/learning method using cross-validation.

Solution: Cross-validation is a useful technique for evaluating the performance and stability of a model or algorithm. Cross-validation can also be used to compare the performance of several different algorithms, or to choose among various models.

In k-fold cross-validation the data is partitioned into k subsets of approximately equal size. Subsequently, each of the k subsets is left out in turn while the remaining $k - 1$ subsets are used to "train" the model (i.e., fit the model or run the algorithm). The omitted subset is used to compute the error rate for the resulting model. If k equals the sample size, then k-fold cross-validation is called "leave-one-out" cross-validation. The overall performance is estimated as the average error rate for the k results.

K-fold cross-validation prediction error for generalized linear models can be computed using the `cv.glm` function from the `boot` package. `cv.glm` requires a matrix or data frame containing the full data (where each row corresponds to a case) as first argument, and an object from a glm model fit as second argument. The arguments K and cost set the number of groups and the function used to evaluate the fit of model. By default, the cost function is the average squared error function. cv.glm returns a list with components K which is the number of groups used, and delta which is a vector of length 2, where the first element is the raw cross-validation estimate of the prediction error, and the second element is an adjusted cross-validation estimate of the prediction error, where the adjustment compensates for the bias introduced by not using leave-one-out cross-validation. If the number of observations in the training set is relatively small, then bias may be introduced and we will overestimate the true prediction error unless it is adjusted.

In the example below, we look at the prediction error of the trees data. Since the default cost is the squared error we prefer smaller values for the prediction error.

```
> library(boot)
> data(trees)
> fit1 <- glm(Volume ~ Girth + Height, data=trees)
> cv.err.loo <- cv.glm(trees, fit1)
> cv.err.loo$delta
[1] 18.2 18.1
> cv.err.5 <- cv.glm(trees, fit1, K=5)  # 5 fold cross validation
> cv.err.5$delta
[1] 19.4 18.7
```

Thus we conclude that the average squared prediction error for the leave-one-out analysis is around 18.2, and we get a slightly larger result of 19.4 when using 5-fold cross-validation. A single K-fold cross-validation result does not necessarily provide an accurate estimate of the prediction error because the split of the data into the K folds may introduce extra variation. Instead we can replicate the cross-validation a number of times and average over the replicates. This is easily done using the sapply function.

```
> res <- sapply(1:200, function(i) {cv.glm(trees, fit1, K=5)$delta[1]})
> mean(res)
[1] 18.4
```

Repeated 5-fold cross-validation yields an estimated prediction error of 18.4 which shows that the single cross-validation result obtained above can be somewhat variable.

If we wish to compare the prediction error between different models, then we can use cross-validation to make sure we are not over-fitting one of the models. Since we are looking at log-transformed variables below we need to define our own cost function such that prediction is measured on the same scale as before. The cost function should take two arguments: the responses used for the test dataset and the expected values computed from the training data.

```
> fit2 <- glm(log(Volume) ~ log(Girth) + log(Height), data=trees)
> mycost <- function(responses, expected) {
+                       mean((exp(responses) - exp(expected))**2) }
> cv2.err.loo <- cv.glm(trees, fit2, mycost)
> cv2.err.loo$delta
[1] 6.98 6.96
```

The leave-one-out prediction error of 6.984 for the model with transformed variables shows a substantial improvement in prediction error compared to the model without transformed variables.

The generic `crossval` function from the `bootstrap` package can be used for general k-fold cross-validation. `crossval` needs a minimum of four arguments: A matrix of explanatory variables, `x`, where each row corresponds to a case, a vector of response values, `y`, and functions `theta.fit` and `theta.predict` which are the function to be cross-validated and the function to compute the predicted values, respectively.

Note that `crossval` converts the `x` object to a matrix before it is passed to `theta.fit` and `theta.predict`. Thus, if `crossval` is used for data frames that contain non-numeric values then the two functions need to convert the variables in the input object to the proper formats before fitting and prediction.

Below we wish to use cross-validation to check for over-fitting when a single-layer neural network is used to predict exercise status from gender, age, and smoking status for the `survey` dataset. We use the `nnet` function from the `nnet` package to fit the neural network.

```
> library(bootstrap)       # For the crossval function
> library(nnet)            # For the nnet function
> library(MASS)            # For the survey data frame
> data(survey)
> cc <- complete.cases(survey$Exer, survey$Sex,
+                        survey$Smoke, survey$Age)
> full <- survey[cc,]      # Keep only the complete cases
> fit <- nnet(Exer ~ Sex + Smoke + Age, data=full,
+               size=10, rang=.05, maxiter=200, trace=FALSE)
> fit
```

```
a 5-10-3 network with 93 weights
inputs: SexMale SmokeNever SmokeOccas SmokeRegul Age
output(s): Exer
options were - softmax modelling
```

Now that we have the fitted model we can compute the confusion matrix and the number of misclassifications (i.e., the off-diagonal elements).

```
> table(full$Exer, predict(fit, type="class"))

       Freq None Some
  Freq   48    0   66
  None    5    3   15
  Some   15    0   83
> # Calculate the number of misclassifications
> sum(predict(fit, type="class") != full$Exer)
[1] 101
```

We can see from the confusion matrix that the neural network at no time predicts that a student never exercises. One-hundred seven out of the 235 students with complete observations were wrongly classified by the model.

We use a "trick" here to circumvent the fact that `crossval` passes the `x` value as matrices since we have both numeric and categorical explanatory variables in our data. Instead we create a vector, `index`, that is used to identify the observations in each group and use the `index` vector inside `theta.fit` and `theta.predict` to extract the correct subsets.

The `theta.fit` function defined below fits the model. It uses the indices passed through `x` to select the subset of the `full` data frame. The response `y` is not really used since we get the correct response with the subset argument to `nnet`. `theta.predict` is also defined below and it returns the predicted class. Again, indices for the test data are passed through `x` and selected from the `full` data frame.

```
> index <- 1:nrow(full) # Work-around to identify obs.
> theta.fit <- function(x,y) {
+   nnet(Exer ~ Sex + Smoke + Age, data=full, size=10,
+        subset=x, rang=.05, maxiter=200, trace=FALSE) }
> theta.predict <- function(fit, x) {
+   predict(fit, full[x,], type="class") }
>
> o <- crossval(index, full$Exer, theta.fit, theta.predict,
+               ngroup=5)
```

```
> table(full$Exer, o$cv.fit)

       Freq None Some
  Freq   62    2   50
  None   15    0    8
  Some   61    1   36
> sum(o$cv.fit != full$Exer, na.rm=TRUE)
[1] 137
```

There were 107 misclassifications for the neural network fit of the full data. Cross-validation showed a slightly larger number of misclassifications, 126, which suggests that we are over-fitting the data with the full dataset.

3.49 CALCULATE POWER OR SAMPLE SIZE FOR SIMPLE DESIGNS

Problem: You want to estimate the number of observations you need to obtain a certain statistical power, or conversely the power you can obtain for a given sample size.

Solution: Sample size and power calculations are typically used to show that under certain conditions, the hypothesis test has a desired chance of detecting an effect if it indeed exists.

Educated sample size and power calculations require that researchers have prior knowledge about the variables they are measuring and the statistical analysis they intend to use, and often the calculations rely on a combination of crude estimates and subjective guesses. In many real situations it may be almost impossible to calculate the necessary sample size (to achieve a certain statistical power) or the statistical power (given a certain sample size) simply because all but the simplest experimental designs require so many guesses and assumptions that may difficult to validate.

R provides functions for sample size calculations for simple designs (one- and two-sample, and paired t tests), one-way analysis of variance, as well as two-sample test for proportions through the `power.t.test`, `power.anova.test`, and `power.prop.test` functions, respectively.

The `power.t.test` function computes power for one- and two-sample t tests. It takes five main arguments: the sample size, `n`, `delta` which is the true difference in mean values between the null hypothesis and the alternative hypothesis, `power` for the statistical power, `sd` for the standard deviation, and `sig.level` for the significance level. If exactly four of the five arguments are specified, then `power.t.test` will calculate the final parameter. Default values for `sig.level` and `sd` are 0.05 and 1, respectively. `power.t.test` assumes a two-sample t test with equal variances and group sizes by default. Setting the `type` option to `"one.sample"` or `"paired"` produces calculations based on one-sample and paired tests.

To calculate the sample size needed in order to use a two-sample t test to detect a mean difference of 2 when the standard deviation within each group is 3.6 with a power of 80% and at a significance level of 5% we type the following.

```
> power.t.test(delta=2, sd=3.6, power=.8, sig.level=0.05)

     Two-sample t test power calculation

              n = 51.8
          delta = 2
             sd = 3.6
      sig.level = 0.05
          power = 0.8
    alternative = two.sided

NOTE: n is number in *each* group
```

Thus we need a total of 104 individuals (52 in each of the two groups) to obtain the desired power of 80% given our prior knowledge/assumption of the true difference in mean values and the standard deviation.

Sometimes, sample size calculations are based on a prior estimate of the effect size, i.e., the ratio of the true mean difference to the standard deviation,

$$\text{ES} = \frac{\mu_1 - \mu_2}{\sigma} = \frac{\delta}{\sigma},$$

where δ is the difference in means and σ is the standard deviation.

If we only know effect size then one of the parameters, say the standard deviation, can be fixed, and we just change the mean difference to obtain the desired effect size. For example, to compute the power of a one-sample t test with 20 individuals where we are looking for an effect size of 0.70 we can use this code.

```
> power.t.test(n=20, delta=.70, sd=1, sig.level=0.05,
+              type="one.sample")

     One-sample t test power calculation
```

```
              n = 20
          delta = 0.7
             sd = 1
      sig.level = 0.05
          power = 0.844
    alternative = two.sided
```

The statistical power of this experiment is 76%.

The `power.prop.test` function computes the power or sample size for two-sample comparisons of proportions. `power.prop.test` has five main arguments: the sample size in each group, `n`, the probabilities of success in the two groups, `p1` and `p2`, as well as the significance level, `sig.level` and the power, `power`. Exactly one of the five parameters should be passed as `NULL`, and that parameter will be calculated based on the other four.

The following code computes the necessary sample size to obtain a power of 80% when testing equality of two proportions at a 5% significance level, where the true probabilities of success are 43% and 75%.

```
> power.prop.test(p1=0.43, p2=0.75, power=.8, sig.level=0.05)

     Two-sample comparison of proportions power calculation

              n = 35.9
             p1 = 0.43
             p2 = 0.75
      sig.level = 0.05
          power = 0.8
    alternative = two.sided

NOTE: n is number in *each* group
```

Consequently, 36 individuals are needed in each of the two groups.

Finally, the `power.anova.test` can be used to compute power for balanced one-way analysis of variance designs. Like the two previous functions, `power.anova.test` needs input on all but one of the six parameters: `groups` (the number of groups or categories), `n` (the number of observations per group), the between group variance (argument `between.var`), the variance within groups (`within.var`), the significance level (`sig.level`), and `power`. Note that it is the between and within group *variances* and not standard deviations that should be specified!

For example, if we have prior information on the group means and

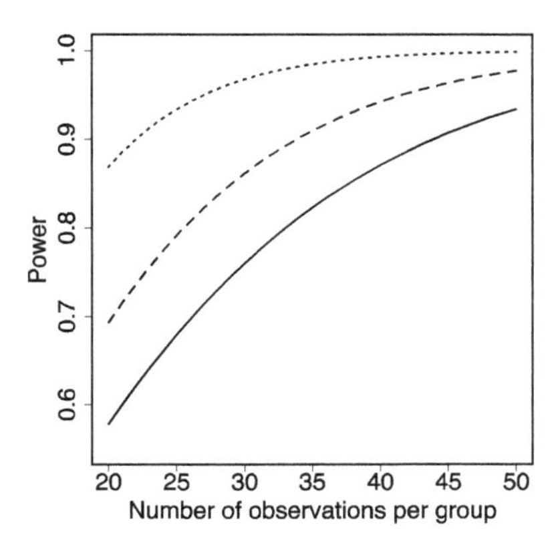

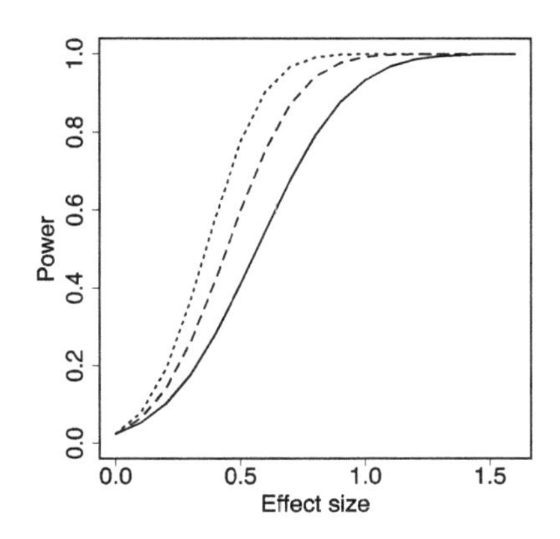

Figure 3.27: Power plots for a t test as a function of sample size (left panel) and effect size (right panel). Three different effect sizes are shown in the left plot: 0.7 (solid line), 0.8 (dashed line), and 1.0 (dotted line), while three different group sample sizes are shown in the right plot: 25 (solid line), 40 (dashed line), and 60 (dotted line).

the variance within groups then we can calculate the necessary sample size as follows:

```
> groupmeans <- c(4.5, 8, 6.2, 7.0)
> power.anova.test(groups = length(groupmeans),
+                  between.var=var(groupmeans),
+                  within.var=5.8, power=.80, sig.level=0.05)

     Balanced one-way analysis of variance power calculation

         groups = 4
              n = 10.7
    between.var = 2.19
     within.var = 5.8
      sig.level = 0.05
          power = 0.8

NOTE: n is number in each group
```

So we would need a total of 44 individuals (11 in each of the four groups) to obtain the desired power.

Power considerations are often used for research proposals and grant applications and in those situations it is often relevant to show how sensitive the power is to the choice of sample or effect size. The code below produces the two plots shown in Figure 3.27 where the power from a two-sample t test is plotted against the sample size (left plot) for different effect sizes and against the effect size for fixed group sample sizes (right plot).

```
> n <- seq(20, 50)          # Consider sample sizes from 20 to 50
> plot(n, power.t.test(n, delta=.7)$power, type="l",
+      ylim=c(0.55,1), ylab="Power",
+      xlab="Number of observations per group")
> lines(n, power.t.test(n, delta=.8)$power, lty=2)
> lines(n, power.t.test(n, delta=1)$power, lty=3)
> delta <- seq(0, 1.6, .1)  # Consider effect sizes from 0 to 1.6
> plot(delta, power.t.test(n=25, delta=delta)$power,
+      type="l", ylim=c(0,1), xlab="Effect size",
+      ylab="Power")
> lines(delta, power.t.test(n=40, delta=delta)$power, lty=2)
> lines(delta, power.t.test(n=60, delta=delta)$power, lty=3)
```

The left-hand plot from Figure 3.27 shows that the power may be somewhat sensitive to the actual sample size — a small change in sample size can have a huge impact on the power. Likewise, the right-hand plot shows that minor changes in the sample size can have a huge impact on the effects sizes that we are able to detect with a given power.

See also: The `MESS` package provides functions for power calculations for various exact tests and an improvement of `power.t.test` to handle groups with different sizes and variances. The `pwr` package provides additional functions for power calculations for specific designs and tests.

NON-PARAMETRIC METHODS

3.50 TEST MEDIAN WITH WILCOXON'S SIGNED RANK TEST

Problem: You want to use a non-parametric method to test if a single sample has a median value, μ.

Solution: Wilcoxon's signed rank sum test is a non-parametric test that is used to test the median of a sample of ordinal or quantitative data. The test can also be used to compare two-sample paired data by taking differences of the paired observations and testing if the median of the differences equals zero.

The function `wilcox.test` computes Wilcoxon's signed rank test statistic and computes the corresponding p-value, and the function handles both one-sample and paired data. The one-sample situation occurs by default if only a single vector is given as input and the paired situation if two vectors are given and the option paired=TRUE is set.

For both the one-sample and the paired data a Wilcoxon signed rank test examines if the distribution of the data has a median value μ, and the value of μ is set by the option mu which defaults to zero.

```
> x <- rbinom(50, size=1, prob=.6)  # Generate random data with 60% 1s
> x
 [1] 0 0 0 1 1 0 0 0 0 0 1 0 1 1 1 1 0 1 0 1 0 1 0 0 1 0 0 0 1 1
[31] 1 1 0 1 1 0 1 1 0 1 1 1 0 0 1 0 0 0 0 1
> wilcox.test(x, mu=0.5)            # Test median equal to 0.5

Wilcoxon signed rank test with continuity correction

data:  x
V = 600, p-value = 0.8
alternative hypothesis: true location is not equal to 0.5
```

Here we get a test statistic of 612 with a corresponding p-value of 0.7816 when we test if the median of the sample x is 0.5. Hence, we fail to reject the null hypothesis that the median could be 0.5.

We can see that this is equivalent to subtracting the desired value of μ first and then use wilcox.test to test for a median value of zero.

```
> x2 <- x - 0.5                     # Deduct 0.5 from all observations
> wilcox.test(x2)                   # Make the same test with median 0

Wilcoxon signed rank test with continuity correction

data:  x2
V = 600, p-value = 0.8
alternative hypothesis: true location is not equal to 0
```

Paired observations are tested whether they are identical (i.e., if the location of their differences is equal to zero) by setting the paired=TRUE argument.

```
> sample1 <- rbinom(24, size=4, p=0.5)
> sample1
 [1] 3 3 2 2 0 1 2 3 1 3 3 3 3 2 3 3 3 0 2 0 2 2 4 2
> sample2 <- sample1 + rbinom(24, size=3, p=0.5) - 1
> sample2
 [1] 2 4 3 3 0 2 1 5 1 2 4 5 4 3 3 2 2 0 2 0 2 4 4 3
> wilcox.test(sample1, sample2, paired=TRUE)
```

```
Wilcoxon signed rank test with continuity correction

data:  sample1 and sample2
V = 40, p-value = 0.07
alternative hypothesis: true location shift is not equal to 0
```

Here we conclude that the two samples have the same location since we fail to reject the null hypothesis ($p = 0.07273$).

`wilcox.test` uses a normal approximation to calculate p-values unless the sample contains less than 50 values or the option `exact=TRUE` is specified. Also, `wilcox.test` cannot compute exact tests in the presence of ties. However, the `coin` package provides the function `wilcoxsign_-test` which handles exact tests even when tied values are present. The argument `distribution=exact()` should be used with `wilcoxsign_-test` to ensure that exact test probabilities are computed.

For both the one-sample and the paired case the model formula used with `wilcoxsign_test` should be specified as `x ~ y`. Here, `x` and `y` are the two numeric vectors (for the paired case) or the numeric vector `x` and a vector of fixed values for μ with the same length as `x` (for the one-sample case).

For the one-sample case we first generate a vector of values under the null hypothesis.

```
> library(coin)
> mu <- rep(.5, length(x))
> wilcoxsign_test(x ~ mu, distribution=exact())

Exact Wilcoxon-Pratt Signed-Rank Test

data:  y by
 x (pos, neg)
 stratified by block
Z = -0.3, p-value = 0.9
alternative hypothesis: true mu is not equal to 0
```

The exact test yields the same conclusions as above, and we fail to reject the null hypothesis that $\mu = 0.5$ ($p = 0.8877$).

In the paired case we get a p-value of 0.1086 and again the exact test provides the same conclusion as the asymptotic test seen above.

```
> wilcoxsign_test(sample1 ~ sample2, distribution=exact())

Exact Wilcoxon-Pratt Signed-Rank Test
```

```
data:  y by
 x (pos, neg)
 stratified by block
Z = -2, p-value = 0.1
alternative hypothesis: true mu is not equal to 0
```

See also: See Problem 3.51 for a non-parametric rank test for two independent samples and Problem 3.52 for a non-parametric rank test for several independent samples.

3.51 USE MANN–WHITNEY'S TEST TO COMPARE TWO GROUPS

Problem: You want to use a non-parametric test to compare if two independent groups of sampled data have equally large values.

Solution: The Mann–Whitney U test (also called the Mann–Whitney–Wilcoxon or Wilcoxon rank-sum test) is a non-parametric test that is used to compare the location shift between two independent groups of ordinal or quantitative data. The test can be employed when the normality assumption of a two-sample t-test is not met.

The function `wilcox.test` calculates the Mann–Whitney U test statistic and tests the hypothesis that the difference in location between the two groups is zero. Input to `wilcox.test` can either be a formula or two vectors representing the values from the two groups.

In the example below we use the `ToothGrowth` data to examine if the supplement type (orange juice or ascorbic acid) of vitamin C influences the length of teeth in guinea pigs.

```
> data(ToothGrowth)
> wilcox.test(len ~ supp, data=ToothGrowth)

Wilcoxon rank sum test with continuity correction

data:  len by supp
W = 600, p-value = 0.06
alternative hypothesis: true location shift is not equal to 0
```

We obtain a test statistic of 575.5 which yields a p-value of 0.06449. Thus we (barely) fail to reject the null hypothesis that the two groups have equally large values.

`wilcox.test` uses a normal approximation to calculate p-values unless the samples contain less than 50 values or the option `exact=TRUE` is given. Also, `wilcox.test` cannot compute exact p-values when ties are present in the data. However, the `coin` package provides the function `wilcox_test` which handles exact tests even when tied values are present. The option `distribution=exact()` should be used with `wilcox_test` to ensure that exact test probabilities are computed.

```
> library(coin)
> wilcox_test(len ~ supp, data=ToothGrowth, distribution=exact())

Exact Wilcoxon-Mann-Whitney Test

data:  len by supp (OJ, VC)
Z = 2, p-value = 0.06
alternative hypothesis: true mu is not equal to 0
```

The conclusion from the exact test is virtually identical to the result from the asymptotic test shown above. We get a p-value of 0.06366 and fail to reject the hypothesis that the two groups have equally large values.

See also: Use Wilcoxon's signed rank sum test (Problem 3.50) to test if *paired* two-sample data are identical. Problem 3.52 extends the Mann–Whitney U test to more than two groups.

3.52 COMPARE GROUPS USING KRUSKAL–WALLIS' TEST

Problem: You want to compare three or more independent groups of sampled data using a non-parametric test.

Solution: The Kruskal–Wallis test is a non-parametric test that can be used to compare two or more independent groups of ordinal or quantitative data when the normality assumption of one-way analysis of variance is not met.

For the Kruskal–Wallis test, the null hypothesis assumes that the samples come from identical populations, and the test is similar to a one-way analysis of variance with the data replaced by their ranks. It is an extension of the non-parametric Mann–Whitney U test (see Problem 3.51) to three or more groups.

In the example below we use the `ToothGrowth` data to examine if the dose (0.5, 1, or 2 mg) of vitamin C influences the length of teeth in guinea pigs. `kruskal.test` computes the Kruskal–Wallis test and accepts input either as a list of vectors, where each vector contains data from the k groups that are compared, or as a model formula. We use the model formula approach below.

```
> data(ToothGrowth)
> # Use only data on ascorbic acid
> teeth <- subset(ToothGrowth, supp=="VC")
> kruskal.test(len ~ factor(dose), data=teeth)

Kruskal-Wallis rank sum test

data:  len by factor(dose)
Kruskal-Wallis chi-squared = 30, df = 2, p-value = 4e-06
```

The p-value for Kruskal–Wallis' test is practically zero so for these data we reject the null hypothesis that the teeth lengths are identical for all three doses.

When the Kruskal–Wallis test is significant, it indicates that at least one of the groups is different from at least one of the other groups. The `kruskalmc` function from the `pgirmess` package can be used to make pairwise multiple comparisons after a significant Kruskal–Wallis test to determine which of the groups is different.

```
> library(pgirmess)
> kruskalmc(len ~ factor(dose), data=teeth)
Multiple comparison test after Kruskal-Wallis
p.value: 0.05
Comparisons
      obs.dif critical.dif difference
0.5-1    10.3         9.43       TRUE
0.5-2    19.7         9.43       TRUE
1-2       9.4         9.43      FALSE
```

The output from `kruskalmc` shows that the differences are found between dose 0.5 and the other two doses and that dose 1 and dose 2 are not significantly different (but almost since the observed difference of 9.4 is close to the critical threshold of 9.42).

Note that `kruskal.test` uses an asymptotic normal distribution to calculate the p-value. Alternatively, the `kruskal_test` from the `coin` package accepts the same input as `kruskal.test` but can compute approximate p-values using Monte Carlo resampling. The argument `distribution=approximate()` ensures that `kruskal_test` computes approximate test probabilities instead of asymptotic probabilities. The number of Monte Carlo replications is 1000 by default but the B option to the approximate function changes the number of replications. For example, 9999 Monte Carlo replications are obtained with distribution=approximate(B=9999).

```
> library(coin)
> kruskal_test(len ~ factor(dose), data=teeth,
+              distribution=approximate(B=9999))

Approximative Kruskal-Wallis Test

data:  len by factor(dose) (0.5, 1, 2)
chi-squared = 30, p-value <2e-16
```

Again, we reject the null hypothesis that the teeth lengths are identical for all three doses.

See also: Problem 3.6 explains how to make one-way analysis of variance. The oneway.test function can test equality of means without assuming equal variances in each group. See Problem 3.51 for the two-group Mann–Whitney test.

3.53 COMPARE GROUPS USING FRIEDMAN'S TEST FOR A TWO-WAY DESIGN

Problem: You want to use a non-parametric test to compare the distribution of observations from k groups when there are measurements from b blocks for each group.

Solution: Friedman's test is a non-parametric test used to compare the distribution of observations from k groups from a balanced block design without replications; i.e., there is exactly one observation for each combination of group and block. Friedman's test is a non-parametric alternative for analysis of variances with repeated measurements (two-way analysis of variance) where the same parameter has been measured under different conditions (the groups) on the same "subjects" (the blocks) and where the assumption of normality may be violated.

Friedman's test is implemented in R by the friedman.test function, and the function tests the null hypothesis that — apart from the effect of the blocks — the location parameter is the same for each of the groups. The input to friedman.test can either be three vectors corresponding to the response values, the grouping factor, and the blocking factor, respectively, or the input can be a model formula of the form response ~ groups | blocks. We use the model formula approach below.

The cabbage data from the isdals package lists the yield of cabbage from four different treatments each grown on four different fields (the blocks). We are only interested in comparing the treatments but want to account for the fact that some of the observations are collected from the same fields.

```
> library(isdals)
> data(cabbage)
> friedman.test(yield ~ method | field, data=cabbage)

Friedman rank sum test

data:  yield and method and field
Friedman chi-squared = 10, df = 3, p-value = 0.01
```

Friedman's rank sum test rejects the null hypothesis of equal location parameter and we conclude that there are significant differences in the yields among the four treatment methods (p-value of 0.0112). If we look at the individual ranks (or the original values), then we see that treatment method "K" has substantially lower yield than the other three treatment methods.

friedman.test uses an asymptotic normal distribution to calculate the p-value. The coin package provides the function friedman_test which can compute approximate p-values using Monte Carlo resampling. Setting the distribution=approximate() argument to the friedman_-test function ensures that approximate test probabilities are computed instead of asymptotic probabilities. The number of Monte Carlo replications is 1000 by default, but it can be changed by setting the B argument to the approximate function; e.g., to use 9999 Monte Carlo replications we set distribution=approximate(B=9999). friedman_test uses the same input as friedman.test except that the "block variable" must be converted to a factor manually if it is not already a factor.

```
> library(coin)
> friedman_test(yield ~ method | factor(field),
+               data=cabbage, distribution=approximate(B=9999))

Approximative Friedman Test

data:  yield by
 method (A, C, K, N)
 stratified by factor(field)
chi-squared = 10, p-value = 9e-04
```

We reach the same conclusion from the Monte Carlo approximation of the p-value as we did from the asymptotic p-value.

See also: Use Cochran's test for this type of situation but with binary responses. Cochran's test can be calculated using the function `cochran.test` from the `outlier` package.

SURVIVAL ANALYSIS

3.54 FIT A KAPLAN–MEIER SURVIVAL CURVE TO EVENT HISTORY DATA

Problem: You want to fit a Kaplan–Meier survival curve to event history data.

Solution: Data that measure lifetime or the length of time until the occurrence of an event are called failure time, event history, or survival time data. For example, the response variable could be the lifetime of light bulbs or the survival time of heart transplant patients. A special consideration with survival time data is that there often is right-censoring present; i.e., we may not know the time of event for all individuals, but we may know that the event *had not* occurred at the given point in time. Thus each response observation consists of two variables: the observed time and a censoring indicator, which tells if the observed time point is due to an event or if it has been censored and we only know that the event has not happened yet. The purpose of Kaplan–Meier survival analysis is to estimate and compare the underlying distribution of the failure time variable among k groups of individuals.

The `survfit` function from the `survival` package computes Kaplan–Meier survival estimates and can be used to create Kaplan–Meier survival curves for right-censored data. The input to `survfit` should be a model formula where the response is a survival object, and where the right-hand side of the formula is a factor that defines the groups. Confidence limits are computed as the `conf.int` argument is set and the default value is 0.95 corresponding to a 95% confidence interval.

Before `survfit` can be used, we need to create the survival object needed as response variable. Survival objects are created with the `Surv` function, and for traditional right-censored data, `Surv` should be called with two arguments: the first argument, `time`, is a vector of survival times and the second argument, `event`, is a vector that indicates if the corresponding time measurement corresponds to an event or if it is censored. The status indicator should be coded as either 0/1 ("1" indicates an uncensored observation and "0" a censored observation), TRUE/FALSE (FALSE means the time is censored), or 1/2 (where "1" means censored).

The ovarian dataset from the survival package contains observations on survival time for two different treatments for ovarian cancer. The variables futime, fustat, and rx, represent the observed time, censoring status, and treatment group, respectively. In the following example, we compare the survival distributions for the two treatment groups.

```
> library(survival)
> data(ovarian)
> head(ovarian)
  futime fustat  age resid.ds rx ecog.ps
1     59      1 72.3        2  1       1
2    115      1 74.5        2  1       1
3    156      1 66.5        2  1       2
4    421      0 53.4        2  2       1
5    431      1 50.3        2  1       1
6    448      0 56.4        1  1       2
> # Fit Kaplan-Meier curves for each treatment
> model <- survfit(Surv(futime, fustat) ~ rx, data=ovarian)
> summary(model)
Call: survfit(formula = Surv(futime, fustat) ~ rx, data = ovarian)

                rx=1
 time n.risk n.event survival std.err lower 95% CI upper 95% CI
   59     13       1    0.923  0.0739        0.789        1.000
  115     12       1    0.846  0.1001        0.671        1.000
  156     11       1    0.769  0.1169        0.571        1.000
  268     10       1    0.692  0.1280        0.482        0.995
  329      9       1    0.615  0.1349        0.400        0.946
  431      8       1    0.538  0.1383        0.326        0.891
  638      5       1    0.431  0.1467        0.221        0.840

                rx=2
 time n.risk n.event survival std.err lower 95% CI upper 95% CI
  353     13       1    0.923  0.0739        0.789        1.000
  365     12       1    0.846  0.1001        0.671        1.000
  464      9       1    0.752  0.1256        0.542        1.000
  475      8       1    0.658  0.1407        0.433        1.000
  563      7       1    0.564  0.1488        0.336        0.946
```

The summary output from the fitted Kaplan–Meier curves shows information on the number of observations under risk and events for each group as well as the estimated survival probabilities and their pointwise confidence limits.

Using plot on the fitted model produces a plot of Kaplan–Meier

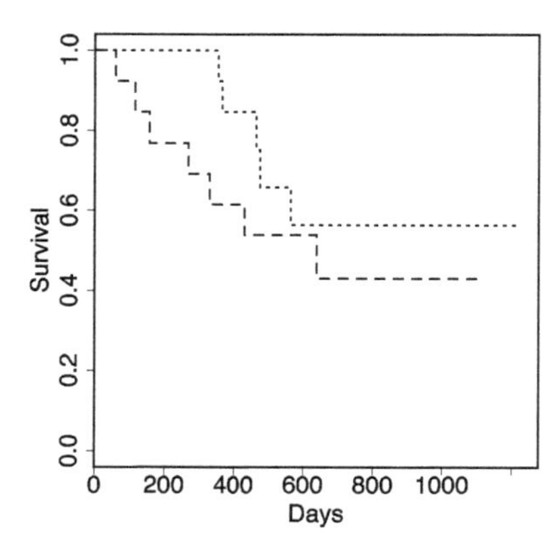

Figure 3.28: Kaplan–Meier survival curves for treatment group 1 (dashed line) and group 2 (dotted line) for the `ovarian` data. Censoring times are shown with a "+".

survival curves. The `mark.time` option to `plot` is set to `TRUE` by default and then censoring times that are not event times are marked on the survival curves. Setting `mark.time` to `FALSE` turns off the marks, and setting it to a numeric vector places a mark at every specified time point. Figure 3.28 shows the Kaplan–Meier survival curves for the two treatment groups.

```
> plot(model, lty=2:3, xlab="Days", ylab="Survival") # Plot them
```

A Kaplan–Meier survival curve for a single group can be fitted by using `~ 1` as the right-hand side of the model formula and in that situation a 95% confidence interval for the curve will be shown when the survival curve is plotted.

The log-rank test tests if there is any difference in survival distribution among the different groups. The `survdiff` function computes the log rank test statistic and computes the corresponding p-value for the test of no difference among groups. `survdiff` takes the same input as `survfit`. The `rho` argument is a scalar parameter that determines the type of test computed by `survdiff`. The default value of 0 produces the log-rank test whereas `rho=1` changes the test so it is equivalent to the Peto and Peto modification of the Gehan–Wilcoxon test.

```
> survdiff(Surv(futime, fustat) ~ rx, data=ovarian)
Call:
survdiff(formula = Surv(futime, fustat) ~ rx, data = ovarian)
```

```
      N Observed Expected (O-E)^2/E (O-E)^2/V
rx=1 13        7     5.23     0.596      1.06
rx=2 13        5     6.77     0.461      1.06

 Chisq= 1.1  on 1 degrees of freedom, p= 0.303
```

We get a chi-square test statistic of 1.1 and a corresponding p-value of 0.303 and conclude from the log-rank test that there is no difference in survival times between the two treatment groups.

When the sample size is rather limited, it might be dangerous to rely on asymptotic tests. Instead, the `logrank_test` function from the `coin` package can be used to compute exact log-rank tests when argument `distribution="exact"` is given. In the following code we need to specify that the grouping variable, `rx`, should be considered a factor since `logrank_test` does not automatically convert the variable to a factor.

```
> library(coin)
> logrank_test(Surv(futime, fustat) ~ factor(rx),
+              data=ovarian, distribution="exact")

Exact Two-Sample Logrank Test

data:  Surv(futime, fustat) by factor(rx) (1, 2)
Z = -1, p-value = 0.3
alternative hypothesis: true theta is not equal to 1
```

The p-value from the exact log-rank test is almost identical to the asymptotic result obtained by `survdiff` and we again conclude that there is no difference in survival time between the two treatments.

See also: The `km.ci` function from the `km.ci` package computes pointwise and simultaneous confidence intervals for the Kaplan-Meier estimator.

3.55 FIT A COX REGRESSION MODEL (PROPORTIONAL HAZARDS MODEL)

Problem: You want to fit a Cox regression model (proportional hazards model) to event history data.

Solution: Problem 3.54 showed how the Kaplan–Meier estimator could be used to estimate the survival function for time-to-event data. Time-to-event data can be analyzed with Cox's regression model (also known as proportional hazards regression model) where one or more explanatory variables may affect the time to the occurrence of the event of interest. Just as in Problem 3.54, right-censoring is often present; i.e., we may not know the exact event time for all individuals. Instead we know that the event *had not* occurred at a given time point and the outcome is essentially a combination of two pieces of information: the observed time and a censoring indicator which states whether the observation time corresponds to an actual event of a censoring time point.

Cox's proportional hazard model is widely used for survival analysis because it does not make any assumptions concerning the shape of the underlying survival distribution. The proportional hazards model assumes that the hazard rate depends multiplicatively on the explanatory variables but requires no assumptions about the shape of the underlying baseline hazard function.

Cox's proportional hazard model can be fitted with the `coxph` function from the `survival` package. The first argument to `coxph` is a model formula, where the explanatory variables are specified on the right-hand side of the formula and where the response is a survival object created by the `Surv` function just as for Kaplan-Meier estimation (see Problem 3.54). The `method` option determines how ties are handled. The default method is `efron` and other alternatives are `breslow` and `exact`.

The `ovarian` dataset from the `survival` package contains observations on survival time for two different treatments of ovarian cancer. The variables `futime`, `fustat`, `rx`, and `resid.ds` represent the observed time, censoring status, treatment group, and a measure of the health condition after chemotherapy, respectively. In the following example, we start with a model with an interaction between age and treatment such that age is allowed to influence the hazard rate differently for the two treatment groups.

```
> library(survival)
> data(ovarian)
> # Make sure rx is a factor
> ovarian$rx <- factor(ovarian$rx, levels=1:2, labels=c("A", "B"))
> ovarian$resid.ds <- factor(ovarian$resid.ds)
> head(ovarian)
  futime fustat  age resid.ds rx ecog.ps
1     59      1 72.3        2  A       1
2    115      1 74.5        2  A       1
3    156      1 66.5        2  A       2
4    421      0 53.4        2  B       1
5    431      1 50.3        2  A       1
```

```
6    448       0 56.4           1  A         2
> model <- coxph(Surv(futime, fustat) ~ rx*age + resid.ds, data=ovarian)
> model
Call:
coxph(formula = Surv(futime, fustat) ~ rx * age + resid.ds,
      data = ovarian)

               coef exp(coef)  se(coef)      z     p
rxB       -9.68e+00  6.24e-05  9.03e+00 -1.07 0.284
age        1.15e-01  1.12e+00  4.67e-02  2.47 0.013
resid.ds2  7.15e-01  2.04e+00  7.60e-01  0.94 0.347
rxB:age    1.49e-01  1.16e+00  1.50e-01  0.99 0.322

Likelihood ratio test=17.9  on 4 df, p=0.0013
n= 26, number of events= 12
```

The fitted model shows the estimated coefficients and the exponentiated coefficients found in the second column for each estimate are the estimated hazard ratios; i.e., the multiplicative effects on the hazard. Thus, the hazard ratio for each additional year for the first treatment is 1.14, while it is $1.14 \times 1.16 = 1.32$ for the other treatment.

The `drop1` (with option `test="Chisq"` to get p-values) and `summary` functions produce likelihood ratio tests for the explanatory variables and parameter estimates, respectively.

```
> drop1(model, test="Chisq")
Single term deletions

Model:
Surv(futime, fustat) ~ rx * age + resid.ds
         Df  AIC   LRT Pr(>Chi)
<none>      60.1
resid.ds  1 59.0 0.932     0.33
rx:age    1 59.2 1.116     0.29
> model2 <- coxph(Surv(futime, fustat) ~ rx + age + resid.ds,
+                  data=ovarian)
> summary(model2)
Call:
coxph(formula = Surv(futime, fustat) ~ rx + age + resid.ds,
      data = ovarian)

  n= 26, number of events= 12

             coef exp(coef) se(coef)      z Pr(>|z|)
rxB       -0.8489    0.4279   0.6392 -1.33   0.1842
age        0.1285    1.1372   0.0473  2.72   0.0066 **
resid.ds2  0.6964    2.0065   0.7585  0.92   0.3586
```

```
---
Signif. codes:  0 '***' 0.001 '**' 0.01 '*' 0.05 '.' 0.1 ' ' 1

           exp(coef) exp(-coef) lower .95 upper .95
rxB            0.428      2.337     0.122      1.50
age            1.137      0.879     1.036      1.25
resid.ds2      2.006      0.498     0.454      8.87

Concordance= 0.812  (se = 0.091 )
Rsquare= 0.475   (max possible= 0.932 )
Likelihood ratio test= 16.8  on 3 df,   p=0.000789
Wald test            = 14.6  on 3 df,   p=0.00216
Score (logrank) test = 20.8  on 3 df,   p=0.000118
```

We find no significant interaction between treatment group and age ($p = 0.3021$), so a model without interaction is specified. The Wald tests in the final model show that age is significant ($p = 0.00141$) but treatment is not ($p = 0.20337$), so treatment could be removed from the model. For age we have an exponentiated estimate of 1.158 so each additional year in age increases the hazard of death by 15.8% (95% confidence interval of 1.059 – 1.268). The health condition status '2' shows a hazard ratio of 2.006 (relative to group '1') but that effect is not significant ($p = 0.3586$).

`summary` also prints out the overall likelihood ratio, Wald and score tests for the hypothesis that all parameters are zero.

The proportionality assumption of the Cox regression model can be checked using the `cox.zph` function from `survival`. `cox.zph` tests the proportional-hazards assumption for each covariate by comparing the (Schoenfeld) residuals to a transformation of time. The default transformation of the survival times is based on the Kaplan–Meier estimate, but other possibilities are `rank` and `identity` which can be set using the `transform` argument.

```
> validate <- cox.zph(model)
> validate
              rho   chisq     p
rxB        0.1260 0.17702 0.674
age       -0.0221 0.00785 0.929
resid.ds2 -0.2586 0.90244 0.342
rxB:age   -0.1043 0.12071 0.728
GLOBAL         NA 2.28638 0.683
```

The proportionality assumption does not appear to be violated for any of the covariates since all p-values are large. The global test is an overall test of proportionality for all variables while the first three lines list tests for the individual explanatory variables.

The Cox model can be stratified according to a categorical explanatory variable by using the `strata` function in the model formula. Stratifying by a covariate enables us to adjust for the variable without estimating its effect on the outcome, and without requiring that the covariate satisfy the proportional hazards assumption since there is a separate underlying baseline hazard for each level of the categorical variable used for stratification.

We can use the following model if we wish to stratify the `ovarian` dataset according to post-treatment health condition.

```
> model4 <- coxph(Surv(futime, fustat) ~ rx + strata(resid.ds),
+                 data=ovarian)
> summary(model4)
Call:
coxph(formula = Surv(futime, fustat) ~ rx + strata(resid.ds),
    data = ovarian)

  n= 26, number of events= 12

       coef exp(coef) se(coef)     z Pr(>|z|)
rxB -0.663     0.515    0.595 -1.11     0.27

    exp(coef) exp(-coef) lower .95 upper .95
rxB     0.515       1.94      0.16      1.66

Concordance= 0.623  (se = 0.108 )
Rsquare= 0.047   (max possible= 0.878 )
Likelihood ratio test= 1.26  on 1 df,   p=0.261
Wald test            = 1.24  on 1 df,   p=0.266
Score (logrank) test = 1.28  on 1 df,   p=0.258
```

If the results from the stratified analysis are compared to the previous analysis we find that the treatment estimate has changed slightly, but that it is still far from significant. Health condition is included differently in the model than before since we now allow the baseline hazard to be different for the two conditions so we do not get any estimated coefficients for `resid.ds` since that is now part of the two baseline hazard functions.

See also: Problem 3.56 shows an example of a proportional hazards model with time-varying covariates. The `survfit` function can be used on the object returned from `coxph` to estimate and plot the baseline survival curves. However, `survfit` function estimates by default at the mean values of the covariates which may not be meaningful. See the help page for `survfit.coxph` — in particular the `newdata` option — for information on how to use sensible values for the covariates.

The `cph` function in package `rms` extends `coxph` to allow for interval time-dependent covariates, time-dependent strata, and repeated events. The `cumres` function in package `gof` computes goodness-of-fit statistics for the Cox proportional hazards model (see Problem 3.33).

3.56 FIT A COX REGRESSION MODEL (PROPORTIONAL HAZARDS MODEL) WITH TIME-VARYING COVARIATES

Problem: You want to fit a Cox regression model (proportional hazards model) with time-varying covariates to event history data.

Solution: Cox regression models with time-dependent covariates are an extension of the Cox regression model where some of the explanatory variables are allowed to change values over time. This could be the case for, say, humans where for example exercise, food intake, or weight measurements change over time.

The `coxph` function from the `survival` package was used to fit Cox regression models with time-constant covariates in Problem 3.55, but `coxph` can also accommodate time-varying covariates. Time-varying covariates are handled by splitting the observation time for an individual into many smaller time intervals, where the covariates are constant, when the survival object is created with the `Surv` function.

To accommodate time-varying covariates we must supply both the start time and the end time of each interval together with the event status indicator; each individual will then be part of the at-risk population with the given explanatory covariates at each interval. For example, if we have an individual who dies at time 600 days after operation, who have had treatment 1 for the first 100 days, treatment 2 for the next 300 days, and then treatment 1 again, then that person should enter the data frame with three lines as shown in the following example data:

```
start  end  sex treatment dead
0      100  M   1            0
100    400  M   2            0
400    600  M   1            1
```

Hence we should make sure that the follow-up of each individual is split according to covariate-constant intervals with one interval per row in the data frame and make sure that the event status vector is coded correctly.

We illustrate the use of time-varying explanatory covariates using the `jasa1` data from the `survival` package. The `jasa1` data frame contains information on survival of patients on the waiting list for the Stanford heart transplant program. People are followed before and after receiving the transplant so the transplant explanatory variable can change over time. The `jasa1` data frame has already split up individuals in time-constant intervals as can be seen below; note that the person with `id` equal to 4 has two records:

```
> library(survival)
> data(heart)
> head(jasa1)
    id start stop event transplant    age  year surgery
1    1     0   49     1          0 -17.16 0.123       0
2    2     0    5     1          0   3.84 0.255       0
102  3     0   15     1          1   6.30 0.266       0
3    4     0   35     0          0  -7.74 0.490       0
103  4    35   38     1          1  -7.74 0.490       0
4    5     0   17     1          0 -27.21 0.608       0
> model <- coxph(Surv(start, stop, event) ~
+                transplant + year + year*transplant,
+                data=jasa1)
> summary(model)
Call:
coxph(formula = Surv(start, stop, event) ~ transplant + year +
    year * transplant, data = jasa1)

  n= 170, number of events= 75

                  coef exp(coef) se(coef)     z Pr(>|z|)
transplant      -0.287     0.751    0.514 -0.56    0.577
year            -0.266     0.767    0.105 -2.53    0.012 *
transplant:year  0.137     1.147    0.141  0.98    0.330
---
Signif. codes:  0 '***' 0.001 '**' 0.01 '*' 0.05 '.' 0.1 ' ' 1

                exp(coef) exp(-coef) lower .95 upper .95
transplant          0.751      1.332     0.274     2.056
year                0.767      1.304     0.624     0.942
transplant:year     1.147      0.872     0.870     1.512

Concordance= 0.611  (se = 0.037 )
Rsquare= 0.05   (max possible= 0.97 )
Likelihood ratio test= 8.64  on 3 df,   p=0.0344
Wald test            = 8.16  on 3 df,   p=0.0428
```

```
Score (logrank) test = 8.46  on 3 df,   p=0.0374
```

Based on this model (without any further model reductions), we see that the log-hazard is decreasing as the *date of acceptance* (`year`) increases for both the non-transplant and the transplant group (estimates of -0.2657 and $-0.2657 + 0.1374 = -0.1283$ per year, respectively). At year 0 the transplanted have a log-hazard that is 0.2868 smaller than the non-transplanted, but this difference diminishes by year and the two groups have identical mortality at year $0.287/0.137 = 2.095$, and after that the transplanted have a higher mortality than the non-transplanted. The hazard (mortality) is decreasing as the date of acceptance (`year`) increases for both the non-transplant and the transplant group (hazard ratio estimates of 0.767 and $0.767 \times 1.147 = 0.880$ per year, respectively). There seems to be a positive effect of transplant since the corresponding estimate is negative and hence decreases the hazard (hazard ratio is 0.75). However, that effect is not statistically significant.

Just as for time-constant covariates, `cox.zph` can be used to test the proportional hazards assumption and the `strata` function can be used to stratify the model. Here we see that none of the covariates reject the proportional hazards assumption so that assumption appears to be valid.

```
> cox.zph(model)
                    rho  chisq     p
transplant      -0.0490 0.2001 0.655
year             0.0117 0.0111 0.916
transplant:year  0.0370 0.1102 0.740
GLOBAL               NA 0.4484 0.930
```

It should be noted, however, that the implicit assumption is that the underlying hazard is a function of time since *admittance* to the waiting list, and is independent of the time since transplant. This assumption is *not* tested by the `cox.zph` function, but requires a more informed analysis of data as in Problem 3.15.

The `survfit` function can be used to generate survival curves from the Cox model with time-varying covariates. Using the `survfit` on a fitted `coxph` object will produce a survival curve that corresponds to a person with covariates equal to the *mean* of each of the covariates in the dataset — rarely of any relevance. Therefore, `survfit` should *always* use the `newdata` argument, where the exact covariates for a hypothetical patient is given when the predicted survival curve is computed.

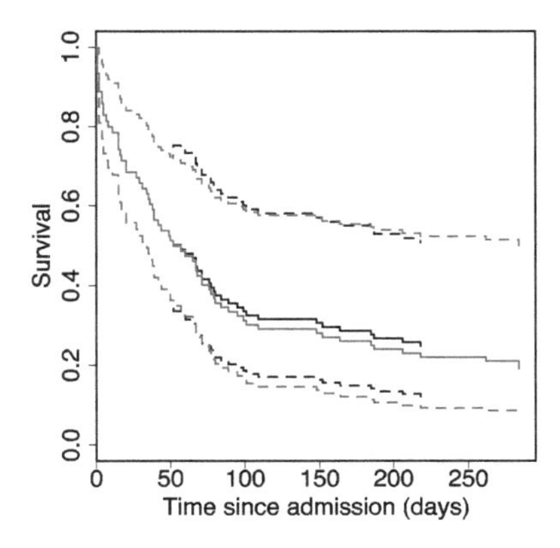

Figure 3.29: Survival curve from a Cox regression model with time-varying covariates. The hypothetical persosn has a transplant at 50 days since admission (black) or no transplant (blue). Solid lines are the survival probability, and dashed lines represent the 95% confidence intervals.

Let us consider a hypothetical person who is admitted at time 0, has a transplant at 50 days after admission, and is followed a maximum of 250 days. If the person was admitted at November 1st, 1968, then we set `year=1` because year has be standardized to start at November 1st, 1967. Since we have time-varying covariates (the transplantation) we include a variable, `id`, in the data frame, and set the `id` argument for `survfit` to ensure that the information from multiple rows with the same id are considered to be from the same person. We want to compare our hypothetical person who does not receive a transplant and is followed until day 300.

```
> nd <- data.frame(start=c(0, 50, 0), stop=c(50, 250, 300), event=1,
+                  transplant=c(0, 1, 0), year=1, id=c(1, 1, 2))
> nd
  start stop event transplant year id
1     0   50     1          0    1  1
2    50  250     1          1    1  1
3     0  300     1          0    1  2
> plot(survfit(model, newdata=nd, id=id),
+      xlab="Time since admission (days)", ylab="Survival",
+      col=c("black", "blue"), conf.int=TRUE)
```

The result is seen in Figure 3.29, where we can clearly see the effect of transplantation at 50 days on the width of the confidence interval for the person receiving the transplant. After day 50 the survival curves for the two persons separate and the person in the transplant group has a higher chance of survival. However, the large confidence intervals shown on the survival curves match the non-significant effect of transplantation.

See also: See the `timecox` function in the `timereg` package for Cox regression models with time-varying parameters.

MISCELLANEOUS

3.57 CORRECT P-VALUES FOR MULTIPLE TESTING

Problem: You have made several simultaneous statistical tests and want to correct the p-values for multiple testing.

Solution: The multiple testing problem refers to the situation where multiple tests or statistical inferences are undertaken simultaneously. Each test has a false positive rate (the type I error rate) which is equal to the threshold set for statistical significance, generally 5%. However, when more than one test is done then the overall type I error rate (i.e., the probability that at least one of the test results is a false positive) is much greater than 5%.

Multiple testing corrections adjust p-values derived from multiple statistical tests in order to control the overall type I error rate and reduce the occurrence of false positives.

The `p.adjust` function takes a vector of p-values as input and returns a vector of p-values adjusted for multiple comparisons. The `method` argument sets the correction method, and the values `"holm"` for Holm's correction (the default), `"bonferroni"` for Bonferroni correction, and `"fdr"` (for false discovery rate) are among the possibilities. The Holm and Bonferroni methods both seek to control the overall false positive rate with Holm's method being less conservative than the Bonferroni method. The `fdr` method controls the false discovery rate which is a less stringent criterion than the overall false positive rate.

We illustrate the `p.adjust` function with a typical microarray gene expression experiment found as the `superroot2` data frame in the `MESS` package. The expression levels of 21,500 different genes are examined and for each gene we wish to determine if there is any difference in gene expression between two plant types (mutant and wild type) after correcting for dye color and microarray. A standard analysis of variance model is used to test the hypothesis of no difference in gene expressions between the two types of plants for each of the 21,500 genes. The `sapply` function is combined with anova and lm to calculate the 21,500 p-values. We would expect 1075 genes to be deemed "significant" by chance with a standard p-value cut-off of 0.05.

```
> library(MESS)
> data(superroot2)
> geneid <- unique(superroot2$gene)    # Get vector of gene names
> # Analyze each gene separately and extract the p-value
> pval <- sapply(1:21500, function(i) {
+               anova(lm(log(signal) ~ array + color + plant,
+                        data=superroot2,
+                        subset=(gene==geneid[i])))[3,5]})
```

Now that we have computed all the p-values we can look at the smallest and adjust them for multiple testing using Holm's correction.

```
> head(sort(pval), 8)                    # Show smallest p-values
[1] 1.05e-05 1.78e-05 2.88e-05 3.30e-05 3.41e-05 3.44e-05
[7] 3.50e-05 4.08e-05
> head(p.adjust(sort(pval)), 8)          # Holm correction
[1] 0.225 0.383 0.619 0.709 0.733 0.739 0.753 0.876
```

The results from p.adjust show that the smallest p-value is 0.2254 after correction with the Holm method, which suggests than none of the 21,500 appear to be significantly differentially expressed when analyzed using analysis of variance.

Below we control the false discovery rate and get slightly smaller p-values than for Holm's method but there are still no significant gene expressions after correcting for the number of tests.

```
> head(p.adjust(sort(pval), method="fdr"), 8) # FDR
[1] 0.108 0.108 0.108 0.108 0.108 0.108 0.108 0.110
```

Note that implicit in all multiple testing procedures for adjusting p-values is the assumption that the distribution of the p-values is correct. If this assumption is violated, then robust methods such as resampling methods should be used to compute the p-values before adjusting for the number of tests.

See also: The multcomp package allows for simultaneous multiple comparisons of parameters whose estimates are generally correlated.

3.58 USE A BOX–COX TRANSFORMATION TO MAKE NON-NORMALLY DISTRIBUTED DATA APPROXIMATELY NORMAL

Problem: You want to find the optimal power transformation of a vector of positive values to ensure that the transformed values appear to approximately follow a normal distribution.

Solution: The Box–Cox transformation is a family of power transformations that can be used to transform non-normally distributed positive quantitative data to approximately follow a normal distribution. The Box–Cox transformation is defined as

$$y(\lambda) = \begin{cases} \frac{y^\lambda - 1}{\lambda} & \text{if } \lambda \neq 0 \\ \log(y) & \text{if } \lambda = 0 \end{cases},$$

where λ is the transformation parameter. λ-values of $0, 0.5$, and 1 correspond to a logarithmic transformation, a square-root transformation, and no transformation, respectively.

Box–Cox transformations are often used in combination with linear models when the normality assumption of the residuals appears to be violated. In those situations, the aim of the Box–Cox transformation is to identify the optimal value of the power transformation parameter for the response variable such that the transformation "best" normalizes the data and ensures that the assumptions for linear model hold.

The `boxcox` function from the `MASS` package applies the Box–Cox transformation to linear models, and the input can be either a model formula or a previously fitted `lm` model object. The `lambda` option is a numerical vector that sets the values of the transformation parameter, λ, to examine. It defaults to the range from -2 to 2 in steps of 0.1. `plotit` is a logical argument that defaults to `TRUE` and it determines if the profile log-likelihood is plotted for the parameter of the Box–Cox transformation.

We use the cherry tree data, `trees`, to find a good transformation when we model the tree volume as a function of log height and log girth. Figure 3.30 shows the resulting plot.

```
> library(MASS)
> data(trees)
> result <- boxcox(Volume ~ log(Height) + log(Girth), data=trees,
+                  lambda=seq(-0.5, 0.6, length=13))
```

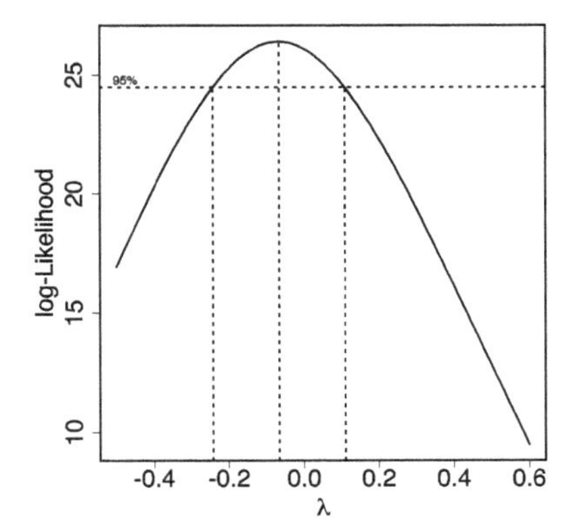

Figure 3.30: Profile log-likelihoods for the parameter of the Box–Cox power transformation for the `trees` data.

The highest profile log-likelihood is attained for $\lambda = -0.06667$ and we also see from the profile log-likelihood plot in Figure 3.30 that the 95% confidence interval for the λ parameter includes zero. Thus we conclude that a log transform of the response variable would be reasonable for this model for the cherry tree data.

`boxcox` returns a list of two vectors: `x` is a vector of λ values and `y` is the corresponding vector of profile log-likelihood values. We can use these vectors to extract the lambda with the highest profile log-likelihood.

```
> result$x[which.max(result$y)]
[1] -0.0667
```

The Box–Cox transformation is defined only for positive data values. This should not pose any problem because an *ad hoc* solution would be to add a constant to all observations if the data contains one or more negative values.

If you wish to find the Box–Cox transformation for a single vector of values then we use a linear model with no explanatory variables as shown below where we find that a square root transformation ($\lambda = 0.5$) is not unreasonable.

```
> y <- rnorm(100) + 5          # Generate data
> y2 <- y**2                   # Square the results
> result <- boxcox(y2 ~ 1, plotit=FALSE)
> result$x[which.max(result$y)]
[1] 0.5
```

See also: See Problems 3.33 and 4.21 for model validation of linear models. Problem 3.9 gives an example of linear models.

CHAPTER 4

Graphics

R is rich in graphical functions and it is very easy to produce high-quality graphics. Base R plotting uses an ink-on-paper approach where high-level graphics functions (e.g., `plot`, `hist`, and `boxplot`) initialize a plot or create an entire plot while low-level graphics functions (e.g., `lines`, `title`, and `legend`) add to the current plot. The ink-on-paper approach means that once something is added to the plot it cannot be removed, and any desired customizations to the plot must be included in the call(s) to the graphics function(s) that produce the plot. The base R graphics have been extended with several packages which extend or redefine graphics, most notably the `lattice` package which produces trellis graphics, and `ggplot2` which produces graphics using graphical grammar. Both packages are highly useful and well worth learning, but they are beyond the scope of this book. Interested readers are referred to the excellent books by Sarkar (2008) and Wickham (2009).

Most of the high-level plotting functions can directly customize many of the graphical parameters through common arguments. If graphical parameters are set as arguments to a plotting function, then the changes are only in effect for the given call to that function. Alternatively, graphical parameters can be set through the `par` function in which case they will be in effect for the rest of the R session or until they are changed again.

The help page for `par` shows a detailed list of all graphical parameters. Some of the most commonly used graphics parameters accepted by the majority of plotting functions are:

- `xlab` and `ylab`, which set the character strings for the x and y axis labels, respectively.

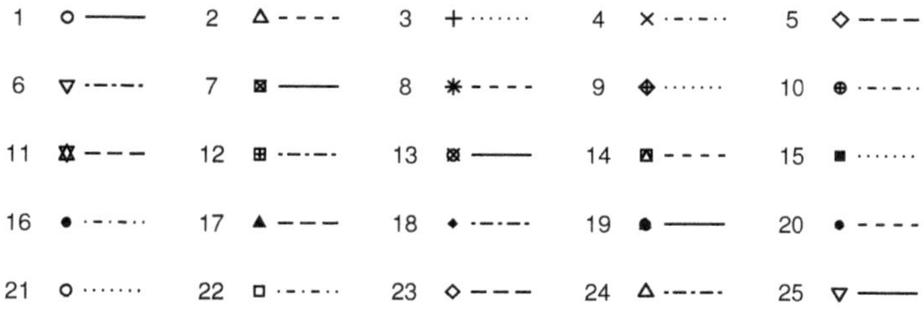

Figure 4.1: Examples of plotting symbols and line types in R plots.

- `col` and `bg`, which change the plotting color and background plotting color (see Problem 4.2).
- `pch`, which changes the plotting symbol for points.
- `lty` and `lwd`, which change the line type and line width when plotting lines. Default line width is 1.
- `main`, which is a character string for the plot title.
- `xlim` and `ylim` set the range of the x and y axes, respectively, and each accepts a pair of numbers for the lower and upper limit of the axis.
- `las`, which sets the style of the axis tick mark labels and should be an integer from 0–3. The value results in tick mark labels that are parallel to the axis (the default), always horizontal, always perpendicular to the axis, and always vertical, respectively.
- `log`, which is used to request logarithmic axes. It can take the character strings `"x"`, `"y"`, and `"xy"` which refer to logarithmic x axis, logarithmic y axis, and double-logarithmic axes, respectively.

The plotting symbol, `pch`, and line types, `lty`, take integer values as shown in Figure 4.1, but the plotting character can also be specified directly; e.g., `pch="a"` which uses the character `a` as plot symbol. Note that for plotting symbols 21 through 25 it is possible to set different border colors (using the `col` option) and fill colors (using the `bg` option). R will recycle graphical values so it is easy to change, for example, plotting symbols and colors for different points.

Text and symbol size are controlled through the cex family of arguments. cex sets the amount of scaling that is used when plotting symbols and text relative to the default. cex defaults to 1, and a value of 1.5 means that symbols and text should be 50% larger, a value of 0.5 is 50% smaller, etc. The options cex.axis, cex.lab, cex.main, and cex.sub are used to set the magnification for axis annotations, labels, main title, and sub-title, respectively, relative to the cex value.

The following code illustrates how graphic parameters are customized and changed. The output is not shown.

```
> x <- 1:5                        # Generate data
> y <- x^2                        # Generate data
> plot(x, y, pch=3:7)             # Use symbols 3-7 for the points
> plot(x, y, pch=c("a", "b"))     # Use symbols a b a b a
> plot(x, y, col="red")           # Each symbol is red
> plot(x, y, pch=c("a", "b"),     # Symbols are a b a b a in colors
+      col=c("red", "blue", "green")) # red blue green red blue
> # Plot circles with a thick red border and blue fill color
> plot(x, y, pch=21, lwd=3, col="red", bg="blue")
> plot(x, y, cex=seq(1,2,.25)) # Symbols with increasing size
> plot(x, y, type="l", lty=2)     # Line with line type 2
> plot(x, y, type="l", lwd=3)     # Line with line width 3
> par(lwd=3, pch=16)              # Line width 3 and symbol 16
> par(cex.lab=2, cex.main=3)      # Large labels and larger title
> plot(x, y, xlim=c(0, 10))       # x axis goes from 0 to 10
> plot(x, y, log="y")             # Logarithmic y axis
> plot(x, y, las=1)               # Rotate numbers on y axis
```

The text function adds a text string at the coordinate given by the x and y arguments. The argument labels should be a character vector or expression that contains the text to be written. Text can be written to one of the four margins of a plot using the mtext function. The first argument to mtext should be a character vector or expression containing the text to write. The side argument determines on which margin the text is printed: 1 is bottom, 2 is left, 3 is top (the default), and 4 is right. Both text and mtext have additional arguments that change the text adjustment and give more control over the exact text placement. Note that when adding code to the margins of a plot it may be necessary to force R to include more space in the margin. This is done with the mar option to the par function (see Problem 4.25 for an example).

```
> text(1, 2, "Where's Wally?")  # Add text to existing plot
> mtext("I'm out here", side=4) # and text to the right margin
```

Table 4.1: Examples of formats possible with text-drawing functions

R code	Interpretation	Example
`x[i]`	subscript	x_i
`x^2`	superscript	x^2
`x%+-%y`	plus/minus	$x \pm y$
`x%*%y`	times	$x \times y$
`x%.%y`	center dot	$x \cdot y$
`x!=y`	not equal	$x \neq y$
`x<=y`	less than or equal	$x \leq y$
`alpha` .. `omega`	Greek letters	α .. ω
`hat(x)`	hat	$\hat{x}$
`bar(x)`	bar	$\bar{x}$
`x%->%y`	right arrow	$x \rightarrow y$
`sqrt(x)`	square root	$\sqrt{x}$

4.1 USE GREEK LETTERS AND FORMULAS IN GRAPHS

Problem: You want to use Greek letters or equations in titles, labels, or text in a plot.

Solution: Greek letters, super- or subscript, and mathematical equations are easy to include in titles, labels, or text-drawing functions. If the text argument to one of the functions that draw text is an expression, then the argument is formatted according to TEX-like rules.

Table 4.1 lists examples of the formatting that can be used in text-drawing functions. Run `demo(plotmath)` to see a thorough demonstration of the possibilities with expressions in R text.

The `paste` function can be used inside the `expression` function to paste regular text and mathematical expressions together. The following lines of code show some of the possibilities, and the resulting output can be seen in Figure 4.2.

```
> x <- seq(0, 5, .1)
> # Plot sine and square root curve and add x axis label
> plot(x, sin(x), type="l", ylim=c(-1, 2.5),
+      xlab=expression(paste("Concentration ", mu[i])))
> lines(x, sqrt(x), lty=2)
> title(expression(paste("This looks like ", Gamma,
+       rho, epsilon, epsilon, kappa, " to me")))
> # Place equations at specific positions
```

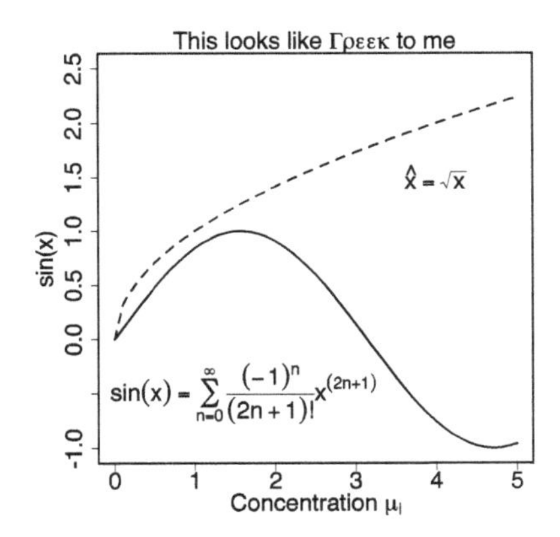

Figure 4.2: Examples of Greek letters and mathematical expressions in R plots.

```
> text(4, 1.5, expression(hat(x) == sqrt(x)))
> text(1.6, -.5, expression(paste(plain(sin)(x) ==
+       sum(frac((-1)^n, paste((2*n+1), plain("!")))*x^(2*n+1),
+       n==0, infinity))))
```

At the moment expressions do not work for axis labels on plots created by the `persp` function.

See also: The help page for `plotmath` and `demo(plotmath)` give a complete list of all possible mathematical annotations in R.

4.2 SET COLORS IN R GRAPHICS

Problem: You want to use a special color in your R graphics.

Solution: Colors can be specified in R graphics using either a character string that defines the color name or a hexadecimal character string that gives the red, green, and blue intensities. If colors are given as hexadecimal strings, then they should have the form `#rrggbb`, where `rr`, `gg`, `bb` are hexadecimal numbers giving the red, green, and blue intensities, respectively.

The `colors` function lists all the predefined colors found in R. The graphical output from the following function calls is not shown.

```
> colors()
  [1] "white"                    "aliceblue"
```

```
  [3] "antiquewhite"           "antiquewhite1"
  [5] "antiquewhite2"          "antiquewhite3"
  [7] "antiquewhite4"          "aquamarine"
  [9] "aquamarine1"            "aquamarine2"
...
[649] "wheat3"                 "wheat4"
[651] "whitesmoke"             "yellow"
[653] "yellow1"                "yellow2"
[655] "yellow3"                "yellow4"
[657] "yellowgreen"
> plot(1:10, col="cyan")          # Plot using cyan color
> plot(1:10, col="#cc0076")       # Plot using purplish color
```

Color specification can also be made with the `rgb` or `hsv` functions. The `rgb` function returns a hexadecimal character string and takes three numerical arguments, `red`, `green`, and `blue`, that give the intensities of the three colors. The `maxColorValue` argument defaults to 1, but can be set to, say, 255, to let the individual color intensities represent 8-bit numbers.

`rgb` can also generate transparent colors with the `alpha` argument. `alpha` takes a number between zero and `maxColorValue` and gives the degree of transparency. Zero means fully transparent while the maximum value means opaque. Transparent colors are supported only on some devices.

```
> rgb(10, 20, 30, maxColorValue=255)
[1] "#0A141E"
> rgb(10, 20, 30, maxColorValue=100)
[1] "#1A334D"
> mycolor <- rgb(10, 20, 30, alpha=50, maxColorValue=100)
> plot(1:10, col=mycolor)   # Use the new colour
```

`hsv` works like `rgb` except it specifies colors using the hue-saturation-value color principle.

See also: Problem 4.3 shows how to specify color palettes.

4.3 SET COLOR PALETTES IN R GRAPHICS

Problem: You want to view or set a complete color palette.

Solution: Colors can be specified in R graphics using character strings as shown in Problem 4.2. Colors can also be specified through color palettes, which is the vector of colors that are used when the color argument in function calls has a numeric index.

The `palette` function views or sets the color palette.

```
> palette()                  # Show the current palette
[1] "black"   "red"      "green3"  "blue"     "cyan"     "magenta"
[7] "yellow"  "gray"
```

If no arguments are given to `palette` then the function lists the current palette. If the character string `"default"` is given as argument, then palette will be set to the default palette, and if a character vector of colors is given then the `palette` function will set the current palette to the vector of colors.

```
> palette(c("red", "#1A334D", "black", rgb(.1, .8, 0)))
> palette()
[1] "red"     "#1A334D" "black"   "#1ACC00"
> palette("default")    # Use the default palette
```

Full-color palettes can be generated directly without having to specify each color separately by using the `RColorBrewer` package. The `RColorBrewer` package contains a single function, `brewer.pal`, which makes color palettes from ColorBrewer available in R. `brewer.pal` takes two arguments, `n` which is the number of different colors in the palette, and `name` which is a character string which should match one of the available palette names. The available palette names can be seen in the `brewer.pal.info` data frame. The output from `brewer.pal` can be used directly as input to the relevant plotting function through, for example, the `col` or `bg` arguments. Alternatively, the palette returned by `brewer.pal` can be used with `palette` to set the global palette.

```
> library(RColorBrewer)
> head(brewer.pal.info, n=10)
         maxcolors category colorblind
BrBG            11      div       TRUE
PiYG            11      div       TRUE
PRGn            11      div       TRUE
PuOr            11      div       TRUE
RdBu            11      div       TRUE
RdGy            11      div      FALSE
RdYlBu          11      div       TRUE
RdYlGn          11      div      FALSE
Spectral        11      div      FALSE
Accent           8     qual      FALSE
> brewer.pal(n=5, name="Spectral")       # 5 colors from Spectral
[1] "#D7191C" "#FDAE61" "#FFFFBF" "#ABDDA4" "#2B83BA"
```

```
> brewer.pal(n=8, name="Accent")         # 8 colors from Accent
[1] "#7FC97F" "#BEAED4" "#FDC086" "#FFFF99" "#386CB0" "#F0027F"
[7] "#BF5B17" "#666666"
> # Set colors locally for a single plotting call
> plot(1:8, col=brewer.pal(n=8, name="Reds"), pch=16, cex=4)
> palette(brewer.pal(n=8, name="Greens")) # Set global palette
> plot(1:8, col=1:8, pch=16, cex=4)       # Plot using colors
```

The examples on the help page for `brewer.cal` should be run on a computer to appreciate the beautiful full-color palettes.

See also: The web page `http://colorbrewer.org` has more information on ColorBrewer palettes. R has several built-in functions such as `gray`, `rainbow`, `terrain.colors`, and `heat.colors` that return color palettes with pre-specified hues.

HIGH-LEVEL PLOTS

4.4 CREATE A SCATTER PLOT

Problem: You wish to create a scatter plot to show the relationship between two quantitative variables.

Solution: Scatter plots are essential for data presentation and explanatory data analysis and are crucial in order to reveal relationships or associations between two numeric variables.

The versatile `plot` function produces scatter plots when two numerical vectors of the same length are provided as input. (If only a single numerical vector is provided then the values are considered the y values and the x values are set to be equidistant.) Plot labels for the x and y axes are set with the `xlab` and `ylab` options, respectively, and the `main` argument can be used to set the plot title. The plotting symbol and color are set with the `pch` and `col` options, respectively.

```
> data(airquality)
> plot(airquality$Temp, airquality$Ozone,
+      xlab="Temperature", ylab="Ozone",
+      main="Airquality relationship?")
```

The `pairs` function produces a matrix of scatter plots where all possible scatter plots between two variables are shown. `pairs` require either a matrix or data frame as input or a model formula where the terms in the formula are used to select the variables plotted with `pairs`.

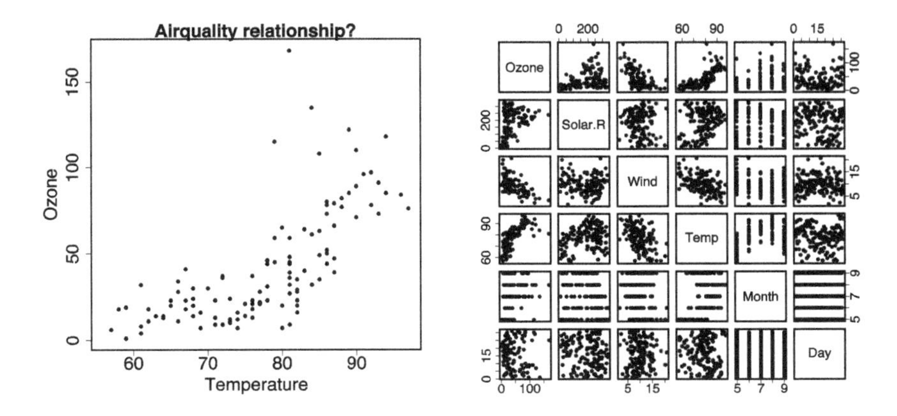

Figure 4.3: Single scatter plot (left) and all possible scatter plots (right) for the `airquality` dataset.

For example to see all possible scatter plots for the `airquality` dataset we use the code below, which first produces a single scatter plot of ozone against temperature and then subsequently plots all possible pairwise scatter plots. The results are shown in Figure 4.3.

```
> pairs(airquality)
```

The `pairs` plot shows the relationship between any pair of variables twice. For example, the upper right plot in the right panel of Figure 4.3 shows ozone level against day number, while the lower left scatter plot in the right panel of Figure 4.3 shows day number against ozone level. It is possible to customize `pairs` such that different plots are shown below, on, and above the diagonal, by supplying `pairs` with the function(s) to generate the necessary plots. The arguments `lower.panel` and `upper.panel` should be functions which accept parameters `x`, `y`, and ... while the diagonal function given to argument `diag.panel` should accept parameters `x` and .... Common to all three functions is that they must not start a new plot but should plot within a given coordinate system. Hence, functions such as `plot`, `boxplot`, and `qqnorm` cannot be used directly for panel functions.

In the following code we use three different panel functions to make the `pairs` plot seen in Figure 4.4. Histograms and empirical density curves are plotted in the diagonal with the `panel.hist` function from the `MESS` package. That package also contains the `panel.r2` function, which is a slight modification of the `panel.cor` function defined on the

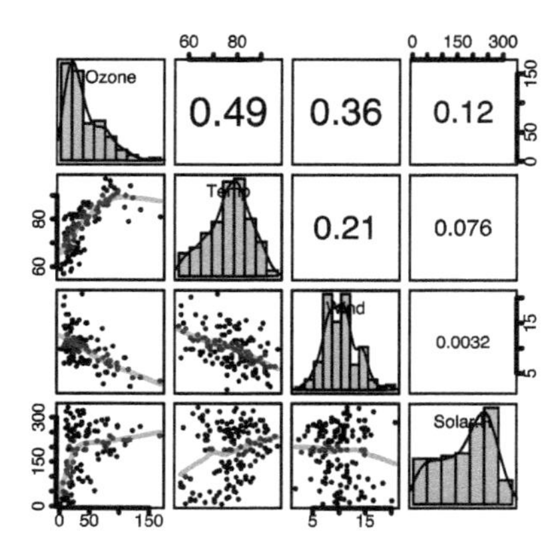

Figure 4.4: Modified `pairs` plot for (some of) the variables in the `airquality` dataset. Histograms and density curves are shown in the diagonal. Upper panels show the proportion of explained variation, R^2, between the two variables and lower panels show scatter plots overlayed with a smoothing function.

`pairs` help page, and it calculates and prints the squared correlation, R^2, with text size depending on the proportion of explained variation. Finally, we use the built-in function `panel.smooth` to produce scatter plots overlayed with a smoothing function to show the general trend (warnings may occur when using multiple panel functions with various graphical parameters, but these can safely be disregarded).

```
> library(MESS)
> pairs(~ Ozone + Temp + Wind + Solar.R, data=airquality,
+       lower.panel=panel.smooth, upper.panel=panel.r2,
+       diag.panel=panel.hist, col.smooth="blue", col.bar="blue")
```

See also: Figure 4.1 shows different plotting options while Problem 4.6 gives an example of histograms.

4.5 CREATE A BUBBLE PLOT

Problem: You wish to create a bubble plot to display the relationship between three or four dimensions of data in a scatter plot.

Solution: Bubble plots are useful for presenting three quantitative variables in a scatter plot where the third dimension is represented by the size of the plot symbol.

Bubble plots are drawn with the `symbols` function which accepts

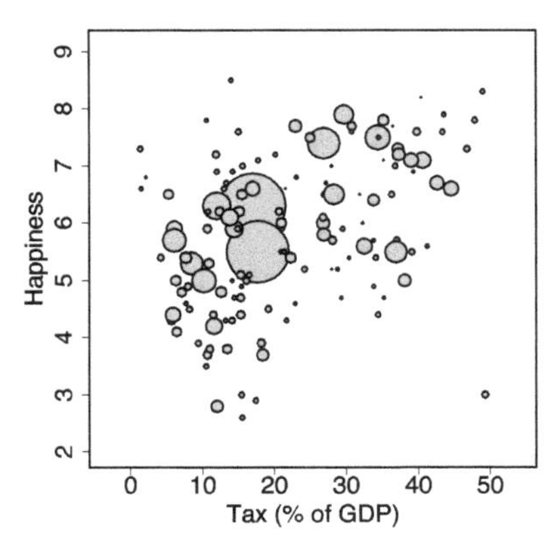

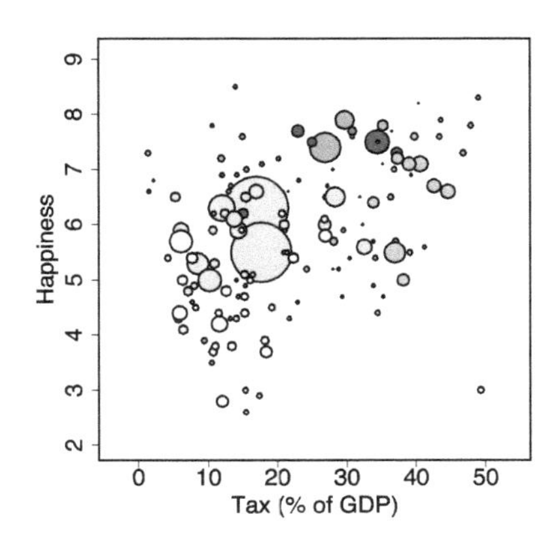

Figure 4.5: Bubble plots showing happiness as a function of taxation and population size (bubble sizes) for 148 countries. The right-hand plot includes a fourth dimension (continent) by adding colors to the bubbles.

two numerical vectors as input for the x and y coordinates in the scatter plot. The third argument, `circles`, requires a numeric vector containing the radii of the plotted circles. By default the plotting symbols are scaled so the largest bubble has a dimension of one inch. The argument `inches=FALSE` prevents this extra scaling of the bubble dimensions.

In the example below we consider a dataset comprising information on happiness, taxation, population size, and continent for 148 countries. We wish to plot happiness as a function of taxation and population size. The `bg` argument sets the color of each circle for clearer visualization.

```
> library(MESS)
> data(happiness)
> with(happiness, symbols(tax, happy, circles=sqrt(population)/8,
+          inches=FALSE, bg="lightgray",
+          xlab="Tax (% of GDP)", ylab="Happiness"))
```

A fourth dimension might be introduced by adding different colors to the bubbles. By default the bubbles are printed in the order of the elements, and this gives problems if the bubbles are overlapping. The solution to this could be to use transparent colors. Below we color the bubbles according to the continent of the country, and the colors are generated with the functions from the `RColorBrewer` package (see Problem 4.3).

```
> library(RColorBrewer)
> newcols <- brewer.pal(6,"Blues")
> with(happiness, symbols(tax, happy, circles=sqrt(population)/8,
+          inches=FALSE, bg=newcols[continent],
+          xlab="Tax (% of GDP)", ylab="Happiness"))
```

See also: The `symbols` function can also be used for plotting other symbols than circles. Squares, rectangles, stars, thermometers, and boxplots are possible by using the corresponding arguments. Problem 4.2 shows how to make transparent colors while Problem 4.3 shows how to generate color palettes.

4.6 CREATE A HISTOGRAM

Problem: You want to summarize the distribution of a quantitative dataset graphically.

Solution: A histogram allows us to graphically summarize the distribution of a quantitative set of observations; e.g., the center, spread, and number of modes in the data.

Histograms and relative frequency histograms are both produced with the `hist` function. By default, the `hist` function automatically groups the quantitative data vector into bins of equal width and produces a frequency histogram. `hist` includes a default title that can be changed with the `main` option and removed altogether with `main=NULL`.

The `LakeHuron` dataset contains information on the annual measurements in feet of the level of Lake Huron in the period from 1875 to 1972. The code below produces a standard frequency histogram as shown in the left plot of Figure 4.6.

```
> data(LakeHuron)
> hist(LakeHuron, main=NULL)  # Create histogram with no title
```

The number of bins is controlled by the `breaks` option to `hist`. If `breaks` is not entered, then R will try to determine a reasonable number of bins. The number of bins is determined by the user if the `breaks` option is set to an integer value, and the exact breakpoints are set when `breaks` is a vector of breakpoint values (note that the breakpoint values should include the entire range of observations).

If the bins are of unequal widths, then `hist` by default produces a relative frequency histogram. Setting the `freq=FALSE` argument forces

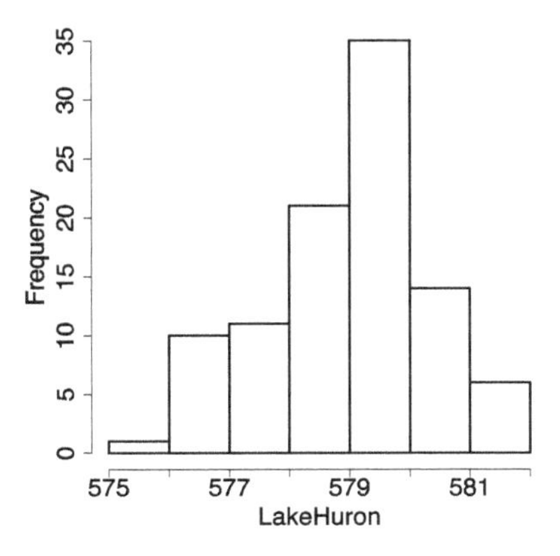

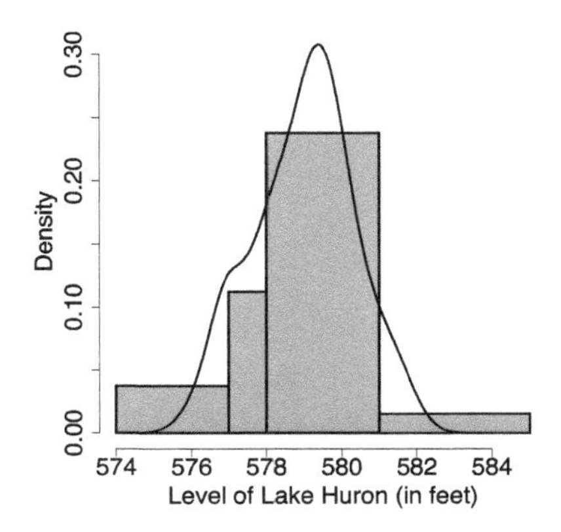

Figure 4.6: Frequency histogram (left plot) and relative frequency histogram with unequal bin size (right plot).

the `hist` function to produce relative frequency histograms even when the bin widths are identical.

The `col`, `density`, and `angle` options are used to set the fill color of the histogram bars, the density and slope of shading lines, respectively. If `density` is set to create shading lines, then the `col` option determines the color of the shading lines.

The `density` function computes the empirical density function and it can be added directly to a relative frequency histogram using the `lines` function.

Below we use the breaks option to specify the number of bins regardless of the size of the dataset, remove the title and make the bars blue, and the result is seen in the right panel of Figure 4.6.

```
> hist(LakeHuron, xlab="Level of Lake Huron (in feet)",
+      freq=FALSE, breaks=c(574, 577, 578, 581, 585),
+      col="blue", main=NULL)
> lines(density(LakeHuron))
```

4.7 CREATE A HANGING ROOTOGRAM

Problem: Create a hanging rootogram to compare an empirical distribution to a theoretical distribution.

Solution: Empirical distributions are often compared to a theoretical distribution graphically by overlaying the histogram of the empirical distribution with the theoretical distribution. It may be difficult to compare an empirical distribution to a theoretical distribution graphically because large frequencies tend to dominate the graph and because it is more difficult to compare curvilinear than linear discrepancies.

The hanging rootogram shifts the histogram bars so they are "hanging" from the theoretical distribution curve in order to visualize deviations from a horizontal line instead of from a curve. Additionally, a square-root scale is used to put less emphasis on the larger frequencies.

The `rootogram` function from the `vcd` package produces hanging rootograms for discrete distributions. `rootogram` requires two arguments: `x` which is a vector or one-way table of observed frequencies, and `fitted` which is a corresponding vector of fitted frequencies.

The `scale` argument defaults to `"sqrt"` but can be set to `"raw"` to prevent the use of the square-root transform of the frequencies. Also, the `type` argument can be set to `"deviation"` to visualize deviations between observed and fitted frequencies instead of hanging frequency bars.

In the example below we consider a classic dataset comprising 6115 families each with 12 children. We assume that the distribution of boys and girls in each family follows a binomial distribution, and estimate the probability of a random child being a boy.

```
> library(vcd)
> families <- c(3, 24, 104, 286, 670, 1033, 1343,
+               1112, 829, 478, 181, 45, 7)
> boys <- 0:12*families                     # Count number of boys
> phat <- sum(boys) / sum(families*12) # Probability of boy
> rootogram(families, 6115*dbinom(0:12, size=12, p=phat))
```

The left-hand plot in Figure 4.7 shows the hanging rootogram. The deviations from the horizontal line appear to be non-random, and there are consistently too many observed families compared to the fitted frequencies under the binomial distribution in the two tails. This suggests that a simple binomial model may not provide a sufficient fit to the data.

Hanging rootograms can also be used to compare a quantitative vector with a continuous distribution. The `rootonorm` function from the `MESS` package compares a histogram from a vector of observed values to a Gaussian distribution. By default, the observed values are compared to a Gaussian distribution with mean and standard deviation that are estimated from the data, but a specific mean and standard deviation can be set using the `mu` and `s` arguments, respectively. Also, the `rootonorm`

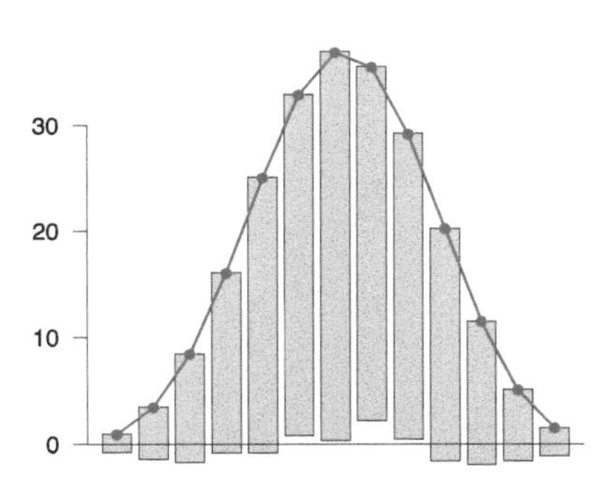

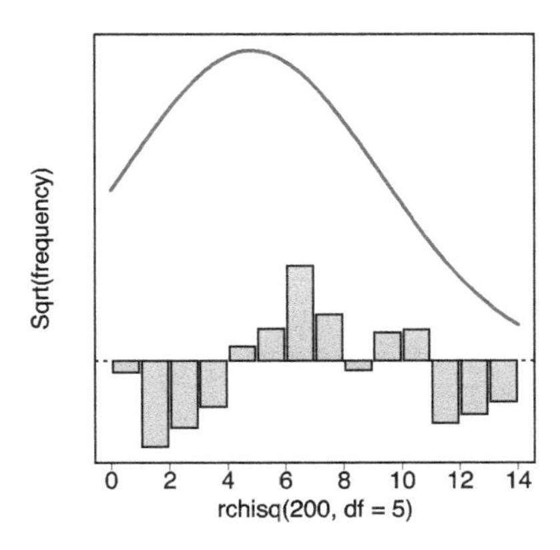

Figure 4.7: Hanging rootogram comparing the observed number of families with $0, \ldots, 12$ boys and the corresponding fitted binomial distribution (left plot), and deviations between a Gaussian distribution and 200 observations from a χ^2_5-distribution (right plot).

function has an optional argument, `breaks`, which either should be a character string `"Sturges"` to use Sturges' algorithm to decide the number of breaks or a positive integer that sets the number of breaks for the histogram.

```
> library(MESS)
> rootonorm(rchisq(200, df=5), type="deviation", breaks=15)
```

The right-hand plot in Figure 4.7 shows the deviations between a hanging rootogram based on 200 observations from a χ^2_5-distribution and a Gaussian distribution with mean and standard deviation estimated from the available data vector.

4.8 CREATE A VIOLIN- OR BOXPLOT

Problem: You wish to make a box-and-whisker plot or violin plot to summarize one or more distributions graphically.

Solution: Boxplots are efficient graphics for examining a single distribution and for making comparisons between several distributions. Boxplots provide a visual summary of several aspects of a distribution, including its central value, general shape, and variability. A boxplot shows the extreme (maximum and minimum values), the lower and upper quartiles, and the median.

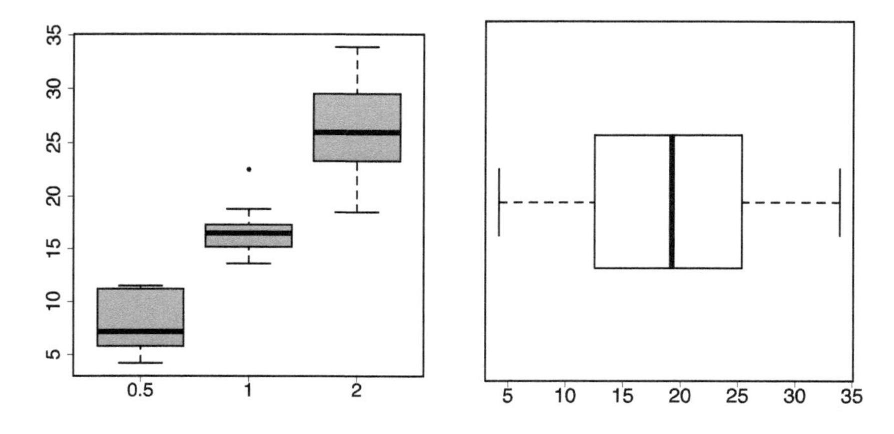

Figure 4.8: Parallel boxplots (left plot) and a single horizontal boxplot where the whiskers extend to the minimum and maximum value (right plot).

Horizontal and vertical boxplots are produced in R by the `boxplot` function and input to `boxplot` can either be a single numeric vector (for a single distribution) or a formula of the form `y ~ group` where `y` is a numeric vector of values and the vector `group` determines the grouping (for parallel boxplots). Parallel boxplots are also created if several numeric vectors are given as arguments to `boxplot`.

By default, R produces a modified boxplot where extreme values are graphed as separate points. Thus, the whiskers extend to the largest data points that are not extreme (extreme observations or outliers are defined as values that are further than 1.5 $\times$ the inter-quartile range away from the lower or upper quartile).

To compare the distribution of teeth lengths for different doses of the `ToothGrowth` data we can use the following code, which produces the left-hand graph of Figure 4.8.

```
> data(ToothGrowth)
> teeth <- subset(ToothGrowth, supp=="VC") # Use only some data
> boxplot(len ~ factor(dose), data=teeth, col="blue")
```

We can see that the distribution corresponding to dose 1 appears to have slightly smaller variance than the other two doses and that there is a single extreme value for the dose 1 observation. The distribution of values for all three doses appears to be fairly symmetric.

The standard boxplot where the whiskers extend to the minimum and maximum value can be obtained by setting the `range` option to zero.

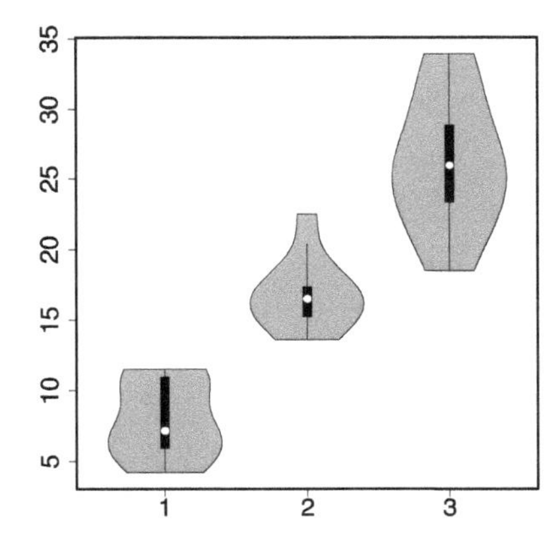

Figure 4.9: Violin plot comparable to the left-hand boxplot in Figure 4.8.

In addition, the `horizontal=TRUE` option ensures that the produced boxplot is horizontal. The following code produces the horizontal boxplot with whiskers extended from minimum to maximum value shown in the right-hand plot of Figure 4.8.

```
> boxplot(ToothGrowth$len, horizontal=TRUE, range=0)
```

Violin plots found in the `vioplot` package can also be useful visual tools to compare distributions. They are essentially similar to a boxplot but show the data distribution density mirrored instead of a box representing the middle 50% of the observations. The current `vioplot` function does not accept a formula argument so multiple vectors are needed as arguments to the function in order to compare the distributions.

Below we extract the three groups to compare and then create the violin plot seen in Figure 4.9.

```
> library(vioplot)
> dose.5 <- teeth$len[teeth$dose==0.5]
> dose1 <- teeth$len[teeth$dose==1.0]
> dose2 <- teeth$len[teeth$dose==2.0]
> vioplot(dose.5, dose1, dose2, col="blue")
```

`vioplot` does not only plot the densities mirrored but also shows the corresponding box plots as the middle of each "violin."

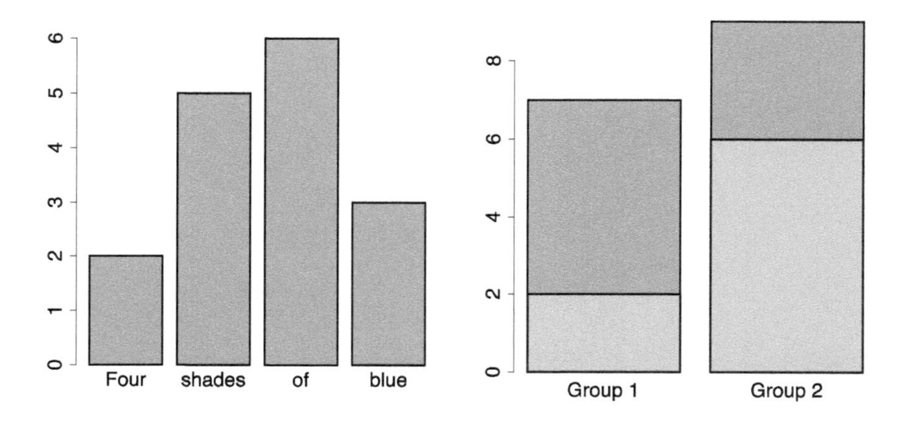

Figure 4.10: Basic bar plots. The left figure is a simple bar plot and the right figure is a stacked bar plot.

4.9 CREATE A BAR PLOT

Problem: You want to present your categorical data or results as a bar plot.

Solution: Categorical data can be visualized as bar plots, which are produced by the `barplot` function in R. The default plot consists of a sequence of bars with heights corresponding to the elements of the vector. If the first argument is a matrix, then each bar is a stacked bar plot where each column of the matrix corresponds to a bar and the matrix values define the height of the elements of each stacked bar. The `names.arg` argument accepts a vector with the names for the columns (see Figure 4.10 to see the output).

```
> barplot(c(2,5,6,3), col="blue",
+         names.arg=c("Four", "shades", "of", "blue"))
> m <- matrix(c(2,5,6,3),2)
> m
     [,1] [,2]
[1,]    2    6
[2,]    5    3
> barplot(m, col=c("lightgray", "blue"),
+         names.arg=c("Group 1", "Group 2"))
```

`barplot` creates stacked bars by default but that can be changed to grouped or juxtaposed bars by setting the `beside=TRUE` argument. The `legend.text` argument constructs a legend for the bar plot. If `legend.text` is set to a vector of strings then these strings will correspond to the legend labels. Alternatively, `legend.text` can be set to `TRUE` in which case the row names of the input matrix are used to create the legend labels. If a vector of character strings is given to the `names` option, then the names are plotted underneath each bar or group of bars. Finally, the `horiz` argument defaults to `FALSE` but can be set to `TRUE` if the bars are to be drawn horizontally.

We need to divide each column in the matrix by the column sum if the stacked bar plot should show relative frequencies. The `prop.table` function computes the relative frequencies of a matrix given as the first argument. The second argument to `prop.table` determines if the proportions are relative to the row sums (`margin=1`) or the column sums (`margin=2`).

As an example, we use the following data from the county of Århus in Denmark where the concentration of the pesticide dichlorobenzamide (also called BAM) in drinking water was examined from two different municipalities within the county. The allowable limit of BAM is $0.10\mu g/l$, and the data are summarized below.

	Concentration (in μg/l)			
Municipality	< 0.01	0.01–0.10	> 0.10	Total
Hadsten	23	12	6	41
Hammel	20	5	9	34

The following example produces the two plots shown in Figure 4.11. We consider the relative frequencies and use the `t` function to transpose the matrix such that frequencies are grouped by municipality instead of concentration.

```
> m <- matrix(c(23, 20, 12, 5, 6, 9), ncol=3)  # Input data
> relfrq <- prop.table(m, margin=1)            # Calc rel. freqs
> relfrq
      [,1]  [,2]  [,3]
[1,] 0.561 0.293 0.146
[2,] 0.588 0.147 0.265
> # Make juxtaposed barplot
> barplot(relfrq,  beside=TRUE,
+         legend.text=c("Hadsten", "Hammel"),
+         names=c("<0.01", "0.01-0.10", ">0.10"))
> # Stacked relative barplot with labels added
> barplot(t(relfrq), names=c("Hadsten", "Hammel"), horiz=TRUE)
```

See also: Problem 4.10 shows how to add error bars to bar plots.

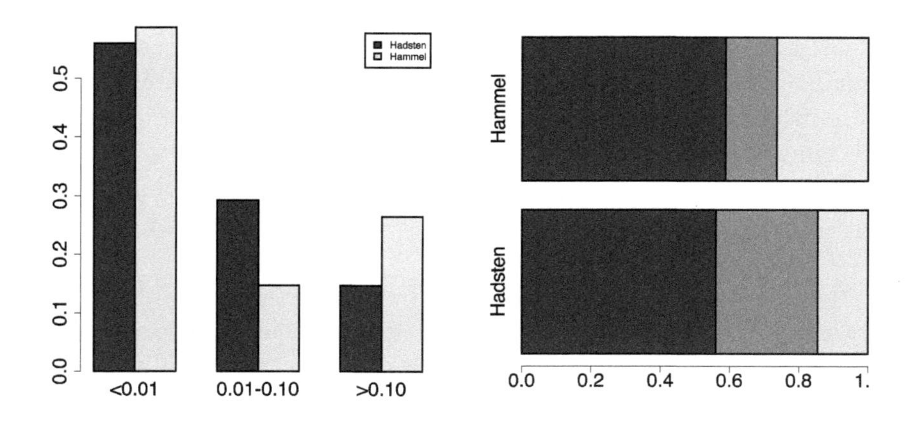

Figure 4.11: Juxtaposed and stacked relative bar plots.

4.10 CREATE A BAR PLOT WITH ERROR BARS

Problem: You want to present your data or results as a bar plot overlayed with error bars.

Solution: The `barplot2` function from the `gplots` package is an enhancement of the standard `barplot` function that can plot error bars for each bar. The function takes the same arguments as `barplot` but in addition the `plot.ci=TRUE` argument forces the barplots to include error bars, and the lower and upper limits for the individual error bars are specified with the `ci.l` and `ci.u` arguments, respectively.

If we want to create a bar plot of the average monthly temperature and corresponding intervals where we can expect to find 95% of the observations of the population (assuming normality and known monthly temperature means) based on the `airquality` dataset, then we can use the following code, where we set the argument `plot.ci=TRUE` to make `barplot2` plot the intervals.

```
> library(gplots)
> data(airquality)
> # Compute means and standard deviations for each month
> mean.values <- by(airquality$Temp, airquality$Month, mean)
> sd.values <- by(airquality$Temp, airquality$Month, sd)
> barplot2(mean.values,
+          xlab="Month", ylab="Temperature (F)",
+          plot.ci=TRUE,
+          ci.u = mean.values+1.96*sd.values,
+          ci.l = mean.values-1.96*sd.values)
```

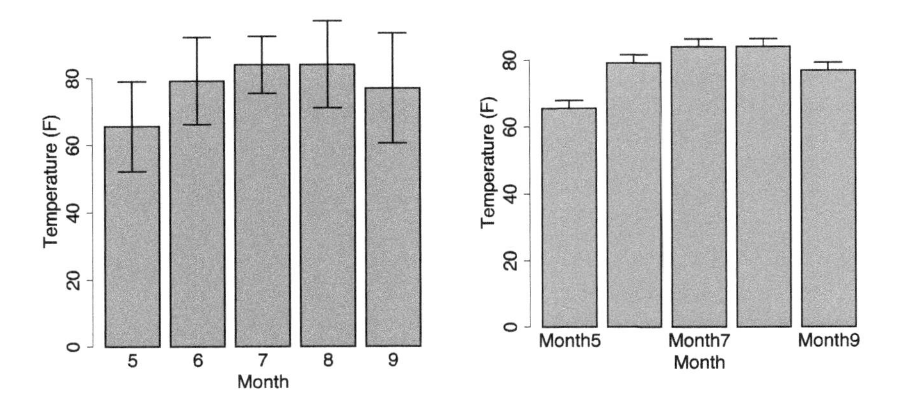

Figure 4.12: Examples of `barplot2` output. The left panel shows the expected intervals for 95% of the observations for each month. The right graph shows (half) the 95% confidence intervals for the mean of each month.

The code above produces the graph seen in the left plot in Figure 4.12. The line color, type, and width of the error bars are controlled with the `ci.color`, `ci.lty`, and `ci.lwd`, respectively, and they accept the usual values related to line plotting.

We can also use the `barplot2` function to show standard error of parameter estimates as shown in the following code. The output is the right-hand graph of Figure 4.12, and by setting the lower confidence interval limit to the same value as the mean value we print only the part of the confidence intervals that is above the bars. We use `coef` and `summary` to extract the estimates and standard errors and the `qt` function to calculate 95% confidence limits directly.

```
> airquality$Month <- factor(airquality$Month)
> model <- lm(Temp ~ Month - 1, data=airquality)
> mean.values <- coef(summary(model))[,1]
> mean.values
Month5 Month6 Month7 Month8 Month9
  65.5   79.1   83.9   84.0   76.9
> sem <- coef(summary(model))[,2]
> scale <- qt(0.975, df=summary(model)$df[2])
> barplot2(mean.values, col="blue",
+          xlab="Month", ylab="Temperature (F)",
+          plot.ci=TRUE,
+          ci.u = mean.values+scale*sem,
+          ci.l = mean.values)
```

See also: The `plotCI` function in the `gplots` package. `plotCI` is used in Problem 4.11.

4.11 CREATE A PLOT WITH ESTIMATES AND CONFIDENCE INTERVALS

Problem: You want to present the results from your analysis graphically with estimates and corresponding confidence intervals.

Solution: Sometimes it is desirable to present the results from an analysis graphically, for example by illustrating parameter estimates and their corresponding confidence intervals. These graphs can be created using standard plotting functions like `plot` and `arrows`, but the `plotCI` function from `gplots` package combines the necessary steps into a single function.

`plotCI` requires one or two arguments, `x` and `y` which define the coordinates for the center of the error bars. The `uiw` and `liw` arguments set the widths of the upper/right and lower/left error bar, respectively, and if `liw` is not specified it defaults to the same value as for `uiw`.

The `OrchardSprays` data frame contains the results from an experiment undertaken to assess the potency of various constituents of orchard sprays in repelling honeybees. We use a one-way analysis of variance to compare the treatment means and we fit the model without intercept so we directly obtain the group means and standard errors from R.

```
> library(gplots)
> data(OrchardSprays)
> fit <- lm(decrease ~ treatment - 1, data=OrchardSprays)
> summary(fit)

Call:
lm(formula = decrease ~ treatment - 1, data = OrchardSprays)

Residuals:
   Min     1Q Median     3Q    Max
-49.00  -9.50  -1.63   3.81  58.75

Coefficients:
           Estimate Std. Error t value Pr(>|t|)
treatmentA     4.62       7.25    0.64  0.52631
treatmentB     7.62       7.25    1.05  0.29766
treatmentC    25.25       7.25    3.48  0.00098 ***
treatmentD    35.00       7.25    4.83  1.1e-05 ***
treatmentE    63.13       7.25    8.70  5.5e-12 ***
```

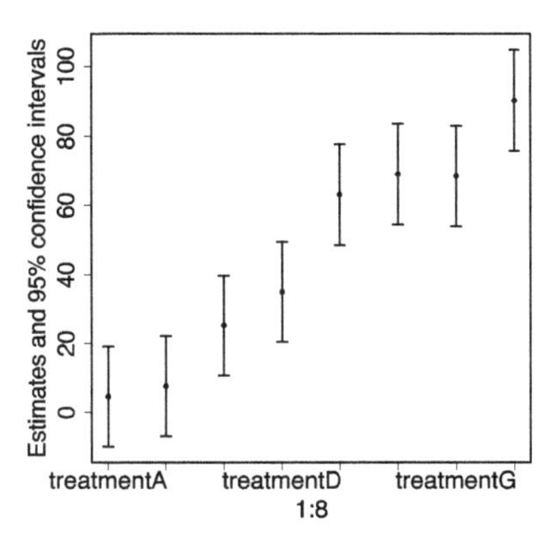

Figure 4.13: Example of `plotCI` output.

```
treatmentF    69.00         7.25     9.51  2.7e-13 ***
treatmentG    68.50         7.25     9.44  3.5e-13 ***
treatmentH    90.25         7.25    12.44  < 2e-16 ***
---
Signif. codes:  0 '***' 0.001 '**' 0.01 '*' 0.05 '.' 0.1 ' ' 1

Residual standard error: 20.5 on 56 degrees of freedom
Multiple R-squared:  0.889,Adjusted R-squared:  0.873
F-statistic: 55.9 on 8 and 56 DF,  p-value: <2e-16
```

The output from `plotCI` can be seen in Figure 4.13. In the following code we suppress the printing of the x-axis when calling `plotCI` since we would like to add the actual coefficient names directly to the graph. Hence we extract the names of the coefficients from the fitted object and add them manually using the `axis` function.

```
> plotCI(coef(summary(fit))[,1],
+        uiw=qt(.975, df=56)*coef(summary(fit))[,2],
+        ylab="Estimates and 95% confidence intervals", xaxt="n")
> axis(1, 1:length(rownames(coef(summary(fit)))),
+      rownames(coef(summary(fit))))
```

See also: The `plotmeans` function from the `gplots` package provides a wrapper to `plotCI` to plot group means and confidence intervals.

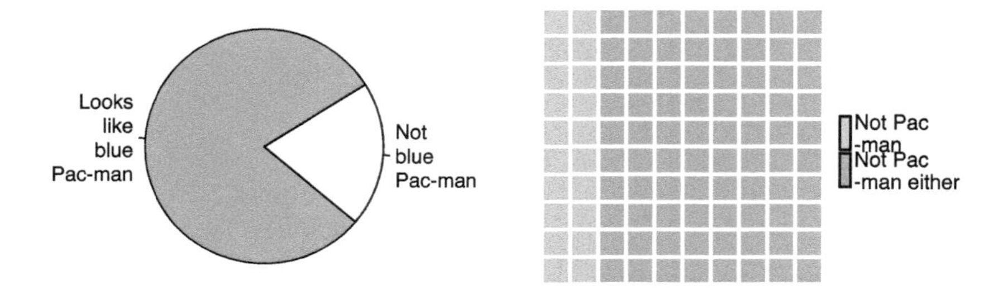

Figure 4.14: Pie (left) and waffle (right) charts.

4.12 CREATE A PIE CHART

Problem: You want to present your categorical data or results as a pie chart.

Solution: It is generally recommended to avoid presenting data as pie charts since it is argued that humans are bad at evaluating areas and because the information from most pie charts often is better represented as numbers (see the book by Tufte (2009) for arguments why most pie charts might be a bad idea).

The `pie` function in R can be used to generate pie charts. Its first argument should be a vector of values that are used to define the relative size of the individual slices of the pie. Additional arguments are `col`, `label`, and `radius` which are vectors that set the colors, and labels of the individual slices, and the circle radius, respectively. Normally, the first cut of the first slice is along the x-axis, but we can change that with the `init.angle` argument as shown below.

```
> pie(c(20, 80), init.angle=-40,
+     radius=.6,
+     col=c("white", "blue"),
+     label=c("Not\nblue\nPac-man",
+             "Looks\nlike\nblue\nPac-man"))
```

An alternative to pie charts is "square pie charts" also known as waffle charts. Results from waffle charts should be easier to compare and evaluate. The `waffle` function from the `waffle` package. It accepts a named numeric vector as input and produces a single square per integer.

By default, the `waffle` function produces ten rows of squares but that can be changed with the `rows` argument. The `colors` option sets the colors of the areas. The results are shown in the right-hand chart of Figure 4.14.

```
> library(waffle)
> areas <- c(`Not Pac\n-man`=20,
+            `Not Pac\n-man either`=80)
> waffle(areas, colors=c("white", "blue"))
```

See also: The `pie3D` function from the `plotrix` package can produce 3D pie charts with optional "exploding" slices.

4.13 CREATE A PYRAMID PLOT

Problem: You want to create a pyramid (opposed horizontal bar) plot to display two groups opposite each other.

Solution: The `pyramid.plot` function from the `plotrix` package plots a pyramid plot (opposed horizontal bar plot) typically used to illustrate population pyramids.

`pyramid.plot` takes two arguments, `lx` and `rx`, each of which should be either a vector, matrix, or data frame of the same length. The values contained in the `lx` and `rx` arguments determine the bar lengths of the left and right part of the plot, respectively. If `lx` and `rx` are both matrices or data frames then `pyramid.plot` produces opposed stacked bar plots with the first column of the matrix/data frame plotted innermost for each bar.

The `labels` option sets the labels for the categories represented by each pair of bars, and it should be a vector containing a character string for each `lx` or `rx` value, even if empty. Different labels for the left and right bars can be used if `labels` is a matrix or data frame (in which case the first two columns are used for the labels). The options `lxcol` and `rxcol` can be used to set the colors of the left and right bars, respectively. If a vector of colors is provided, then `pyramid.plot` will cycle through them from the bottom of the pyramid and upwards. `pyramid.plot` returns the vector of graphics margins that was used when the function was called and it has no impact on the output.

In this example we plot a population pyramid using 10-year age intervals for the Danish population based on the Danish registry on January 1, 2011. The number of men and women are both normalized

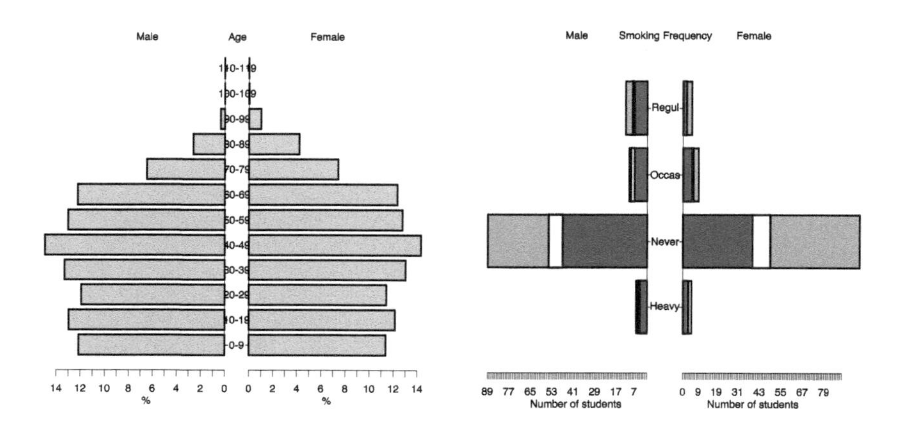

Figure 4.15: Examples of `pyramid.plot` output.

to get the percentage of men (and women) at each age group. The output is shown in the left plot of Figure 4.15.

```
> library(plotrix)
> men <- c(334700, 357834, 328545, 367201, 411689,
+          359363, 336841, 178422, 72296, 9552, 139, 0)
> women <- c(318662, 340357, 320462, 365180, 401521,
+            357405, 345884, 208057, 117843, 27914, 761, 0)
> groups <- paste(seq(0, 110, 10), "-", seq(9, 119, 10), sep="")
> groups                                 # Show group labels
 [1] "0-9"     "10-19"   "20-29"   "30-39"   "40-49"   "50-59"
 [7] "60-69"   "70-79"   "80-89"   "90-99"   "100-109" "110-119"
> men <- 100*men/sum(men)                # Normalize men
> women <- 100*women/sum(women)          # and women
> pyramid.plot(men, women, labels=groups,
+              lxcol="lightgray", rxcol="lightgray")
[1] 4.7 4.7 0.1 0.1
```

Below we use the `survey` data frame from the `MASS` package to compare the male and female student frequencies of exercise for different groups of smoking status.

```
> library(MASS)
> data(survey)
> # Tabulate exercise and smoking status by gender
> result <- by(survey, survey$Sex,
+              function(x) {table(x$Smoke, x$Exer)})
> result
survey$Sex: Female
```

```
      Freq None Some
  Heavy    3    0    2
  Never   39   10   50
  Occas    5    1    3
  Regul    2    0    3
-----------------------------------------
survey$Sex: Male

      Freq None Some
  Heavy    4    1    1
  Never   47    8   34
  Occas    7    2    1
  Regul    7    1    4
```

The default options for `pyramid.plot` are set up to create population pyramids. However, the function can also be used for other opposed bar plots with a little tweaking of the optional arguments. The `gap` option sets the width between the opposing bars and it may need to be increased if the labels are wider than the age groups typically used for population pyramids. The `unit` option sets the labels on the x axis (default is "%"). Finally, the `top.labels` option is a character vector of length three that sets the top left, middle, and right labels, respectively. Its default values are `Male`, `Age`, and `Female`. The output is seen in the right plot of Figure 4.15.

```
> pyramid.plot(lx=result$Male, rx=result$Female,
+              labels=row.names(result$Male),
+              gap=10, unit="Number of students",
+              top.labels=c("Male","Smoking Frequency","Female"),
+              lxcol=c("darkblue", "white", "lightblue"),
+              rxcol=c("darkblue", "white", "lightblue"))
```

4.14 PLOT MULTIPLE SERIES

Problem: You want to create a graph where multiple series (points or lines or both) are plotted.

Solution: The `matplot` function can be viewed as an extension of the `plot` function which plots multiple series of points or lines simultaneously. `matplot` takes two arguments, `x` and `y` each of which can be vectors or matrices with the same number of elements/rows.

The first column of `x` is plotted against the first column of `y`, the second column of `x` against the second column of `y`, etc. `matplot` will

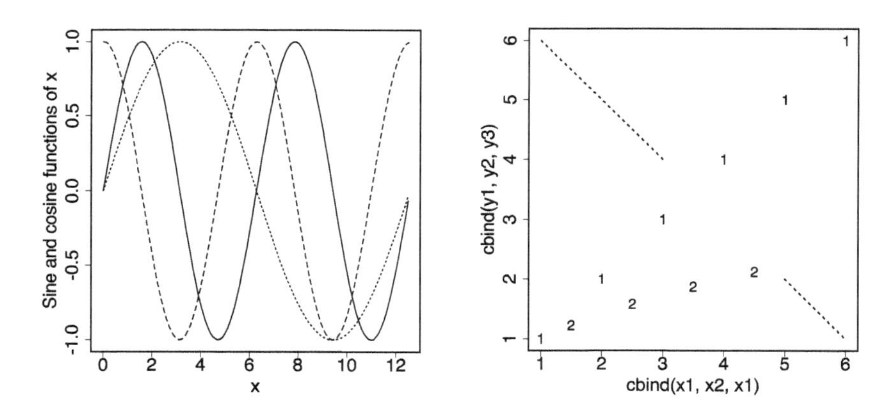

Figure 4.16: Example of `matplot` output.

cycle through the columns if one matrix has fewer columns than the other.

The left panel of Figure 4.16 is created by the following two lines:

```
> x <- seq(0, 4*pi, .1)
> matplot(x, cbind(sin(x), cos(x), sin(x/2)), type="l",
+         ylab="Sine and cosine functions of x")
```

`matplot` takes the same graphical parameters as `plot`. In particular, `type` can be used to set the plot type (`type="l"` for lines and `type="p"` for points) and the parameters `pch`, `lty`, and `lwd` are used to set the plotting symbol, line type and line width, respectively. As always, these parameters are recycled if their length is shorter than the number of series plotted.

Missing values are not plotted and can be used to create series with a different number of observations as shown below and in the right-hand panel of Figure 4.16.

```
> x1 <- 1:6
> y1 <- x1
> x2 <- c(1.5, 2.5, 3.5, 4.5, NA, NA)
> y2 <- sqrt(x2)
> y3 <- 7-x1
> y3[4] <- NA
> y3
[1]  6  5  4 NA  2  1
> matplot(cbind(x1, x2, x1), cbind(y1, y2, y3),
+         type=c("p", "p", "l"), lwd=3)
```

If only one of x and y is specified, then x will be set to the vector 1:n, and the other vector will act as y.

See also: matlines and matpoints are matrix plot versions of the lines and points functions, respectively. matlines adds lines to an existing plot while matpoints adds points.

4.15 MAKE A 2D SURFACE PLOT

Problem: You want to create a 2D contour plot of a surface.

Solution: R has several built-in functions that produce two-dimensional plots of a surface. The contour function displays isolines from a matrix z, where the elements of z are interpreted as heights with respect to the xy plane.

The following code makes a contour plot for the function $\sin(x \cdot y)$ evaluated over the rectangle $[0, 3] \times [0, 4]$. The result is shown in the upper left panel of Figure 4.17.

```
> x <- seq(0, 3, .2)     # Set grid points for x
> y <- seq(0, 4, .1)     # Set grid points for y
> # Compute function at all grid points
> z <- outer(x, y, FUN=function(xx,yy) { sin(xx*yy) })
> contour(x, y, z)       # Make contour plot
```

The location of the grid lines given by the x and y parameters must be in ascending order and they are set to be equidistant on the interval from 0 to 1 if they are not specified. The z matrix is interpreted as a table of $f(x_i, y_j)$ values, so the x and y axes in the plot produced by contour corresponds to the rows and columns of z, respectively.

The number of contour isoline levels is set by the nlevels options (default is 10) which produce equidistant isolines. Alternatively, specific isolines can be set by supplying a vector of desired isolines to the levels option. This is shown below and the result is in the upper right panel of Figure 4.17.

```
> contour(x, y, z, levels=c(-1, -.3, 0, .1, .2, .3, .8, 1))
```

Plain contours are not very striking and often the surface is more easily envisioned if colors are used to fill the spaces between the contours. The filled.contour function creates a contour plot where the areas are filled in solid color and adds a color map that shows the relationship between the colors and the isolines. This type of plot is sometimes also

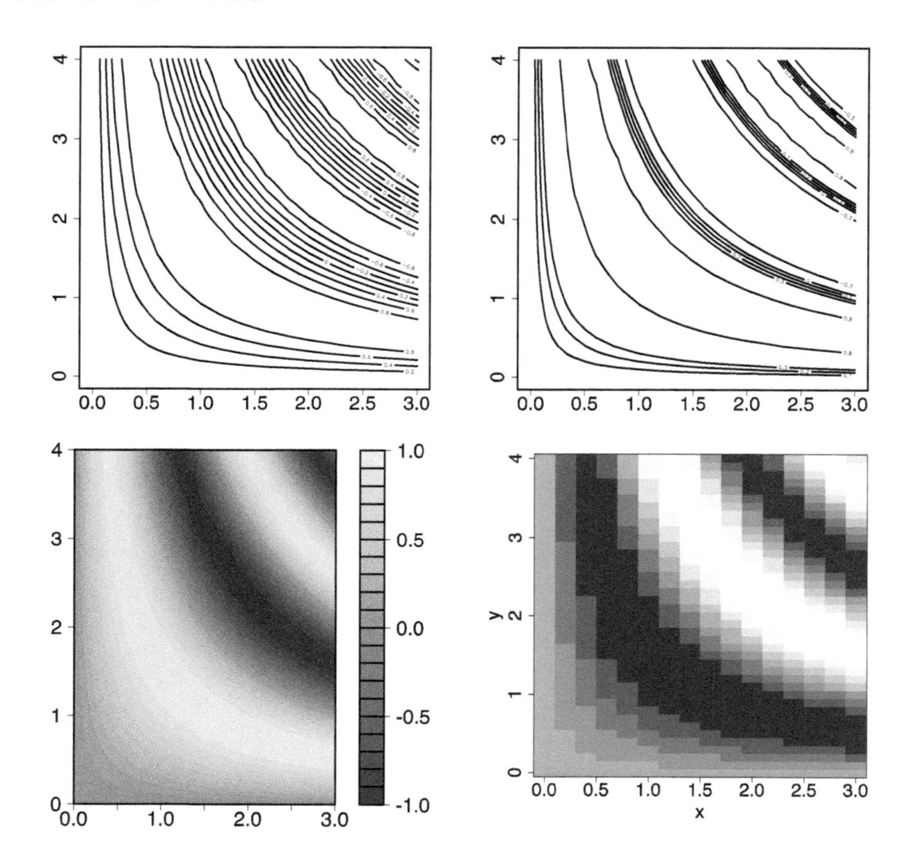

Figure 4.17: Examples of `contour`, `filled.contour`, and `image` output. The upper panels show `contour`, the lower left plot is `filled.contour`, and the lower right panel is `image`.

called a "level plot." The `color.palette` option to `filled.contour` sets the color scheme.

Finally, you can use the `image` function to plot your matrix without it being contoured. `image` needs slightly different input than `contour` and `filled.contour`. If the length of `x` equals `nrow(z)+1` then the values of `x` are interpreted as the boundaries between the cells. Alternatively, if the length of `x` equals `nrow(z)` then the values of `x` specify the midpoints of the cells. The same applies to `y`. Also, for `image` the option to set the color scheme is `col` so the following code produces the output shown in the two lower plots of Figure 4.17. The `RColorBrewer` package is used to generate the blue palette of colors.

```
> filled.contour(x, y, z, color.palette=grey.colors)
> library(RColorBrewer)
> blue.colors <- brewer.pal(9, "Blues")
> image(x, y, z, col=blue.colors)
```

See also: The `contourplot` from the `lattice` package uses formula notation to specify contour plots, i.e., `z ~ x*y`. The functions `rainbow`, `terrain.colors`, `heat.colors`, and `cm.colors` are used to create color palettes with different color schemes. See Problem 4.3 for more information on `RColorBrewer`.

4.16 MAKE A 3D SURFACE PLOT

Problem: You want to create a 3D graph of a surface.

Solution: The `persp` function draws perspective plots of surfaces represented by a matrix z, where the elements of z are interpreted as "heights" with respect to the xy plane.

The following code creates a perspective plot for the function $\sin(x \cdot y)$ evaluated over the rectangle $[0, 3] \times [0, 4]$. The result is shown in the upper left panel of Figure 4.18.

```
> x <- seq(0, 3, .2)          # Set grid for x dimension
> y <- seq(0, 4, .1)          # Set grid for y dimension
> z <- outer(x, y, FUN=function(xx,yy) { sin(xx*yy) })
> persp(x, y, z)
```

The `z` matrix is interpreted as a table of $f(x_i, y_j)$ values, so the x and y axes in the plot correspond to the rows and columns of `z`, respectively. The function arguments `phi` and `theta` rotate the viewing angle of the surface in the azimuthal and colatitude directions, respectively. The plot can often be substantially improved by setting a reasonable view point using these two options. Thus, with the default values of `phi` and `theta`, the top left corner of the matrix is displayed at the left-hand side, closest to the user.

Three additional arguments are worth mentioning here: `col`, `shade`, and `ticktype`, which set the surface color, the degree of shading, and the tick marks along the axes, respectively. `col` is the color of the surface facets, and can be set to a single color (which is recycled), or a longer vector of colors. `shade` should be a number between 0 and 1 which

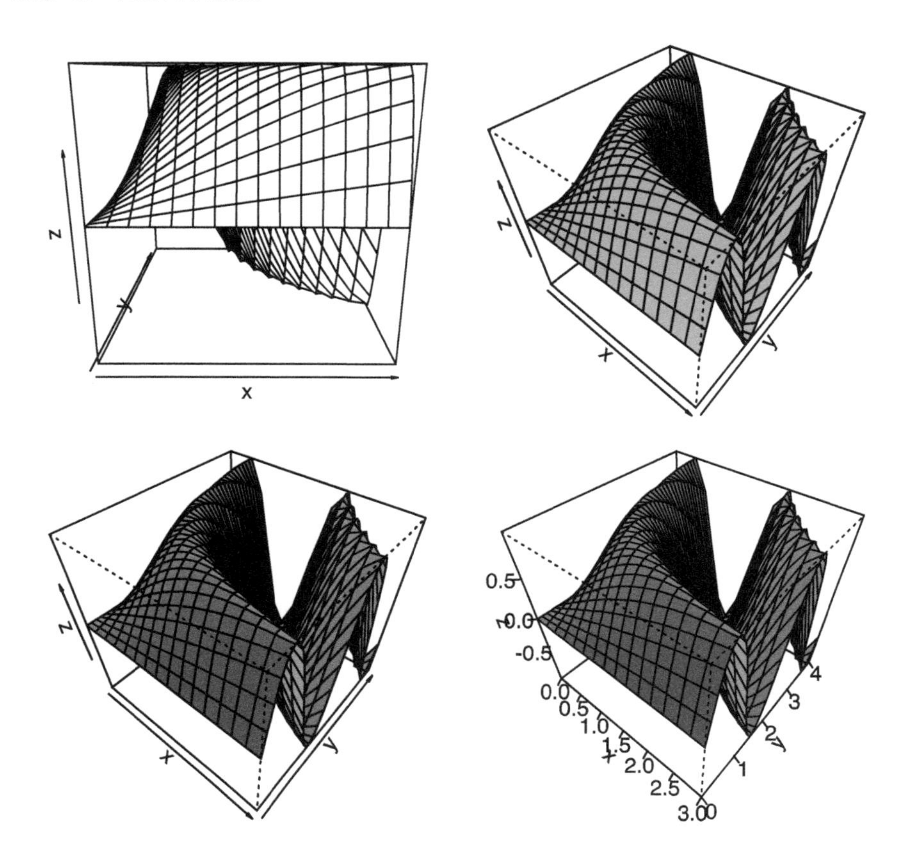

Figure 4.18: Examples of `persp` output. The upper right sets rotate the plot and set the color to light gray. The lower left graph adds shading to the plot and the lower right provides tick marks.

sets the level of shading, and `ticktype` can be set to the character string `"detailed"` to provide tick marks along the axes. The default for `ticktype` is `"simple"` which just draws arrows parallel to the axes.

These options are used below and the output is seen in Figure 4.18.

```
> persp(x, y, z, theta=40, phi=45, col="blue")
> persp(x, y, z, theta=40, phi=45, col="blue", shade=.3)
> persp(x, y, z, theta=40, phi=45, col="blue", shade=.3,
+        ticktype="detailed")
```

See also: The help page of `persp` gives an example of how to make the surface colors correspond to the values of the `z` matrix. The `wireframe` function from the `lattice` package can also produce surface plots. Problem 4.31 uses the persp3d function from the rgl package to create interactive 3D surface plots.

4.17 PLOT A 3D SCATTER PLOT

Problem: You want to make a 3D scatter plot.

Solution: Visualizing a three-dimensional point cloud in just two dimensions can be difficult but a few functions exist that try to accomplish that. Here we will look at the scatterplot3d function from the scatterplot3d package for plotting three-dimensional scatter plots.

scatterplot3d works like plot but takes three arguments, x, y, and z, which define the three coordinates of the points. The y and z options can be omitted if x is data frame with numeric variables cols, rows, and value, *or* if x is a data frame that contains the variables x, y, and z.

The color of the points/lines can be given by the col option as for most graphics. Alternatively, the highlight.3d option can be set to TRUE to override the color option, and makes sure that the points are drawn in different colors depending on their depth (i.e., the value of y coordinate).

The relationship between volume, height, and girth for the trees data frame is plotted by the following code using the scatterplot3d function. The output is shown as the upper left plot of Figure 4.19.

```
> library(scatterplot3d)
> data(trees)
> scatterplot3d(trees$Girth, trees$Height, trees$Volume,
+               xlab="Girth", ylab="Height", zlab="Volume")
```

Just as for plot, the type option sets the type of plot: p for points (the default), h for vertical lines, and l for lines. The vertical lines are especially useful for indicating the "depth" of a three-dimensional point since both the "height" and the position on the $x - y$ plane is shown (the upper left plot of Figure 4.19).

```
> scatterplot3d(trees$Girth, trees$Height, trees$Volume,
+               type="h", color=grey(31:1/40),
+               xlab="Girth", ylab="Height", zlab="Volume")
```

The scale.y and angle arguments set the depth of the plotting box

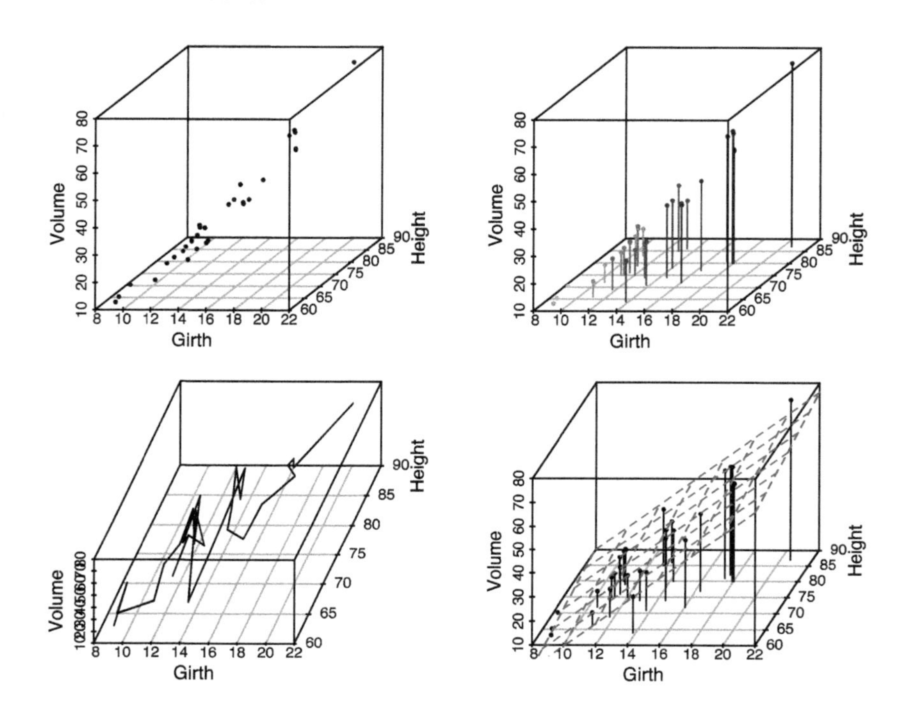

Figure 4.19: Examples of `scatterplot3d` output for the `trees` data. The upper left plot is the default 3D plot, while the upper right plot shows the use of vertical lines (`type="h"`). The lower left plot shows how `scale.y` and `angle` changes the plot, and the lower right plots adds a 3D plane.

(relative to the width and height), and the angle between the x and y axis, respectively. Default values are 1 and 40.

The object returned by `scatterplot3d` provides the two functions, `points3d` and `plane3d`, which are used to add graphics to the current 3D plot. `points3d` adds points or lines (much like the `points` and `lines` functions) while `plane3d` draws a plane into the existing plot. Both functions include information on the graphical setup and plot dimensions to ensure that the new graphical elements are correctly added. `points3d` works just like `points` and requires three numeric vectors as input. `plane3d` either needs options `Intercept`, `x.coef`, and `y.coef` for the three parameters that define the plane in 3D, or the resulting object from an `lm` call with two explanatory variables. The first explanatory variable is taken to be the x axis while the other variable is the y axis.

In the code below we first order trees according to their volume and then plot curve from lowest to highest volume. After that we add the fitted linear plane to the points. The code produces the two lower plots shown in Figure 4.19.

```
> o <- order(trees$Volume)
> scatterplot3d(trees$Girth[o], trees$Height[o], trees$Volume[o],
+               type="l", scale.y=3, angle=75,
+               xlab="Girth", ylab="Height", zlab="Volume")
> myplot <- scatterplot3d(trees$Girth, trees$Height, trees$Volume,
+                         type="h", angle=60,
+                         xlab="Girth", ylab="Height", zlab="Volume")
> model <- lm(Volume ~ Girth + Height, data=trees)   # Fit a model
> myplot$plane3d(model, col="blue")                  # Add fitted plane
```

See also: The `plot3d` from the `rgl` package produces interactive 3D plots (see Rule 4.31).

4.18 CREATE A HEAT MAP PLOT

Problem: You wish to create a heat map plot to visualize data from a two-dimensional array.

Solution: Heat map plots are used to visualize data from a two-dimensional array using colors to indicate the values in the array. In R, the function `heatmap` plots a heat map, and it requires a numeric matrix as its first argument. There are two important arguments to `heatmap`. The `distfun` argument determines which function is used to calculate distances between the rows of the matrix and it defaults to the function `dist`. The other important argument is `hclustfun`, which sets the function that is used to compute the hierarchical clustering of the rows and columns (it defaults to the function `hclust` which computes hierarchical clusterings).

`heatmap` automatically produces row and column dendrograms but the arguments `Rowv` and `Colv` can be set to `NA` to suppress the row and column dendrograms, respectively. Finally, the `scale` argument indicates if the matrix values should be scaled. Its default value is `row` if the matrix is not symmetric, and `none` otherwise, which corresponds to values being centered and scaled in the row direction or not at all, respectively. If

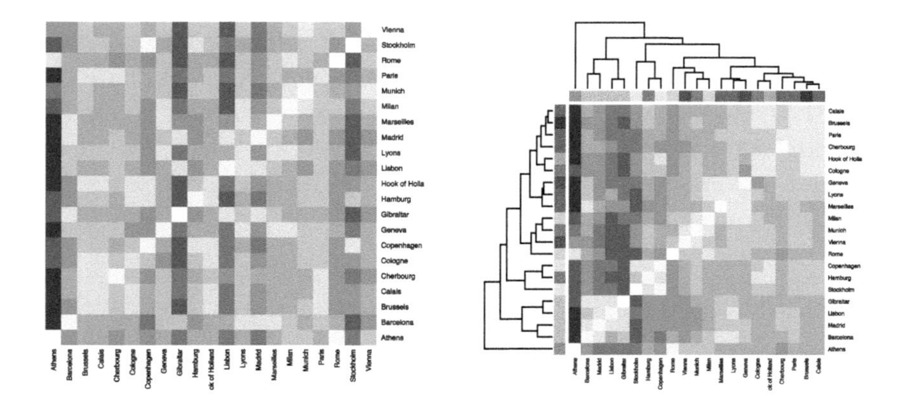

Figure 4.20: Examples of `heatmap` output. The right-hand plot has added vertical and horizontal side bars.

`scale="column"` then the values are centered and scaled in the column direction.

The `eurodist` data frame contains distances between major European cities and the distances are used below to illustrate the `heatmap` function. The `RColorBrewer` package is used to generate the blue palette of colors, and the output is seen in the left panel of Figure 4.20.

```
> data(eurodist)
> x <- as.matrix(eurodist)
>
> library(RColorBrewer)
> blue.colors <- brewer.pal(9, "Blues")
> heatmap(x, Rowv=NA, Colv=NA, col=blue.colors)
```

Notice how the row and column names of the matrix `x` are automatically included in the plot to make it easier to identify the individual rows and columns.

The color scheme of the heat map plot can be set with the `col` option. By default it uses heat colors obtained from the `heat.colors` function, but that can be changed to give other color schemes. The color of the horizontal/vertical side bars that annotate the columns and rows of the heat map are controlled by the arguments `ColSideColors` and `RowSideColors`, respectively. This can be used to indicate previously known groupings by color code as shown below.

In the following, we would like to add side bars that indicate which cities are from the same areas. We do that by providing a vector of colors corresponding to the areas for the individual city names. The output is seen in the right-hand graph of Figure 4.20.

```
> colnames(x)
 [1] "Athens"           "Barcelona"        "Brussels"
 [4] "Calais"           "Cherbourg"        "Cologne"
 [7] "Copenhagen"       "Geneva"           "Gibraltar"
[10] "Hamburg"          "Hook of Holland" "Lisbon"
[13] "Lyons"            "Madrid"           "Marseilles"
[16] "Milan"            "Munich"           "Paris"
[19] "Rome"             "Stockholm"        "Vienna"
> # Add a color bar to the heatmap
> areas <- c("Greece", "Iberia", "Belgium", "France", "France",
+            "Germany", "Scandinavia", "Central Europe", "Iberia",
+            "Germany", "Holland", "Iberia", "France", "Iberia",
+            "France",  "Italy", "Germany", "France", "Italy",
+            "Scandinavia", "Central Europe")
> f.areas <- factor(areas)
> area.colors <- gray.colors(nlevels(f.areas))
> heatmap(x, col=blue.colors, ColSideColors = area.colors[f.areas],
+         RowSideColors = area.colors[f.areas])
```

See also: The heatmap.2 function from the gplots package extends the standard heatmap function by adding several extra options including the possibility to see a color key for the values in the heatmap. The functions rainbow, terrain.colors, and cm.colors are some of the possible choices for color schemes. For red/green heat maps that are commonly used to plot microarray expression data you can use the redgreen function from the gplots packages to generate the color scheme. See Problem 4.3 for more information on RColorBrewer.

4.19 PLOT A CORRELATION MATRIX

Problem: You have a covariance matrix or a matrix of correlations and you want to visualize it graphically.

Solution: A correlation plot displays correlations between the variables in a dataset or between the estimates from a fitted model. Correlations can be depicted as ellipses and the degree of correlation is indicated by the shape and/or color of the ellipses. A variable is always perfectly correlated with itself (so the diagonal of the correlation plot will by definition always consist of straight lines), and perfect circles indicate no correlation.

The plotcorr function from the ellipse package plots a correlation matrix using ellipses for each entry in a correlation matrix. `plotcorr` expects a matrix with values between -1 and 1 as its first argument. If the `numbers` argument is set to `TRUE` then the actual correlation coefficients are rounded (and multiplied by 10) and printed instead of the ellipses. The `type` option determines if the `full` (the default), `lower` or `upper` triangular matrix is plotted. The `diag=FALSE` argument suppresses printing of the diagonal elements (when `type="full"`) while `diag=TRUE` adds the diagonal elements (when `type="upper"` or when `type="lower"`).

Below, we plot the correlations between different observations for the `airquality` data. The `cor` function only works on complete cases so we remove all rows with missing variables.

```
> library(ellipse)
> data(airquality)
> complete <- airquality[complete.cases(airquality),]
> plotcorr(cor(complete), col="blue")
```

The following example overlays two correlation plots so the actual correlation coefficients are printed in the upper triangular matrix and the ellipses are printed in the lower triangular matrix. We construct the correlation plot from the estimates from the linear model. Here, the `vcov` function is used in combination with `cov2cor` to extract the variance-covariance matrix of the estimated parameters from a linear model and convert it to a correlation matrix.

```
> library(isdals)
> data(fev)
> model <- lm(FEV ~ Ht + I(Ht^2) + Gender + Smoke + Age,
+              data=fev)
> corr2 <- cov2cor(vcov(model))      # Correlation matrix
> plotcorr(corr2,  type="lower", diag=TRUE, , col="blue")
> par(new=TRUE)                      # Keep 1st plot and add numbers
> plotcorr(corr2, type="upper", diag=TRUE, numbers=TRUE, cex=1.6)
```

The left plot of Figure 4.21 shows the correlation between the measurements and shows that, for example, wind and ozone are negatively correlated, while temperature and ozone are positively correlated. In the right plot of Figure 4.21 we see that all variables but height are uncorrelated. The strong negative correlation between height and height squared and the intercept is caused by collinearity in the data.

See also: Colors can also be added to the ellipses to help differentiate between the magnitudes of correlation. This is done with the `col` option. See the help page for an example of how to add color palettes.

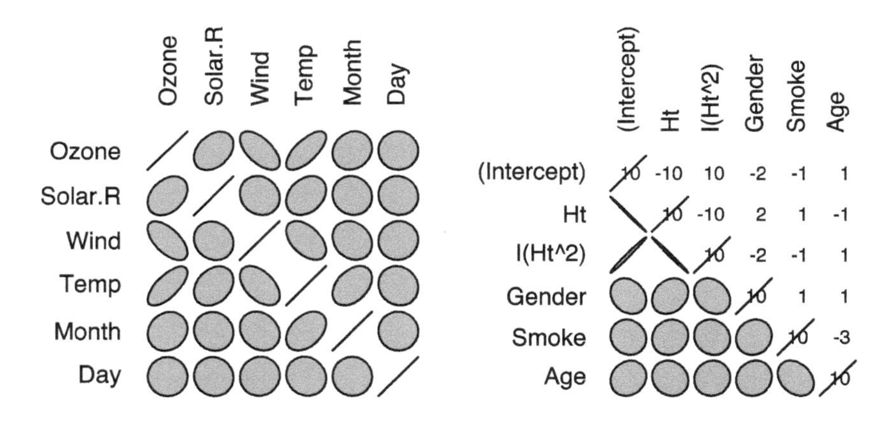

Figure 4.21: Correlation plot for `airquality` measurements (left graph) and for parameter estimates from the `fev` dataset.

4.20 MAKE A QUANTILE-QUANTILE PLOT

Problem: You wish to make a quantile-quantile (q-q plot) to graphically compare the distribution of two datasets or to compare the distribution of one dataset to a known distribution.

Solution: Quantile-quantile plots (q-q plots) are used to compare the distribution function of two datasets by plotting the quantiles of one dataset against the quantiles of the other dataset. Shifts in location, scale, or changes in symmetry can be seen from the q-q plot, and the points on the q-q plot should fall approximately along a 45-degree line if the two datasets come from the same distributions.

The q-q plot is often used for model validation to check if the sample observations come from a specific distribution by comparing the quantiles of the dataset with the theoretical quantiles from the hypothesized distribution. In particular, checking for normality is done by comparing the sample quantiles to the expected theoretical quantiles from a standardized normal distribution. If the points in the normal q-q plot are on a straight line, then the sample observations appear to be from a normal distribution with a standard deviation corresponding to the slope of the points, and a mean that corresponds to the intercept of the points.

The `qqplot` function compares the distribution of two data samples and requires two quantitative vectors as input arguments — one for each dataset. The two vectors do not have to have the same length. The normal q-q plot is implemented by the `qqnorm` function which requires a

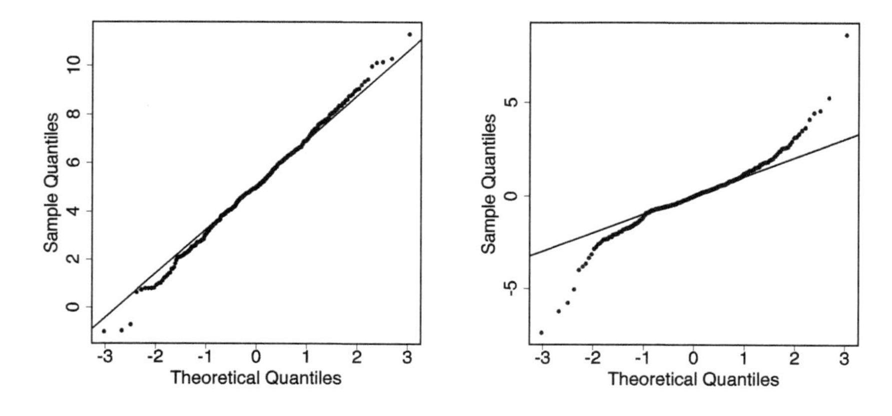

Figure 4.22: Normal quantile-quantile plots for data from a normal distribution (left) and from a t distribution (right).

single numeric vector of quantitative observations as input. The `qqline` adds a comparison line that passes through the first and third quartiles, and it requires the same numeric vector of observations as input as was used for generating the q-q plot with `qqnorm`.

If we observe substantial departures from a straight line — either in the tails, the middle or both — then we conclude that the fitted normal model is not appropriate for the data. An "inverse S" shape of the points indicate that the sample distribution has heavier tails than the normal distribution used for comparison while an "S" shape suggests lighter tails. If both tails are above the comparison line, then this is an indication that the observations are skewed to the right. If both tails are below the comparison line, then the distribution of the observations is left-skewed.

We illustrate the use of normal q-q plots below for two situations: when data are from the normal distribution and from the t distribution (which has the same form as the standardized normal distribution but has heavier tails). The result is shown in Figure 4.22.

```
> x <- rnorm(400, mean=5, sd=2)    # Normal data
> y <- rt(400, df=4)               # Data from t distribution
> qqnorm(x)                        # Create normal q-q plot
> qqline(x)                        # Add comparison line
> qqnorm(y)                        # Create normal q-q plot
> qqline(y)                        # Add comparison line
```

The points in the left plot of Figure 4.22 follow a straight line so nothing suggests that they do not come from a normal distribution (roughly with a mean of 5 and a standard deviation of 2 as seen by the intercept and slope of the line). The right plot in Figure 4.22 suggests that the observations come from a distribution with heavier tails than the normal distribution since the points roughly follow an "inverse S" shape.

4.21 GRAPHICAL MODEL VALIDATION FOR LINEAR MODELS

Problem: You wish to validate a linear normal model graphically to verify if the underlying assumptions appear to be fulfilled.

Solution: Linear normal models play a major role in statistical analysis and model validation is important for all analyses. Here we will focus on graphical methods for linear model validation since they illustrate a broad range of aspects of the relationship between the model and the data. Diagnostic plots provide a way to check for heteroscedasticity, normality, and influential observations.

Linear models are fitted in R using the `lm` function and the generic `plot` function (which calls the `plot.lm` function for an `lm` object) provides several standard plots for model validation.

The `which` argument specifies which model diagnostics plots are produced. There are six possible plots: 1) a plot of residuals against fitted values; 2) a scale-location plot of $\sqrt{|\text{residuals}|}$ against fitted values; 3) a normal q-q plot; 4) a plot of Cook's distances versus row labels; 5) a plot of residuals against leverages; and 6) a plot of Cook's distances against leverage/(1-leverage), and by default plots 1, 2, 3, and 5 are produced.

We use the cherry tree data, `trees`, to illustrate the use of `plot` for graphical model validation. The output is seen in Figure 4.23.

```
> data(trees)
> result <- lm(Volume ~ Height + Girth, data=trees)
> plot(result, which=1:6)
```

The six plots produced by `plot` can be used as follows:

Residuals vs. Fitted. The residuals should be randomly and uniformly scattered about zero and they should be independent of the fitted (predicted) values which can be checked on the "Residual vs. Fitted Values" plot. A lack of uniform scatter around zero suggests that there is still some structure left in the data that is not captured by the model and that a more complex model and/or a

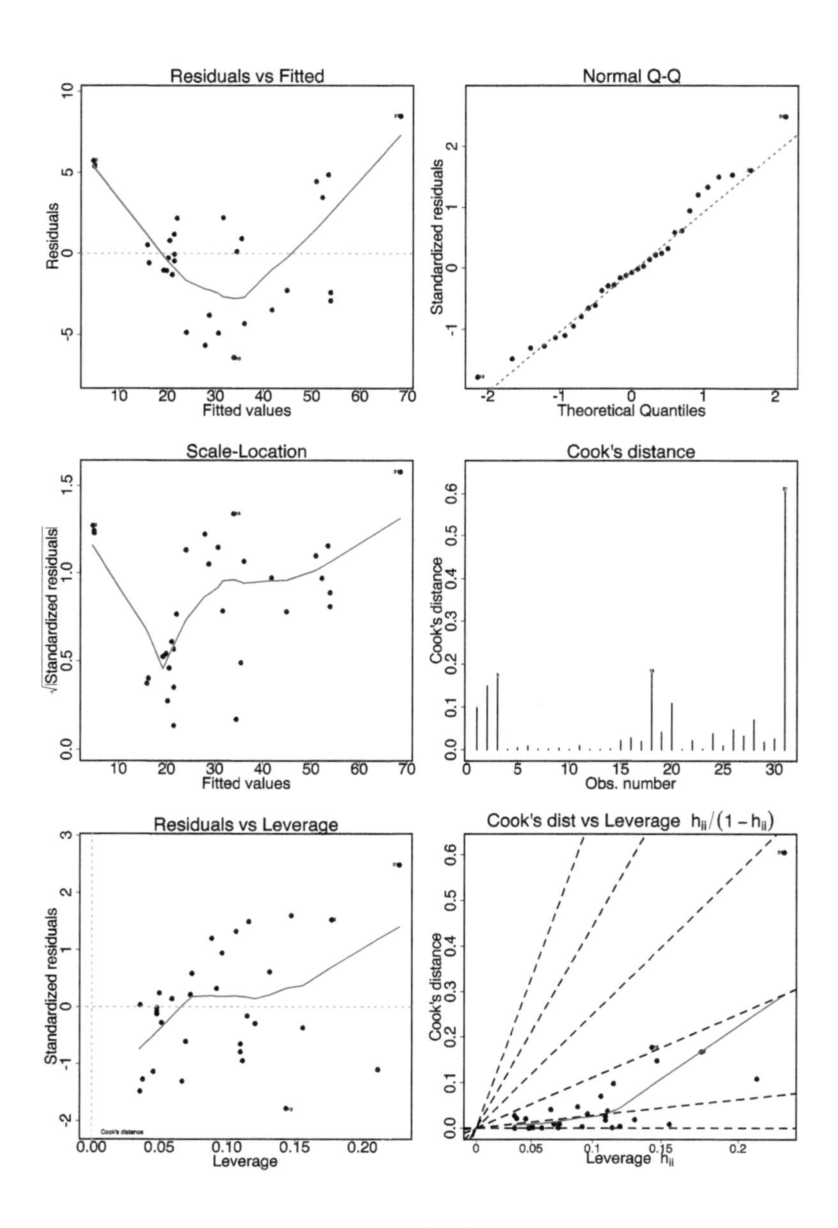

Figure 4.23: Model diagnostic plots for fit of a linear model to cherry tree data. The plots are (row-wise from top): 1) residuals vs. fitted, 2) a scale-location plot, 3) q-q plot, 4) Cook's distance, 5) residuals against leverages, and 6) Cook's distances against leverage/(1-leverage).

transformation of some of the variables may be relevant. The solid curve shows a smoothed average level of the residuals.

Normal Q-Q. The normal quantile-quantile plot compares the distribution of the standardized residuals with the expected theoretical quantiles from a standardized normal distribution, and provides information on whether the residuals are realistic to come from a normal distribution. See Problem 4.20 for details about the q-q plot.

Scale-Location. The scale-location plot can be used to check for variance homogeneity and in particular to see if the variance is changing with the mean fitted value. The size of the square root of the absolute residuals should be approximately constant if there is variance homogeneity, and the solid curve indicates a smoothed average of the square root of the absolute standardized residuals.

Cook's Distance. Cook's distance is a measure of the influence of a single observation and can be used to identify observations that are highly influential and/or may be potential outliers. Cook's distance measures how much the fitted values change when each observation is left out of the analysis; substantial changes to the fitted values mean that the fit is dominated by that observation. The "Cook's distance" plot graphs Cook's distance for each observation. As a rough rule-of-thumb any observation with a value of 1 or more is particularly worth checking for validity as is any observation with a Cook's distance substantially larger than other observations, but any single observation with a Cook's distance vastly different from the other values is of potential interest.

Residuals vs. Leverage. Observations with both high leverage and large residuals can have a huge impact on the fitted model and the corresponding estimates. The "Residuals vs. Leverages" plot can identify potential outliers, observations with high leverage, or a combination of both. Ideally, the standardized residuals should be scattered around zero when compared to the leverage. The smoothed average of the standardized residuals is indicated by the solid line and contour levels of Cook's distance are shown by dashed lines.

The "Residuals vs. Leverages" plot contains the same information as the Cook's distance plot, but at the same time enables the viewer to determine if influential points are caused by large residuals, large leverages, or both.

Cook's Distance vs. Leverage. In the "Cook's distance vs. Leverage" plot the "Residuals vs. Leverage" plot has been turned around. Instead of plotting the standardized residuals against the leverage and contours for Cook's distance, it plots Cook's distance against the leverage and contours for standardized residuals.

The "Residuals vs. Fitted" plot from Figure 4.23 suggests that there is some kind of uncaptured structure in the data that is not captured by the multiple linear relationship between the volume and the girth and height. The standardized residuals follow a straight line in the normal q-q plot which suggests that the standardized residuals follow a normal distribution, and the scale-location plot is fairly constant and does not suggest variance heterogeneity. None of the observations have particularly high values of Cook's distance although observation 31 has a relatively large value of Cook's distance. This is caused by the fact that the tree girth and height of observation 31 are both larger than any of the other observations which is why the leverage of that observation is relatively high. The "Residuals vs. Leverage" and "Cook's distance vs. Leverage" plots both show that observation 31 is relatively influential but that the standardized residuals generally are fairly constant with increasing leverage.

The standardized residual plot plots the standardized residuals against the fitted values (and usually also against each of the explanatory variables) and it is another common plot used for graphical model validation. Ideally, the standardized residuals should be randomly dispersed around the horizontal axis with approximately 95% of the standardized residuals in the interval from $[-1.96; 1.96]$. We should look for variance homogeneity (even spread of observations for all fitted values), the overall structure (the average should be centered around zero), and possible observations with a very large standardized residual just as described above. Plots of the standardized residuals against each of the explanatory variables serve to check if the explanatory variables enter the model correctly. This is especially useful for quantitative explanatory variables, where any systematic structure in a residual vs. explanatory variable plot suggests that the explanatory variable should enter the model in a more complex way.

We can plot a standardized residual plot directly using the follow-

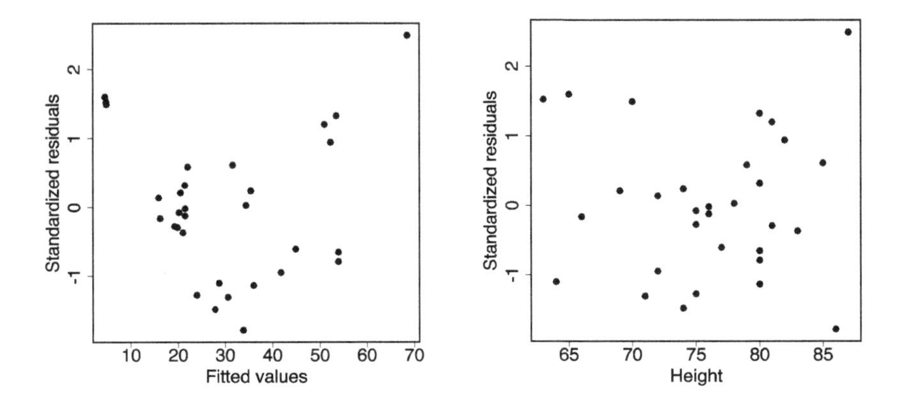

Figure 4.24: Standardized residual plot.

ing code (where we only include a plot for one of the two explanatory variables):

```
> plot(fitted(result), rstandard(result),
+      xlab="Fitted values", ylab="Standardized residuals")
> plot(trees$Height, rstandard(result),
+      xlab="Height", ylab="Standardized residuals")
```

The standardized residual plot in Figure 4.24 shows the same overall result as the "Residuals vs. Fitted" plot from Figure 4.23 above. The standardized residual vs. height plot in Figure 4.24 has the standardized residuals randomly spread around the horizontal axis so there is nothing that suggests that height needs to enter the model in a more complex way than linear.

See also: Problems 3.3–3.9 show examples of linear models. The `MESS` package provides the `residualplot` function which produces residual plots with extra information to help the interpretation.

4.22 CREATE A VENN OR EULER DIAGRAM

Problem: You want to create a Venn or Euler diagram to visualize relationships between groups.

Solution: An Euler diagram is a graphical representation of sets and the relation between these sets. The Venn diagram is a special case of an Euler diagram where all possible relationships between the sets are included. Euler and Venn diagrams are useful tools for visualizing complex hierarchies among sets when the number of sets is not too large.

The venn function from the gplots package draws Venn and Euler diagrams with up to 5 sets. It takes a list of vectors as input, where each vector defines a set and the vector elements define the elements that comprise the set.

The optional argument simplify can be set to TRUE to produce Euler diagrams (i.e., where empty group combinations are omitted when possible).

The following code shows the relationship between three sets where the set elements are given as letters a through p. The output is seen in the left-hand graph of Figure 4.25.

```
> library(gplots)                       # Load the gplots package
> list1 <- c("a", "b", "c", "d", "e", "f", "g", "h")
> list2 <- c("a", "b", "c", "i", "j", "k", "l", "m", "p")
> list3 <- c("a", "b", "d", "e", "i", "j", "k", "n", "o")
> venn(list(Group1=list1, list2=list2, "Set 3"=list3))
```

Alternatively, the venn function from the gplots package accepts a data frame with Boolean variables that define group membership. In the example below we compare the relationship between 4 sets. The data frame rows correspond to all possible elements (one row for each element), while the variables define set membership for each element (TRUE or FALSE). The output is seen in the right-hand graph of Figure 4.25.

```
> list4 <- c("a", "b", "d", "i", "p")     # Add extra group
> lets <- tolower(LETTERS)[1:17]          # Set possible names
> # Create a data frame
> df <- data.frame(g1=lets %in% list1, g2=lets %in% list2,
+                  g3=lets %in% list3, g4=lets %in% list4)
> head(df, 10)
      g1    g2    g3    g4
1   TRUE  TRUE  TRUE  TRUE
2   TRUE  TRUE  TRUE  TRUE
3   TRUE  TRUE FALSE FALSE
4   TRUE FALSE  TRUE  TRUE
5   TRUE FALSE  TRUE FALSE
6   TRUE FALSE FALSE FALSE
7   TRUE FALSE FALSE FALSE
8   TRUE FALSE FALSE FALSE
9  FALSE  TRUE  TRUE  TRUE
10 FALSE  TRUE  TRUE FALSE
> venn(df)
```

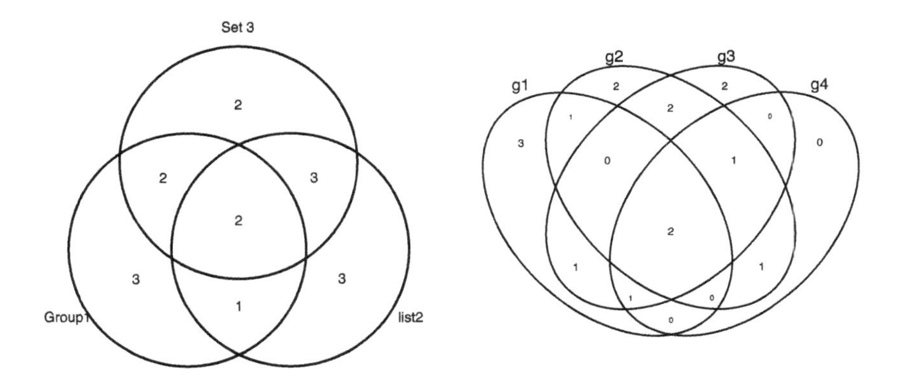

Figure 4.25: Venn diagram for comparison of relationships between elements of three sets (left-hand plot) and four sets (right-hand plot).

See also: The `venneuler` package produces Venn and Euler diagrams where it is possible to tweak some of the graphical parameters and add colors. Set sizes are replaced by weights (i.e., area of the circles comprising the diagram). The `VennDiagram` package can create beautiful diagrams for printing including extensive customization of plot shape and structure.

TWEAKING GRAPHICS

4.23 ADD A BROKEN AXIS TO INDICATE DISCONTINUITY

Problem: You want to include a break mark on your axis to indicate a discontinuity.

Solution: The `axis.break` from the `plotrix` package can add break marks on the axis of an existing plot. There are two different break mark styles (the default `slash`, and `zigzag`) that are set with the `style` option. The axis, position of the break, and relative width of the break are controlled with the `axis`, `breakpos`, and `brw` arguments, respectively.

In the example below we wish to reduce the size of the y axis, so we subtract 40 from the larger values to bring the values closer together.

```
> library(plotrix)
> year <- 2001:2005
> prevalence <- c(10, 15, 12, 60, 65)
> prevalence[prevalence>40] <- prevalence[prevalence>40] - 40
```

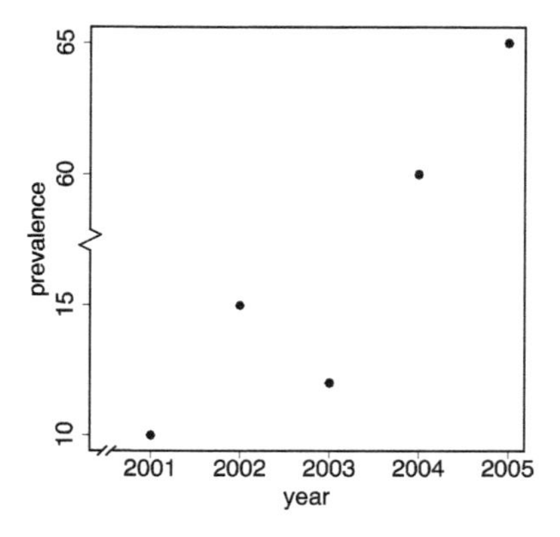

Figure 4.26: Example of `axis.break` output.

```
> prevalence
[1] 10 15 12 20 25
```

Then we plot the data but manually add the y axis and insert the proper labels corresponding to the original scale of the data. Finally, a zigzag break point is positioned at 17.5 on the y axis with a width of 0.05 of the full plot width.

```
> plot(year, prevalence, yaxt="n", xlim=c(2000.5, 2005))
> axis(2, at=seq(10, 25, 5), labels=c(10, 15, 60, 65))
> axis.break()     # Default slash break point on the x axis
> axis.break(axis=2, breakpos=17.5, brw=0.05, style="zigzag")
```

The output is seen in Figure 4.26. By default, `axis.break` creates the break point on the x axis (`axis=1`) close to the y axis. Due to R's default ink-on-paper printing approach the break mark is created by first printing a bit of background color and then subsequently printing the break mark. That also means that if you use a non-default background color for your plots you should use the `bgcol` option in the call to `axis.break` to set the correct background color.

4.24 CREATE A PLOT WITH TWO Y-AXES

Problem: You want to create a plot with two different y-axes.

Solution: While it is easy to create a plot in R that has two different

y axes it is usually not recommended to use two y axes as it is too easy to mislead the viewer.

The general approach to create a plot with two y axes is to superimpose the second graph (with the corresponding axis) on the first plot and then use the axis function to add an extra axis. We superimpose a second plot on an existing graph by preventing R from clearing the graphics device by calling par(new=TRUE) after the first plot. The second plot is then created with no axes, and then finally the second axis is added manually.

The function twoord.plot from the plotrix package automates the process somewhat. As a minimum, twoord.plot should be given four arguments: the x and y values for the first plot and the x and y values for the second plot (corresponding to the options lx, ly, rx, and rx, respectively). The labels on the x axis, left y axis, and right y axis are set by the options xlab, ylab, and rylab, respectively, and special labels on the x axis (like dates) can be set using the xticklab option. Finally, the type argument accepts a vector of character strings indicating the types of plots to make.

The following plot shows monthly Danish soft drink sales and traffic accidents in the same graph, and the output is seen in Figure 4.27.

```
> library(plotrix)
> # Enter data
> softdrinks <- c(19.13, 15.94, 22.96, 28.70, 49.75, 76.53,
+                 70.15, 89.29, 56.12, 44.01, 31.89, 25.51)
> accidents <- c(557, 527, 566, 598, 631, 657,
+                640, 733, 724, 663, 678, 627)
> toword.plot(1:12, softdrinks, 1:12, accidents, xlab="Month",
+             ylab="Soft drink sales (mio. l.)",
+             rylab="Traffic accidents",
+             lcol="black", rcol="blue", type=c("l", "bar"),
+             xtickpos = 1:12,
+             xticklab=month.name[1:12])
```

See also: The doubleYScale function from the latticeExtra package also allows for two y axes.

4.25 ROTATE AXIS LABELS

Problem: You want to create a plot where some of the axis labels are rotated.

Solution: Text can be added to a graph using the text function and we

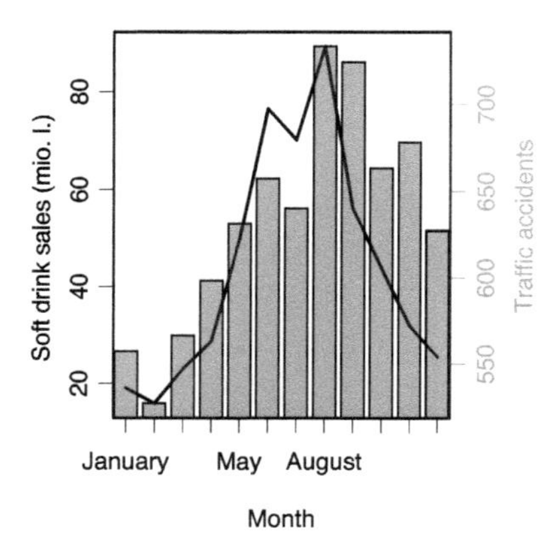

Figure 4.27: Example of plotting two y-axes on a single graph.

will use this function to manually add rotated axis labels. The margin text function, `mtext`, which we would normally use to add text to the margins of the current figure cannot be used for this as it does not support the string rotation argument `srt`.

By default, R, assigns space around the figure region so it has room for most labels. The amount of space can be changed with the `mar` option to the `par` function, and we may need to provide extra space for the rotated labels if they take up more space after rotation.

To add rotated axis labels, we first set the option `xaxt="n"` to suppress the default axis labels when the graph is created. Then we use the `axis` function with option `labels=FALSE` to add an axis with tick marks but without labels before we finally add the label to the plot using `text`.

When `text` is called we use the options `srt` to set the text rotation angle, `adj` to set the horizontal and/or vertical adjustment of the label text, and `xpd=TRUE` to prevent R from clipping the text at the figure region. `adj=1` places the right end of the text at the tick marks. In the code below we use the optional argument `y` to place the labels. `par("usr")[3]` returns the lower coordinate of the plotting region and we use that to identify the location of the x axis. The value 0.1 can then be adjusted to move the labels up or down relative to the x axis.

Before starting the plot we add extra space to make room for the rotated labels, and then we plot without any x-axis.

```
> par(mar = c(7, 4, 1, 1) + 0.1)    # Add extra space
> x <- 1:6
> plot(x, sin(x), xaxt="n", xlab="", pch=1:6)  # No x axis
```

Then we manually create and add the new x axis as well as labels rotated 25° at the default tick marks.

```
> axis(1, labels=FALSE)
> labels <- c("Are", "you", "master", "of", "your", "domain?")
> text(x, y=par("usr")[3] - .1, srt=25, adj=1,
+       labels=labels, xpd=TRUE)
> mtext(1, text="Added label for x-axis", line=5)
```

The output from the above code can be seen in Figure 4.28.

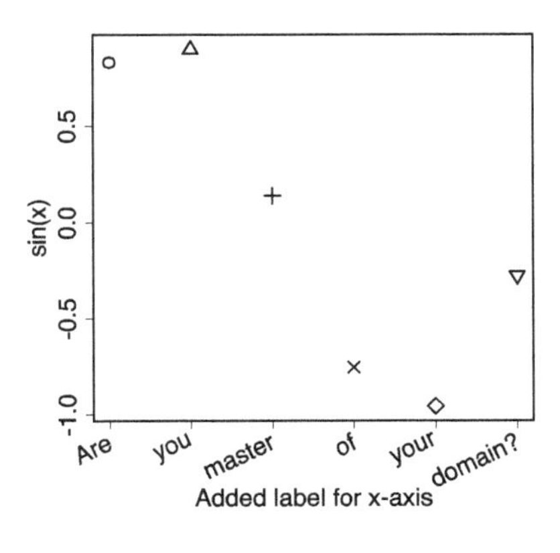

Figure 4.28: Example of rotated labels on the x-axis.

See also: The help page for `par` gives information on all the graphical parameters.

4.26 MULTIPLE PLOTS

Problem: You want to create a figure that consists of multiple sub-figures. This can be used to make complex plot arrangements with plotting regions of varying sizes.

Solution: The `mfrow` and `mfcol` options to `par` can partition the plotting region into a number of equally sized rows and columns defined by the values of `mfrow` or `mfcol`.

The `layout` function allows for more complicated partitions of the plotting area. `layout` uses a matrix as input to imitate the total area of the plotting device and determine how the area is partitioned. The input matrix should contain integer values, and cells with the same value are combined into a single plotting region. Values used to define the plotting

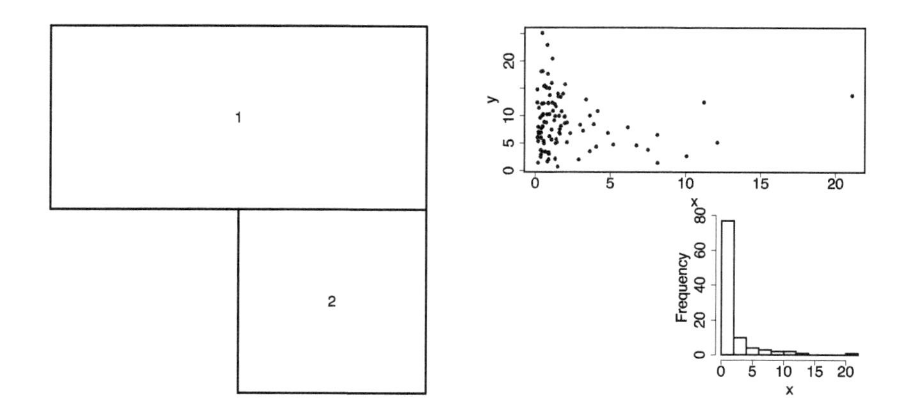

Figure 4.29: Examples of `layout.show` and `layout` output.

regions determine the order in which they are filled. The `layout.show` function shows the current layout and the numbers indicate the order in which plots are added to the graph.

Below we prepare a layout with a plotting region that spans the entire width of the device on top, a blank lower left area, and a second plotting area to the lower right that spans half the device width.

```
> design <- matrix(c(1,1,0,2), 2, 2, byrow=TRUE)
> design                              # Show the layout matrix
     [,1] [,2]
[1,]    1    1
[2,]    0    2
> layout1 <- layout(design)
> layout.show(layout1)
> x <- exp(rnorm(100))                # Simulate data
> y <- (rnorm(100)+3)**2              # Simulate more data
> plot(x, y)                          # Make a scatter plot
> hist(x, main="")                    # and a histogram
```

The output is shown in the two plots found in Figure 4.29. Note how the output from the `layout.show` function matches the matrix used as input to the `layout` function. The first plot spans the entire width of the plotting device by using the value '1' for both values in the first row of the input layout matrix. More complicated patterns can be achieved by clever selection of the matrix elements.

By default, `layout` assumes that all rows have the same height and all columns have the same width. In principle, we could control the exact size of the individual plotting regions by using a matrix with a large number of rows and columns. However, the `widths` and `heights` options allow us to control the relative widths and heights of the columns and rows, respectively. In the following we make a 2 × 2 layout where the top row fills 25% of the vertical plotting region and the right-hand column fills 25% of the horizontal plotting region. The options `widths` and `heights` require vectors with lengths that match the number of columns and number of rows, respectively, and the vector values determine the relative size of each column or row.

```
> layout2 <- layout(matrix(c(1, 0, 2, 3), nrow=2, byrow=TRUE),
+                   widths=c(3, 1),
+                   heights=c(1, 3))
> par(mar=c(4, 4.3, 0, 0) +.1)      # Remove space around plots
> plot(density(x), main="", xlab="")
> plot(x, y)
> boxplot(y)
```

The result is seen in Figure 4.30.

4.27 ADD A LEGEND TO A PLOT

Problem: You have created a graph and want to add a legend to the plot.

Solution: The `legend` function adds a legend to an existing plot. The `x` and `y` arguments determine the position of the (upper left corner of the) legend and the `legend` argument is a character or expression vector that contains the text for the legend.

If any of the usual graphics parameters `lty` or `pch` are supplied, then a line or plotting symbol is placed in front of each line of the legend text, respectively. Likewise, `col` and `lwd` modify the color and the line width, respectively.

The following code plots the CO_2 uptake profile as a function of CO_2 concentration for each of 12 plants from two different areas and adds a legend to the plot. The output is seen in Figure 4.31.

```
> data(CO2)
> matplot(matrix(CO2$conc, ncol=12), matrix(CO2$uptake, ncol=12),
+         type="l", lwd=2, lty=rep(c(1,2), c(6,6)), xlim=c(80, 1100),
+         ylim=c(0, 45), xlab="Concentration", ylab="Uptake",
+         col=rep(c("black", "blue", "black", "blue"), times=rep(3,4)))
> legend(600, 9, legend=c("Quebec - nonchilled", "Quebec - chilled",
+                 "Mississippi - nonchilled", "Mississippi - chilled"),
+        lty=c(1, 1, 2, 2), col=rep(c("black", "blue"), 2))
```

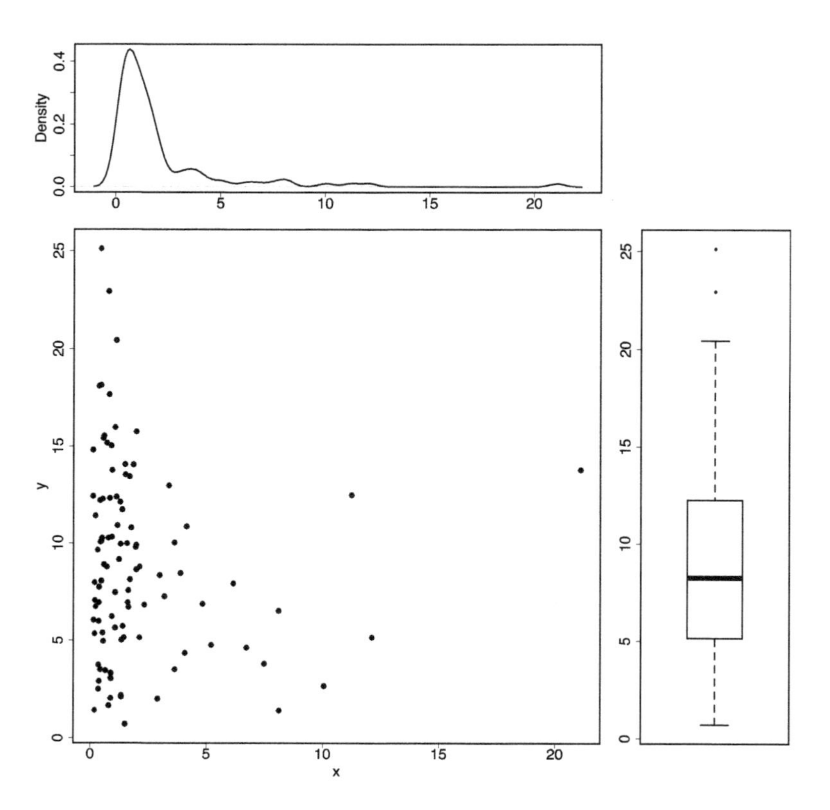

Figure 4.30: Examples of `layout` output where the relative widths of the rows and columns are controlled.

Several other options exist that can change the appearance of `legend` — for example setting `horiz=TRUE` sets the legend horizontally instead of vertically.

See also: Another example of `matplot` can be found in Rule 4.29.

4.28 ADD A TABLE TO A PLOT

Problem: You have created a graphics plot and want to add a table to the plot.

Solution: The `addtable2plot` function from the `plotrix` package adds a table to an existing plot. Data for the table can be a data frame or matrix and the table is added in the same way a legend is added to a plot.

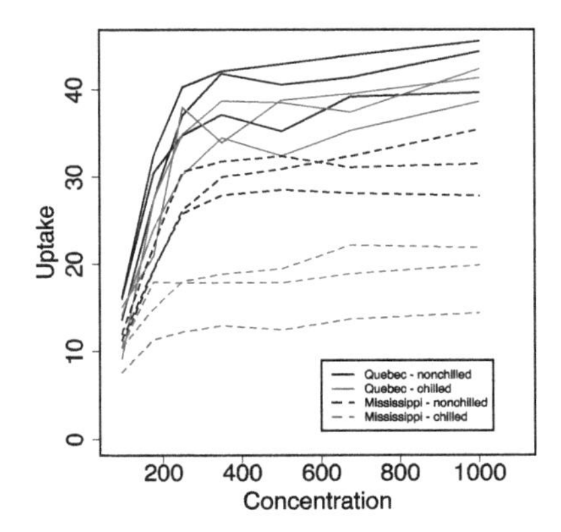

Figure 4.31: Example of `legend` function output.

The coordinates of the lower left corner of the table location are given as the parameters `x` and `y`, and the parameters `bty`, `bg`, and `title` set the box type surrounding the table, the background color, and the title, respectively. These three options work the same way as they do for the `plot` function. If the option `display.rownames` is set to `TRUE` then rownames are displayed in the table. The output from the following code is seen in Figure 4.32.

```
> library(plotrix)
> cats <- matrix(c(53, 115, 17, 502, 410, 14),ncol=2)
> rownames(cats) <- c("Yes", "No", "No info")
> colnames(cats) <- c("Problems", "No problems")
> cats
        Problems No problems
Yes           53         502
No           115         410
No info       17          14
```

We can then calculate the group-wise relative frequencies and plot them as a bar chart.

```
> cattable <-  cats / apply(cats, 1, sum)
> barplot(t(cattable), beside=TRUE, ylim=c(0,1.1))
> addtable2plot(3.0, .8, cats, display.rownames=TRUE, bty="o",
+               bg="blue", title="Cat behavior data", cex=1.8)
```

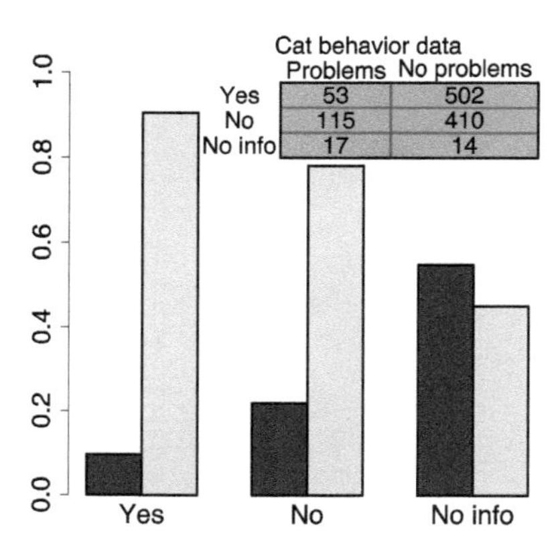

Figure 4.32: Example of `addtable2plot` output.

4.29 LABEL POINTS IN A SCATTER PLOT

Problem: You want to make a graph where you label points in a scatter plot.

Solution: Individual labels can be put on a plot using the `text` function. If you wish to change the plotting symbol and/or color according to a grouping variable, then we can do that directly through the arguments `pch`, `col`, and `bg`, respectively. In particular, we use the grouping variable as index to `pch`, `col`, or `bg` to set the plotting symbol or colors for each group.

The `CO2` data frame contains information on ambient dioxide concentrations and carbon dioxide uptake for two origins of the grass species *Echinochloa crus-galli*. The `Type` variable provides the two origins: Quebec and Mississippi. The following code produces a graph where observations from the two origins are labeled as white or black squares. The output can be seen in the left panel of Figure 4.33.

```
> data(CO2)
> plot(CO2$conc, CO2$uptake, pch=22,
+      bg=c("black", "white")[as.numeric(CO2$Type)],
+      xlab="Concentration", ylab="Uptake")
```

Here we use `Type` as a factor and `as.numeric` returns the index of the levels. Otherwise, we can always start by first converting the grouping variable to a factor and then converting it back using `as.numeric`.

In the `CO2` dataset there is also information on the treatment —

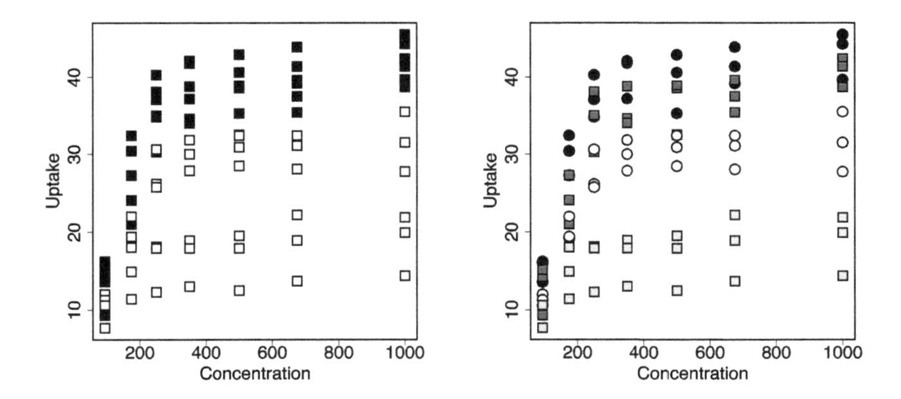

Figure 4.33: Individual symbols, labels, and colors on points.

whether or not that grass was chilled. In the code example below we use the `interaction` function to create a vector that gives combined information on origin and treatment.

```
> origintreat <- interaction(CO2$Treatment, CO2$Type)
> head(origintreat)
[1] nonchilled.Quebec nonchilled.Quebec nonchilled.Quebec
[4] nonchilled.Quebec nonchilled.Quebec nonchilled.Quebec
4 Levels: nonchilled.Quebec ... chilled.Mississippi
> levels(origintreat)
[1] "nonchilled.Quebec"      "chilled.Quebec"
[3] "nonchilled.Mississippi" "chilled.Mississippi"
```

This result is automatically a factor, so we can extract the factor level indices and use those to plot the origin in black, blue, or white and the treatment using either circles (non-chilled) or squares (chilled). The resulting plot is found in the right panel of Figure 4.33.

```
> plotcode <- as.numeric(origintreat)
> plot(CO2$conc, CO2$uptake, cex=2.6,
+      pch=c(21, 22, 21, 22)[plotcode],
+      bg=c("black", "blue", "white", "lightblue")[plotcode],
+      xlab="Concentration", ylab="Uptake")
```

See also: The `ggplot2` package extends the base plotting system in R and can easily change labels according to groups or variables.

4.30 IDENTIFY POINTS IN A SCATTER PLOT

Problem: You want to identify individual points in a scatter plot.

Solution: The `identify` function can be used to interactively identify individual points in an existing graphics device. It requires either the coordinates of the points in the scatter plot as the first two input arguments, `x` and `y`, or simply any R object which defines coordinates like `xy.coords`.

By default, `identify` registers the position of the graphics pointer when the mouse button is pressed and searches for the closest point as defined by the `tolerance` option. If any such point is found on the plot, then the observation number is printed on the plot. Plotting the observation number can be omitted by setting `plot=FALSE`.

`identify` is terminated by pressing either any other mouse button than the first or by pressing the ESC key, depending on the graphics device. When `identify` terminates, it prints the indices of points that were identified while the function was running.

```
> x <- 1:10
> y <- x**2
> plot(x, y)
> identify(x,y)
[1]  3  9 10
```

Observations 3, 9, and 10 were identified in the example above.

Instead of printing the index number for the closest observation, we can use the `labels` option to set the labels to print for each point. For example if we wish to know if a point has an odd or even x value, then we could use the following code

```
> plot(x, y)
> group <- rep(c("Odd", "Even"), 5)
> identify(x,y, labels=group)
[1] 2 3 8
```

The `identify` function is only supported on some graphics screen devices and will do nothing on an unsupported device.

See also: The `locator` function can be used to print the position of the graphics cursor when the mouse button is pressed.

4.31 VISUALIZE POINTS, SHAPES, AND SURFACES IN 3D AND INTERACT WITH THEM IN REAL-TIME

Problem: You want to visualize data in 3D and be able to interactively turn, rotate, and zoom in and out on the plot.

Solution: Three-dimensional images can be difficult to depict in two dimensions but the `rgl` package provides functionality that offers three-dimensional, real-time visualization of points, shapes, and surfaces, and the package allows the user to generate interactive 3D graphics to help visualize the plot.

The `rgl` package uses OpenGL (the Open Graphics Library) for rendering 3D images in real-time and creates a graphics device which the user can interact with. Since the graphics are rendered in real-time it is possible to turn, scroll, rotate, and zoom in and out on the plot.

The `rgl` package has a number of functions for 3D plotting, and we will just illustrate a few of them here. To really appreciate the `rgl` package you should run the code and not just look at the screen dumps shown in Figure 4.34.

The `plot3d` function creates a 3D scatter plot, and takes three numeric vectors (corresponding to the values on the x, y, and z axes, respectively) as the first three arguments. Like `plot`, `plot3d` can plot points (use option `type="s"` for spherical points), lines (option `type="l"`), or vertical lines relative to the x-y-plane (`type="h"`). The `size` option sets the size of the plotted spheres, and `col` is used to set the plotting color.

The following code plots the height, girth, and volume variables from the `trees` dataset, and it is possible to rotate the points to see the nice relationship between the volume and the height and girth. Screen dumps of the output are seen in the upper plots of Figure 4.34.

```
> library(rgl)


Attaching package: 'rgl'
The following object is masked from 'package:plotrix':
        mtext3d

> data(trees)
> plot3d(trees$Height, trees$Girth, trees$Volume,
+        type="s", size=2) # Plot spheres
> plot3d(trees$Height, trees$Girth, trees$Volume,
+        type="h")         # Vertic. lines
```

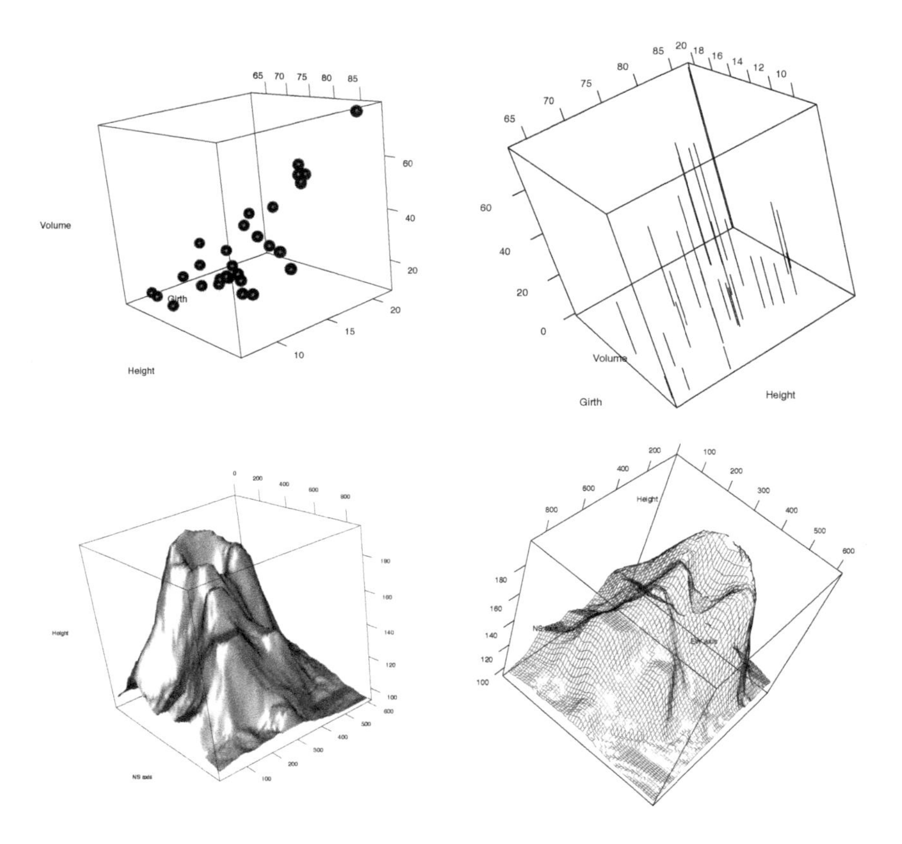

Figure 4.34: Examples of snapshots of output from the `rgl` package. Upper plots are created by `plot3d` while the lower plots are made by `surface3d`.

The `persp3d` function is similar to `persp` and can be used to create 3D surface plots. As usual, coloring of the surface can be set using the `col` option but `persp3d` also has the `front`, `back`, and `smooth` arguments which are used to set the front and back polygon fill mode, and whether shading is added to the rendering, respectively. `front` and `back` defaults to `"fill"` which produces filled polygons — both when viewed from "above" (the front) and "below" (the back). Other possibilities are `"line"` for wire-framed polygon, `"point"` for point polygon, and `"cull"` for hidden polygon. When the `smooth` option is set to `TRUE` then smooth shading is used for the rendering instead of the default flat shading.

The `volcano` dataset contains topographical data about the Maunga Whau volcano and consists of a matrix of heights corresponding to a 10-meter by 10-meter grid. The lower plots in Figure 4.34 show screen dumps from the following code.

```
> data(volcano)
> z <- volcano                # Height
> x <- 10 * (1:nrow(z))      # 10 meter spacing (S to N)
> y <- 10 * (1:ncol(z))      # 10 meter spacing (E to W)
> persp3d(x, y, z, smooth=TRUE, col="lightgray",
+         xlab="NS axis", ylab="EW axis", zlab="Height")
> persp3d(x, y, z, col="darkgray", front="line", back="cull",
+         xlab="NS axis", ylab="EW axis", zlab="Height")
```

Both `plot3d` and `persp3d` have an `add` option, which can be set to `TRUE` if points or surfaces should be added to the current plot instead of creating a new plot. In general, graphics parameters can be set or queried using the `par3d` function, which has the same functionality as the `par` function except it works with the functions in `rgl`.

Graphical animations of 3D plots can be created by the `movie3d` or `play3d` functions. The only difference between the two functions is that `movie3d` saves the animation as a file (for example as an animated GIF file) while `play3d` plays the animation directly on the screen. Both functions work on the current rgl device and for input they require a function that returns a list of viewpoints (readable by `par3d`) used for the animation. The `spin3d` function is an example of a function that provides viewpoints by spinning the current plot. `spin3d` takes two arguments: `axis` which is a vector of length 3 which contains the desired axis of rotation, and `rpm` which sets the rotation speed in rotations per minute. When creating an animated file, the `duration` option should be set to determine the duration in seconds. The output location of the movie is set with the `dir` option.

The following code creates an animated GIF file where the current plot rotates slowly around the z axis.

```
> movie3d(spin3d(axis=c(0,0,1), rpm=3), duration=60, dir=".")
```

See also: The `decorate3d` function is used to add boxes, labels, and axes to plots in `rgl`. Screen shots can be exported from an rgl graphics device using the `rgl.postscript` function which can export in different formats. `rgl.light` can be used to add light sources to the rendering.

WORKING WITH GRAPHICS

4.32 EXPORTING GRAPHICS

Problem: You have made a nice graph and want to export it so you can use it in a manuscript or elsewhere.

Solution: R is extremely useful for producing high-quality graphics. Here we focus on how to save plots as pdf files so they can be included in other documents. R can export to several other graphics formats but we will mainly focus on the pdf file format here because it is easily used under most operating systems.

Base R plotting uses an ink-on-paper approach where each graphics command either creates a new plot or adds to the current plot. Once you have finished a complete plot, you can save the current graphics device to a pdf file using the `dev.copy2pdf` function. If no filename is specified, then a file called `Rplot.pdf` is created in the current directory.

```
> x <- 1:10
> y <- rnorm(10)
> plot(x,y)                            # Create a scatter plot
> title("Very lovely graphics")        # Add title
> dev.copy2pdf(file="lovelygraph.pdf") # Copy graph to pdf file
```

The width and height of the output file created by `dev.copy2pdf` are taken from the current device unless otherwise specified with the `width` or `height` options.

The `dev.print` function copies the graphics contents of the current device to a new device. Hence, we can save the current graphics device in a different format by adding the relevant `device` option when calling `dev.print`. The following lines show how to create other graphical formats. Both png and jpeg files measure the width and height in pixels and the output size needs to be set by specifying at least one of the `width` or `height` arguments

```
> dev.print(file="pngfile.png", device=png, width=480)
cairo_pdf
        2
> dev.print(file="jpgfile.jpeg", device=jpeg,
+           width=480, height=240)
cairo_pdf
        2
```

Note that `dev.copy2pdf` and `dev.print` copy the entire plotting device region *including* the background color. The background is transparent on most screen devices which may result in undesirable surprises if the graphic is saved in a format such as png that actually accommodates transparent colors.

As an alternative to `dev.copy2pdf` and `dev.print`, the plot may be produced directly to a file, without printing it on the screen at all:

```
> pdf("gorgeousgraph.pdf")          # Opens a pdf device
> plot(x,y)                         # Create a scatter plot
> title("Very lovely graphics") # Add title
> dev.off()                         # Close device and file
```

The `pdf` function opens a file as the current graphics device, adds all graphics output to the file until the `dev.off` command closes the file and ends the device. The width and height of the graphics are set by the `width` and `height` options, respectively, and are measured in inches (both default to 7 inches). The files are saved in the working directory unless a path is given in the file name.

See also: The `pdf` function (as well as the other graphical device functions) has additional options that control fonts, paper size, etc. See the help file for more information. The functions `svg`, `png`, `jpeg`, and `bmp` are used to create scalable vector graphics, portable network graphics, joint photographic experts group, and bitmap image file formats, respectively. The `postscript` function creates postscript files and `win.metafile` can be used to create a Windows metafile.

4.33 PRODUCE GRAPHICS IN LATEX-READY FORMAT

Problem: You want to produce graphics that are typeset directly in LaTeX so the plots have a clean, unified look with the rest of the text.

Solution: If you are typesetting text with LaTeX you can use the Ti*k*Z graphics language to typeset R graphics from within LaTeX such that the plots have a clean, unified look with the remaining text.

The `tikz` function from the `tikzDevice` package opens a Ti*k*Z graphics device and creates a file, `Rplots.tex`, that contains the necessary Ti*k*Z code to be included in LaTeX. The default file name is changed with the `file` argument. Graphics output from `tikz` is typeset by LaTeX and consequently will match whatever fonts are currently used in the document.

The following code shows an example of how to create Ti*k*Z output,

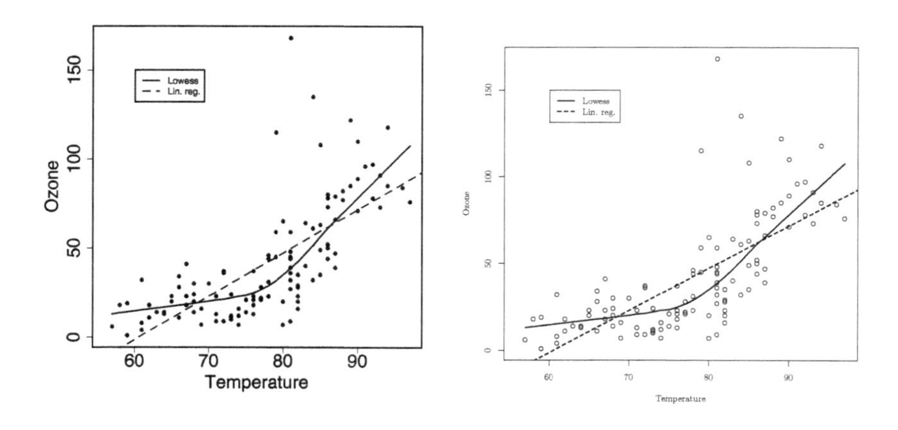

Figure 4.35: Rendering of the same graphics using base R graphics (left) and a Ti*k*Z-device in LaTeX (right).

where the relationship between ozone and temperature is modeled for the `airquality` data frame.

```
> library(tikzDevice)
> data(airquality)
> tikz(file="tikzexample.tex")                    # Open tikz device
> plot(airquality$Temp, airquality$Ozone,
+      xlab="Temperature", ylab="Ozone")          # Make scatter plot
> model <- lm(Ozone ~ Temp, data=airquality)  # Estimate lin. reg.
> abline(model, lwd=3, lty=2)                     # Add line to plot
> # Add a lowess smoothed line to the plot as well
> # lowess requires complete cases only
> cc <- complete.cases(airquality$Temp, airquality$Ozone)
> lines(lowess(airquality$Temp[cc], airquality$Ozone[cc]),
+       lwd=3, lty=1)
> # Add a legend
> legend(60, 150, c("Lowess", "Lin. reg."), lty=1:2, lwd=3)
> dev.off()                                       # Close tikz device
```

The following lines show how to include the file produced by `tikz` in a LaTeX document. Figure 4.35 shows an example of the graphics created by the code above both as regular R graphics and as Ti*k*Z graphics. Alternatively, if the option `standAlone` is set to `TRUE` then the output produced by `tikz` will be wrapped in a complete LaTeX document that can be compiled immediately.

```
\documentclass{article}\usepackage{knitr}
```

```
\usepackage{tikz}           % Necessary for tikz output
\IfFileExists{upquote.sty}{\usepackage{upquote}}{}
\begin{document}
Here goes some text.
\begin{figure}[h]
\begin{center}
\input{tikzexample.tex}     % Import output from tikz
\end{center}
\caption{Simple example}
\label{fig:myfigure}
\end{figure}
Some text after the figure. See Figure~\ref{fig:myfigure}
to see my new integrated plot.
\end{document}
```

Note that it is necessary to have LaTeX and the TikZ package installed to typeset the final graphical output. When `tikzDevice` loads, it searches for the location of a LaTeX compiler in order to be able to query the compiler for string widths and font methods. If no LaTeX executable is found then the path can be set manually with the global `tikzLatex` option, e.g., `options(tikzLatex="/path/to/latex/compiler")`.

4.34 EMBED FONTS IN POSTSCRIPT OR PDF GRAPHICS

Problem: You have made a nice plot in pdf or postscript format and want to embed all fonts to ensure that it looks the same on all machines.

Solution: If you wish to share your postscript or pdf graphics files and want to make sure that they will appear the same on all computers, you must make sure that all the necessary fonts are embedded in the file.

By default R does not embed its fonts when producing pdf files but that is generally not a problem if only the standard default fonts (e.g., Times, Courier, Helvetica) are used since these standard fonts are part of the postscript and pdf formats. However, if you use special fonts and want to make sure that they are included in the postscript or pdf file, then the `embedFonts` function can be used.

As the name suggests, `embedFonts` takes an existing postscript or pdf file and embeds fonts in the file. It takes the name of the existing file as first argument and overwrites the existing file unless the `outfile` option is given to set the name of the new graphics file with all the fonts embedded. `embedFonts` relies on the external program Ghostscript to do the actual conversion, so that program should be installed on the computer and the executable should be in the search path so it can be found by R.

To embed the necessary fonts for the graphics file `myplot.pdf` located in the current working directory, we type:

```
> embedFonts("myplot.pdf")
```

Note that `embedFonts` does *not* embed the standard pdf fonts such as Helvetica and Times in the graphics file since these fonts are copyrighted and cannot be distributed legally without licensing. Since R uses the Helvetica font by default, it may be necessary to use a free font such as "Nimbus Sans" instead to make sure it is embedded. The available pdf fonts can be seen with the `pdfFonts` function.

```
> names(pdfFonts())
 [1] "serif"                 "sans"
 [3] "mono"                  "AvantGarde"
 [5] "Bookman"               "Courier"
 [7] "Helvetica"             "Helvetica-Narrow"
 [9] "NewCenturySchoolbook"  "Palatino"
[11] "Times"                 "URWGothic"
[13] "URWBookman"            "NimbusMon"
[15] "NimbusSan"             "URWHelvetica"
[17] "NimbusSanCond"         "CenturySch"
[19] "URWPalladio"           "NimbusRom"
[21] "URWTimes"              "ArialMT"
[23] "Japan1"                "Japan1HeiMin"
[25] "Japan1GothicBBB"       "Japan1Ryumin"
[27] "Korea1"                "Korea1deb"
[29] "CNS1"                  "GB1"
```

Thus to make sure all fonts are embedded, we can use the following code where the argument `useDingbats` is set to `FALSE` to prevent the use of the copyrighted ZapfDingbats font for creating symbols.

```
> pdf(file="test.pdf", useDingbats=FALSE)
> par(family="NimbusSan")
> plot(1,1, pch=16)
> title("All embedded")
> dev.off()
> embedFonts("test.pdf")
```

Alternatively, the `Cairo` package can create a Cairo graphics device which extends the default R graphics device and which automatically includes fonts in postscript and pdf files. Cairo devices also allow for extra graphical features not found in the base R graphics device.

The `CairoPS` and `CairoPDF` functions initialize new postscript and pdf graphics devices in the same way as the `pdf` and `postscript` functions, respectively. The following code shows that `CairoPDF` is used just like `pdf`.

```
> library(Cairo)
> CairoPDF("mycairoplot.pdf")
> hist(rnorm(100))
> dev.off()
```

See also: The `fonts` argument to the `pdf` function can be used to specify R graphics font family names to include in the pdf file.

CHAPTER 5

R

GETTING INFORMATION

5.1 GETTING HELP

Problem: You have run into a problem and are looking for help about a function or topic.

Solution: There are several ways of getting help in R but sometimes it may be difficult to find the answer to a particular question. Knowledge of the different ways of using the help system greatly improves the chance of locating the desired information.

The `help` function returns the documentation about a function or dataset. It is the primary source of information about functions, their syntax, arguments, and the values returned. The `help.start` function opens a browser window with access to introductory and advanced manuals, FAQs, and reference materials.

Not all of the output from the calls below is shown.

```
> help(mean)        # Get help about the function mean
> help("mean")      # Same as above
> ?mean             # Question mark is the same as help
> help.start()      # Start browser and the general help
```

The `apropos` function lists all the R command names that contain a specific text pattern. Alternatively, `help.search` searches for a match in the name, alias, title, concept, or keyword entries in the help system and installed packages. `help.search` can search for phrases but it can only find phrases that actually appear in the help pages and not general terms. For example, when searching for "logistic regression" there is no reference to the `glm` function, which fits logistic regression models under the handle "generalized linear model."

`RSiteSearch` performs a search of help pages, the R-help mailing list archives, and vignettes and returns the result in a web browser.

```
> apropos("mean")
[1] ".colMeans"     ".rowMeans"     "colMeans"      "kmeans"
[5] "mean"          "mean.Date"     "mean.default"  "mean.difftime"
[9] "mean.POSIXct"  "mean.POSIXlt"  "rowMeans"      "weighted.mean"

> help.search("mean")
Help files with alias or concept or title matching
`mean' using regular expression matching:

gplots::bandplot   Plot x-y Points with Locally Smoothed
                   Mean and Standard Deviation
gplots::plotmeans
                   Plot Group Means and Confidence
                   Intervals
    .                   .
    .                   .
base::DateTimeClasses
                   Date-Time Classes
base::Date         Date Class
base::colSums      Form Row and Column Sums and Means
base::difftime     Time Intervals
base::mean         Arithmetic Mean
boot::sunspot      Annual Mean Sunspot Numbers
lattice::tmd       Tukey Mean-Difference Plot
Matrix::Matrix-class
                   Virtual Class "Matrix" Class of
                   Matrices
Matrix::colSums    Form Row and Column Sums and Means
Matrix::dgeMatrix-class
                   Class "dgeMatrix" of Dense Numeric
                   (S4 Class) Matrices
rpart::meanvar     Mean-Variance Plot for an Rpart
                   Object
stats::kmeans      K-Means Clustering
stats::oneway.test
                   Test for Equal Means in a One-Way
                   Layout
stats::weighted.mean
                   Weighted Arithmetic Mean

> ??mean                 # Same as above
> RSiteSearch("sas2r")   # Search for the string sas2r in online
>                        # help manuals and archived mailing lists
```

See also: The `example` function executes the example code associated with a function and can be used to run the examples. See Problem 5.12 for more information on vignettes.

5.2 FINDING R SOURCE CODE FOR A FUNCTION

Problem: You want to see the R source code for a function.

Solution: The source code for many R functions can be viewed simply by typing the function name at the command prompt. For example, to see the way the standard deviation function `sd` is defined we type

```
> sd
function (x, na.rm = FALSE)
sqrt(var(if (is.vector(x) || is.factor(x)) x else as.double(x),
    na.rm = na.rm))
<bytecode: 0x7fe3f08b0098>
<environment: namespace:stats>
```

However, for generic functions like `print`, `plot`, or `boxplot` where the type of input object influences the way the function works, we need to use the `methods` function to identify which methods are available for the generic function. As shown below, we cannot see the source code for the `boxplot` function directly, but the call to `methods` shows which functions are actually called by `boxplot`.

```
> boxplot
function (x, ...)
UseMethod("boxplot")
<bytecode: 0x7fe3ed4edd58>
<environment: namespace:graphics>
> methods(boxplot)
[1] boxplot.bclust*  boxplot.default  boxplot.formula*
[4] boxplot.matrix   boxplot.n
see '?methods' for accessing help and source code
```

The source code for the individual functions can be obtained by typing the function name directly but that only works for non-hidden methods. Functions that are hidden inside a name space are shown with a trailing asterisk and in those cases we can use the `getAnywhere` function to extract objects within all loaded name spaces.

```
> boxplot.matrix   # See source for non-hidden method
function (x, use.cols = TRUE, ...)
{
    groups <- if (use.cols) {
        split(c(x), rep.int(1L:ncol(x), rep.int(nrow(x), ncol(x))))
    }
    else split(c(x), seq(nrow(x)))
    if (length(nam <- dimnames(x)[[1 + use.cols]]))
        names(groups) <- nam
    invisible(boxplot(groups, ...))
}
<bytecode: 0x7fe3cf19d098>
<environment: namespace:graphics>
> boxplot.formula  # Try to see source for hidden method

Error in eval(expr, envir, enclos): object 'boxplot.formula' not found

> getAnywhere("boxplot.formula")
A single object matching 'boxplot.formula' was found
It was found in the following places
  registered S3 method for boxplot from namespace graphics
  namespace:graphics
with value

function (formula, data = NULL, ..., subset, na.action = NULL)
{
    if (missing(formula) || (length(formula) != 3L))
        stop("'formula' missing or incorrect")
    m <- match.call(expand.dots = FALSE)
    if (is.matrix(eval(m$data, parent.frame())))
        m$data <- as.data.frame(data)
    m$... <- NULL
    m$na.action <- na.action
    m[[1L]] <- quote(stats::model.frame)
    mf <- eval(m, parent.frame())
    response <- attr(attr(mf, "terms"), "response")
    boxplot(split(mf[[response]], mf[-response]), ...)
}
<bytecode: 0x7fe3f7f631c8>
<environment: namespace:graphics>
```

The output from `getAnywhere` tells you that the `boxplot` function is located in the `graphics` name space and gives you the source code.

R has two internal systems for handling generic functions. The older S3 classes (as described above) and the newer internal S4 classes. Things are slightly different for the newer S4 classes where the `showMethods` and findMethods functions are used to list the available methods and to get their source code, respectively.

```
> library(lme4)


Attaching package: 'lme4'
The following object is masked _by_ '.GlobalEnv':
      cbpp
The following objects are masked from 'package:ordinal':
      VarCorr, ranef

> refit           # Try to show the source for the refit function
function (object, newresp, ...)
UseMethod("refit")
<environment: namespace:lme4>
> showMethods("refit")     # Show available methods

Function "refit":
 <not an S4 generic function>
> # List the source code for refit for class type numeric
> findMethods("refit", classes="numeric")
An object of class  "listOfMethods"
named list()
Slot arguments:
[1] "object"  "newdata"
Slot signatures:
list()
Slot generic:
standardGeneric for "refit" defined from package "modeltools"

function (object, newdata, ...)
standardGeneric("refit")
<environment: 0x7fe3f332c9c8>
Methods may be defined for arguments: object, newdata
Use  showMethods("refit")  for currently available ones.
```

5.3 TIMING R COMMANDS AND FUNCTIONS

Problem: You wish to time the running time of commands or functions.

Solution: Computing time becomes an issue when running R commands on large datasets since simple tasks that may take very little time on smaller datasets can become cumbersome once they are scaled to run on massive datasets or when they need to be run a large number of times. Consequently, it becomes crucial to evaluate computing time when we need to analyze large datasets or when we need to repeat a computation a large number of times.

The microbenchmark package provides accurate timing of R expressions. Its main function, microbenchmark, accepts a number of R expressions as input and runs each of them 100 times in order to compute summary statistics on the computing times. The times argument modifies the number of times that the expressions are computed.

In the first example below we create a 10000×12 matrix of random numbers and compute the sum of each of the 12 columns using three different methods: using the apply, matrix multiplication, and colSums functions.

```
> library(microbenchmark)
> m1 <- matrix(rnorm(120000), ncol=12) # Simulate data
> m2 <- rep(1, 10000)                  # Vector of 1s for multiplication
> microbenchmark(apply(m1, 2, sum),    # Compare apply,
+                m2 %*% m1,            # matrix multiplication,
+                colSums(m1))          # and colSums methods
Unit: microseconds
              expr  min   lq mean median   uq   max neval
 apply(m1, 2, sum) 1301 2179 3096   2978 3403 26645   100
         m2 %*% m1  396  479  511    495  513  1665   100
       colSums(m1)  139  170  206    199  216   440   100
```

The output lists summary statistics for the computing time (in microseconds) for each of the three methods: minimum, 25% quantile, median, mean, 75% quantile, and maximum. The colSums function is roughly 10 times faster at computing the column sums than the apply function and about twice as fast as matrix multiplication. It is clearly highly advantageous to use the colSums function to compute the column sums.

The statistics produced by microbenchmark are by default based on 100 runs of the functions. The number of runs can be changed with the times argument.

See also: The microbenchmark package also includes function boxplot that accepts the output from microbenchmark and produces a boxplot of the results.

R PACKAGES

5.4 INSTALLING R PACKAGES

Problem: You want to install an R package.

Solution: Apart from the functions that are available in R, there exists a very large number of R packages that extend the usefulness of R. An R package is essentially a collection of R functions and/or datasets. Assume we want to install the `foreign` package which can be found on CRAN. The command

```
> install.packages("foreign")  # Install the foreign package
```

downloads and installs the package from one of the CRAN mirrors. Installation of a package is only required once, but you should make sure that each installed package is updated.

Packages can be installed globally if R is run with administrator privileges; otherwise, they must be installed locally. A package can be installed in a local directory by setting the `lib` argument to a directory where you have write permissions, for example

```
> install.packages("foreign", lib="C:/my/private/R-packages/")
```

The `.libPaths` function shows the locations where R currently looks for packages. These locations can be used for the `lib` argument to make sure packages are installed in the same directory.

```
> .libPaths()      # Output from a machine running ubuntu
[1] "/home/user_4/ekstrom/R/x86_64-pc-linux-gnu-library/2.10"
[2] "/usr/local/lib/R/site-library"
[3] "/usr/lib/R/site-library"
[4] "/usr/lib/R/library"
```

If you do not have Internet access or want to install a package that is not part of CRAN, then use the `repos` argument for `install.packages` to tell R where the package can be found. If `repos=NULL` then R will install from a local file on the computer.

```
> # Install a package from http://www.some.url.dk/Rfiles/
> install.packages("newRpackage",
+                  repos="www.some.url.dk/Rfiles/")
> # Install the newRpackage package from a local source file
> # located in the current working directory
> install.packages("newRpackage_1.0-1.tar.gz", repos=NULL)
```

The full path to a local package can be given, if it is not located in the current working directory.

Packages may be distributed in source form or in compiled binary form. The default is to use binary packages for Windows and Mac OS X and source packages for unix-like systems. Source packages may contain C/C++/Fortran code that requires the necessary tools to compile them whereas binary packages are platform-specific and generally work right out of the box. `install.packages` consults CRAN for available packages in the default form, but the argument `type="source"` can be set to force R to download source packages.

The functions and datasets in a package cannot be accessed before the package is installed (and loaded in the current R session). `library` loads a package in the current R session.

```
> library("foreign")  # Loads the foreign package
> library("isdals", lib.loc="C:/my/private/R-packages/")
```

The `library` function needs only to be run once in every R session where the functions and datasets from the package are needed.

See also: Problem 5.14 on how to automatically load packages when R starts. Problem 5.5 shows how to update packages. Look at Problem 5.13 if you permanently want R to search a local directory for installed packages so you do not have to specify the `lib` option. In the Windows graphical user interface a package can also be installed from the `Packages -> Install package(s)` menu item. The `Packages` menu item can also be used to set the repositories or to install a downloaded zip package for Windows (use `Packages -> Install package(s) from local zip files`). In the graphical user interface under OS X a local package can be installed from the `Packages & Data -> Package Installer` menu item.

5.5 UPDATE INSTALLED R PACKAGES

Problem: You want to update the installed R packages.

Solution: Use `update.packages` to automatically search the R repositories for newer versions of the installed packages and then download and update the most recent version on-the-fly.

```
> update.packages()
```

You need administrator privileges to update globally installed packages. Locally installed packages are updated if the `lib.loc` option is specified, so to check for new versions of both global and local packages we use the command

```
> update.packages(lib.loc="C:/my/private/R-packages/")
```

Note that Windows may lock the files from a package if the package is currently in use. Thus, under Windows you may have to start a new R session where the packages have not been loaded before you run `update.packages`.

See also: Look at Problem 5.13 if you permanently want R to search a local directory for installed packages so you do not have to specify the `lib.loc` option.

5.6 UNLOAD A PACKAGE

Problem: You want to "unload" an R package that was previously loaded.

Solution: A package that was previously loaded with the `library` function can be unloaded with the `detach` function. The first argument to `detach` should be the string `package:` followed by the name of the package to unload. The option `unload=TRUE` should be set to make R attempt to unload any corresponding namespace.

```
> library(isdals)                            # Load the isdals package
> detach("package:isdals", unload=TRUE) # Unload isdals
```

Note that detaching a package may leave some code behind, so detaching and re-attaching a package is generally inadvisable since some of the components may not be refreshed.

5.7 INSTALL AN R PACKAGE FROM A REPOSITORY

Problem: You want to install an R package from GitHub or BitBucket or SVN.

Solution: A lot of R package development is done using the repository service GitHub and several newer packages and development versions can only be found and installed from there. Here we just focus on GitHub — see the note below for the names of the functions for SVN or BitBucket.

The `install_github` function from the `devtools` package installs R packages directly from GitHub. It needs a string with the username and repository as its first argument in the format `username/repository`. For example, the following commands can be used to install the development version of the MESS package.

```
> library(devtools)
> install_github("ekstroem/MESS")
```

Similarly to the install.packages function from Problem 5.4, the dependencies=TRUE argument can be used with install_github function to try to install any necessary dependencies from CRAN as well.

See also: The install function from the devtools lists all the extra arguments that can be passed to install_github. devtools also provides install_svn for use with Subversion repositories, and install_-bitbucket for installing from BitBucket repositories.

5.8 INSTALL A PACKAGE FROM BIOCONDUCTOR

Problem: You want to install an R package that is part of the BioConductor project.

Solution: The BioConductor project provides tools for the analysis and handling of high-throughput genomic data and contains a large number of R packages that cannot be found on CRAN.

BioConductor packages are most easily installed using the biocLite function from the biocLite.R installation script. We use the source function to read the installation script directly from its location on the Internet, and once it is sourced we have access to the biocLite function.

If biocLite is called without any arguments, then we will automatically install around 20 common packages from the BioConductor project. Alternatively, we can specify the desired package as the first argument.

```
> # Read the installation script directly from BioConductor site
> source("http://bioconductor.org/biocLite.R")
> biocLite("limma")  # Install the limma package
> biocLite()         # Install the common BioConductor packages
```

The update.packages function updates installed packages from the BioConductor project if the correct repository is set with the repos option. The biocLite.R script provides the biocinstallRepos, which returns the correct URL for the BioConductor repository.

```
> source("http://bioconductor.org/biocLite.R")
> update.packages(repos=biocinstallRepos(), ask=FALSE)
```

5.9 LIST THE INSTALLED PACKAGES

Problem: You want to see a list of the installed packages.

Solution: The `library` function lists the installed packages and the locations where each of the packages is installed, when `library` is called without any arguments.

```
> library()

Packages in library '/usr/local/lib/R/site-library':

gplots              Various R programming tools for plotting data
isdals              Provides datasets for Introduction to
                    Statistical Data Analysis for the Life Sciences

Packages in library '/usr/lib/R/site-library':

gdata               Various R programming tools for data
                    manipulation
gtools              Various R programming tools
lme4                Linear mixed-effects models using S4 classes
RODBC               ODBC Database Access
snow                Simple Network of Workstations

Packages in library '/usr/lib/R/library':

base                The R Base Package
boot                Bootstrap R (S-Plus) Functions (Canty)
```

See also: The `installed.packages` function also lists the installed packages and their location and provides additional information on version number, dependencies, licenses, suggested additional packages, etc.

5.10 LIST THE CONTENT OF A PACKAGE

Problem: You want to view the content and the description of an installed package.

Solution: Use the `package` option with the `help` function to obtain information about an installed package and list its content.

```
> help(package=MESS, help_type="text")

Information on package 'MESS'

Description:

Package:       MESS
Type:          Package
Title:         Miscellaneous Esoteric Statistical Scripts
Version:       0.3-10
Date:          2016-02-01
Author:        Claus Ekstrøm <claus@rprimer.dk>
Maintainer:    Claus Ekstrøm <claus@rprimer.dk>
Depends:       geepack,
Imports:       MASS, Matrix, Rcpp, glmnet, kinship2, mvtnorm,
               parallel
LinkingTo:     Rcpp, RcppArmadillo
Suggests:      lme4
Description:   A mixed collection of useful and semi-useful
               diverse statistical functions, some of which may
               even be referenced in The R Primer book.
ByteCompile:   true
License:       GPL-2
RoxygenNote:   5.0.1
Built:         R 3.2.3; x86_64-apple-darwin13.4.0; 2016-02-14
               02:06:50 UTC; unix

Index:

MESS                    Collection of miscellaneous useful and
                        semi-useful functions
QIC.geeglm              Quasi Information Criterion
adaptive.weights        Compute weights for use with adaptive lasso.
auc                     Compute the area under the curve for two
                        vectors.
bdstat                  Danish live births and deaths

...

tracemp                 Fast computation of trace of matrix product
wallyplot.default       Plots a Wally plot
write.xml               Write a data frame in XML format
```

5.11 UNINSTALL AN R PACKAGE

Problem: You want to remove an installed R package.

Solution: The `remove.packages` function uninstalls R packages that were previously installed with `install.packages`, through R's graphical user interface or from the command line. `remove.packages` requires a character vector of package names to remove as input.

```
> remove.packages("emma")   # Uninstall the emma package
```

By default, `remove.packages` tries to uninstall packages from R's default package directory. If the package has been installed in another directory then the `lib` argument must be set with a character string providing the path to the package installation directory. If the `lib` argument is not given then `remove.packages` uses the first location given by the `.libPaths` function as shown in the code above.

```
> # Remove the mystuff packages installed in ~/my/own/R-packages
> remove.packages("mystuff", lib="~/my/own/R-packages/")
```

Note that it may be necessary to run R with administrator privileges to uninstall global packages.

5.12 LIST OR VIEW VIGNETTES

Problem: You want to list the available vignettes or view a specific vignette.

Solution: Vignettes are additional documents that accompany some R packages and provide additional information, user guides, examples or a general introduction to a topic or to the package. They are typically provided as pdf files (often also with the source file included) and it is possible to extract any R code so the user can run and experiment with the code.

The `vignette` function shows or lists available vignettes. If no arguments are given to `vignette` then it lists all vignettes from all *installed* packages. The `all` argument can be set to `FALSE` to only list vignettes from attached packages. If a character string is given as an argument, then the `vignette` function will display the vignette matching the string. The optional `package` argument can be supplied with a character vector to restrict the packages that are searched through. The `edit` function used on the object returned by the `vignette` function shows the R source code from the vignette.

```
> vignette(all=FALSE)  # List all attached vignettes
no vignettes found

Use `vignette(all = TRUE)'
to list the vignettes in all *available* packages.

> vignette(all=TRUE)   # List all installed vignettes
Vignettes in package `coin':

LegoCondInf          A Lego System for Conditional
                     Inference (source, pdf)
MAXtest              Order-restricted Scores Test (source,
                     pdf)
coin                 coin: A Computational Framework for
                     Conditional Inference (source, pdf)
coin_implementation
                     Implementing a Class of Permutation
                     Tests: The coin Package (source, pdf)

...

> library(Matrix)
> vignette(all=FALSE, package="Matrix")
Vignettes in package `Matrix':

Comparisons          Comparisons of Least Squares
                     calculation speeds (source, pdf)
Design-issues        Design Issues in Matrix package
                     Development (source, pdf)
Intro2Matrix         2nd Introduction to the Matrix
                     Package (source, pdf)
Introduction         Introduction to the Matrix Package
                     (source, pdf)
sparseModels         Sparse Model Matrices (source, pdf)


Use `vignette(all = TRUE)'
to list the vignettes in all *available* packages.

> vignette("Intro2Matrix")          # Show vignette
> vign <- vignette("Intro2Matrix")  # Store vignette
> edit(vign)                        # Extract/edit source code
```

The `edit` function starts an editor window where the extracted R code can be viewed and altered.

5.13 PERMANENTLY CHANGE THE DEFAULT DIRECTORY WHERE R INSTALLS PACKAGES

Problem: You want to permanently change the default directory where R installs packages so you do not have to manually specify the desired installation directory every time a new package is to be installed.

Solution: If you do not have global/administrator rights on a computer, then it can be desirable to permanently set up R to install packages in a specific local directory.

The current site-wide directories used for installation can be identified with the `.libPaths` function:

```
> .libPaths()   # Example for Ubuntu installation
[1] "/usr/local/lib/R/site-library" "/usr/lib/R/site-library"
[3] "/usr/lib/R/library"

> .libPaths()   # Example for Windows installation
[1] "C:/Program Files/R/R-3.2.4revised/library"
```

The default location for package installation can be overridden by setting the `R_LIBS_USER` environment variable. The value of an environmental variable can be seen from within R by using the `Sys.getenv` function:

```
> Sys.getenv("R_LIBS_USER")   # On machine running Ubuntu
[1] "~/R/x86_64-pc-linux-gnu-library/3.2"

> Sys.getenv("R_LIBS_USER")   # On machine running Windows XP
[1] "C:\\Documents and Settings\\Claus\\Documents/R/win-library/3.2"

> Sys.getenv("R_LIBS_USER")   # On machine running Mac OS X
[1] "~/Library/R/3.2/library"
```

The default path specified by `R_LIBS_USER` is only used if the corresponding directory actually exists, so in order for the `R_LIBS_USER` path to have any effect, the directory may have to be created manually.

The `R_LIBS_USER` environmental variable is set in the `.Renviron` file and at startup, R searches for an `.Renviron` file in the current directory or in the user's home directory in that order. Thus, if different projects are kept in separate directories, then different environment setups can be used by using a different `.Renviron` file in each directory. Under Windows the `.Renviron` file can be placed in the current directory or in the user's Documents directory.

For example, creating the `.Renviron` file with the following contents adds the path `~/sandbox/Rlocal` to the set of libraries.

```
R_LIBS_USER="~/sandbox/Rlocal"
```

When R is started the specified package library is included among the libraries found by R (provided that the directory exists):

```
> .libPaths()
[1] "/home/claus/sandbox/Rlocal"  "/usr/local/lib/R/site-library"
[3] "/usr/lib/R/site-library"     "/usr/lib/R/library"
```

The `R_LIBS_USER` environmental variable can specify multiple library paths by providing a list of paths separated by colons (semicolons on Windows).

5.14 AUTOMATICALLY LOAD A PACKAGE AT STARTUP

Problem: You want to make sure that a package is loaded automatically when R starts.

Solution: At startup, R searches for a site-wide startup profile file and a user profile file (in that order). Code added to any of these files are run when R launches so we can add the necessary commands in the site-wide profile or user profile to automatically load a package. Code added to the individual user profile file obviously only affects that particular user, while changes to the site-wide file affects all users.

The location of the site-wide and user profile files are determined by the `R_PROFILE` and `R_PROFILE_USER` environment variables, respectively. If `R_PROFILE` is unset, then the default site-wide profile file is `R_HOME/etc/Rprofile.site`, where `R_HOME` points to the R installation directory. R searches for a file `.Rprofile` in the current directory or in the user's home directory in that order if the environment variable `R_PROFILE_USER` is unset.

The following code can be added to, say, the `.Rprofile` file to have the `foreign` package loaded for the current user when R is launched.

```
> # Example of .Rprofile to have the foreign package loaded
> # automatically when R launches
> local({
+         options(defaultPackages=c(getOption("defaultPackages"),
+                                   "foreign"))
+       })
```

Note that we call the getOption function to ensure that the default packages from R are also loaded and then add any additional packages in the call to options. Everything is wrapped in the local function in case any of the packages require any code to be executed when they start to prevent them from cluttering up the workspace.

Use the following commands with the Sys.getenv function to determine the relevant environmental variable or to find the R_HOME folder.

```
> Sys.getenv("R_PROFILE")
[1] ""
> Sys.getenv("R_PROFILE_USER")
[1] ""
> Sys.getenv("R_HOME")  # Output from machine running Ubuntu
[1] "/usr/lib/R"
> Sys.getenv("R_HOME")  # Output from machine running Windows
[1] "C:/Program Files/R/R-3.2.4revised"
```

Hence, to make the foreign load automatically for all users on a machine running Windows XP we would add the necessary lines to the file C:\Program Files\R\R-3.2.4revised\etc\Rprofile.site (where the version number should be replaced with the version of R installed).

See also: See help(Startup) in R to get more information on the R startup process.

THE R WORKSPACE

5.15 LIST OR DELETE OBJECTS

Problem: You wish to list or delete objects in the R workspace.

Solution: The R workspace is your current working environment and includes any user-defined objects: vectors, matrices, data frames, lists, and functions.

The ls function lists the names of the existing objects in the current workspace. rm is used to remove objects.

```
> x <- 1:5
> myfactor <- factor(c("a", "b", "b", "b", "c", "c"))
> ls()                  # List local objects
[1] "hook_output" "myfactor"    "x"
> x                     # Print x
[1] 1 2 3 4 5
```

```
> rm(x)                    # Remove object x
> x                        # Try to print x

Error in eval(expr, envir, enclos): object 'x' not found

> ls()
[1] "hook_output" "myfactor"
```

5.16 CHANGE THE CURRENT WORKING DIRECTORY

Problem: You want to change or identify the current working directory.

Solution: R reads and writes files from the current work directory unless a full path to the files is given. The functions `getwd` and `setwd` are used to get and set the current working directory of R.

```
> getwd()          # Get the current working directory from OS X
[1] "/Users/cld189"
> getwd()          # Get the current working directory from Windows
[1] "C:/R"
> setwd("d:/")     # Set the current working directory to d:/
```

Note that when specifying paths in R you should generally use the forward slash '/' to indicate directories. This is standard on unix-like systems but different from Windows, and using forward slashes makes your R-code portable to all operating systems. For example, R will interpret the path `c:/My Documents/mydata.txt` correctly under Windows. The backslash character '\' functions as an escape character in R character strings and should be double quoted if used in paths; i.e., `c:\\My Documents\\mydata.txt`.

5.17 SAVE AND LOAD WORKSPACES

Problem: You wish to save the current workspace or load a saved workspace.

Solution: The `save.image` function saves the current workspace and `load` is used to load a saved workspace into the current session. If no path is specified in any of the functions, then R saves and looks for the workspace file in the current working directory.

```
> ls()                      # List the current objects
[1] "hook_output"
> x <- 1:5                  # Create a new vector
> ls()                      # List the current objects
[1] "hook_output" "x"
> save.image(".RData")      # Save the workspace to .RData
```

After quitting and restarting R we use `load` to load the workspace.

```
> ls()                      # List the current objects
[1] "hook_output" "x"
> load(".RData")            # Load the saved image
> ls()                      # List the current objects
[1] "hook_output" "x"
> x
[1] 1 2 3 4 5
```

The default file name for `save.image` is `.RData` and you need write permissions for the relevant directory for `save.image` to work.

On startup, R automatically loads a saved workspace if it is found in the current working directory. Likewise, whenever you exit R you are asked if you wish to save the current workspace (to the current working directory).

Note that a saved workspace only contains the saved objects — not the command history or any loaded packages.

See also: Problem 5.16 shows how to change the current working directory. Problem 5.18 shows how to save and restore histories and the `library` function loads an installed package.

5.18 SAVE AND RESTORE HISTORIES

Problem: You want to save (and subsequently restore) the commands you have issued in an R session.

Solution: R allows the user to save (and reload) the complete history of session commands as a text file. This allows the user to save the commands for a later time (and possibly to use them again) and keep the complete command history as documentation.

The `savehistory` function saves the complete R session history. The history is saved as a plain text file with the default name `.Rhistory`. The file name and location can be set with the `file` option and the location is relative to the current working directory. Currently, the `savehistory` function does not work on Mac OS X.

```
> x <- 1:5
> y <- c(2, 3, 3, 1, 2)
> lm(y ~ x)

Call:
lm(formula = y ~ x)

Coefficients:
(Intercept)            x
        2.8         -0.2
> savehistory("myanalysis.txt")
```

The file `myanalysis.txt` now contains the lines

```
x <- 1:5
y <- c(2, 3, 3, 1, 2)
lm(y ~ x)
savehistory("myanalysis.txt")
```

Note that `savehistory` only saves the commands used during the session and not the actual objects that were loaded or created. Thus, you still need to save a copy of any data if the data were read from an external source.

When R is restarted we read in a saved history with the `loadhistory` function.

```
> loadhistory("myanalysis.txt")
> x                              # x is not found

Error in eval(expr, envir, enclos): object 'x' not found
```

`loadhistory` only reads in the list of saved commands but the actual commands are *not* rerun. They are only entered in the list of previous commands.

To get R to read and evaluate all the commands from a file the `source` function should be used. `source` accepts a file name, URL, or connection as argument and reads (and executes) input from that source

```
> source("myanalysis.txt")
> ls()
[1] "hook_output" "x"           "y"
> y
[1] 2 3 3 1 2
```

See also: Problem 5.17 explains how to save the objects in the current workspace.

5.19 INTERACT WITH THE FILE SYSTEM

Problem: You wish to interact with the file system for example to list or rename files or to create new directories.

Solution: The `dir` function (alternatively the `list.files` function) lists the files in the current working directory. The `path` option can be set to change the directory listed, and the `pattern` option can be set to a regular expression such that only files which match the regular expression are listed.

```
> dir()                # List files in the current working directory
 [1] "addtable.r"   "auc.r"        "barplot2.r"   "boxcox.r"
 [5] "break-mark.r" "ci.r"         "cluster.r"    "contour.r"
 [9] "gfx.r"        "graphhelp.r"  "heatmap.r"    "label.r"
[13] "layout.r"     "lda.r"        "legend.r"
> dir(pattern="^b") # List files starting with b
[1] "barplot2.r"   "boxcox.r"     "break-mark.r"
```

R has a large number of functions that provide an interface to the computer's file system like `file.copy`, `file.remove`, `file.rename`, `file.append`, and `file.exists`. The functions have the obvious functionality indicated by their names, and they take one or two character vectors containing file names or paths as input arguments. The `dir.create` function creates a new directory and takes a character vector containing a single path name as input.

```
> file.copy("boxcox.r", "boxcox2.r")      # Copy file to boxcox2.r
[1] TRUE
> file.rename("boxcox2.r", "newrcode.r") # Rename to newrcode.r
[1] TRUE
> file.exists("newrcode.r")              # Check if file exists
[1] TRUE
> file.exists("boxcox2.r")               # Old file is gone
[1] FALSE
> file.remove("newrcode.r")              # Delete the file
[1] TRUE
> dir.create("newdir")                   # Make new directory
```

The `basename` and `dirname` functions extract the base (i.e., all but the part after the final path separator) of a file name or path and the path up to the last path separator, respectively.

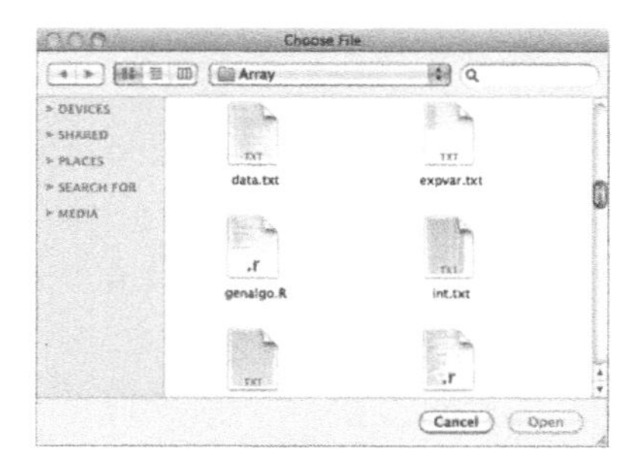

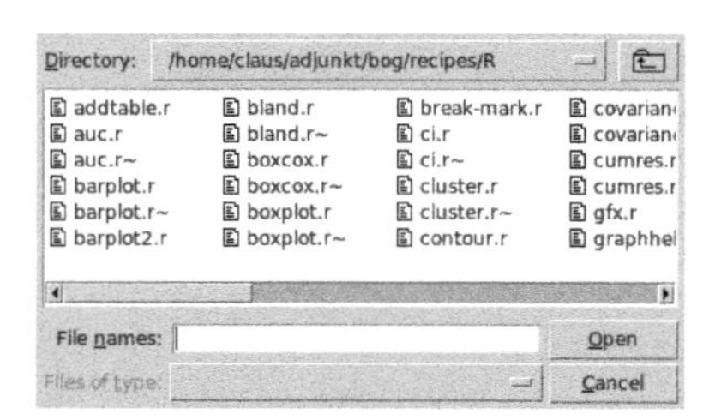

Figure 5.1: File selection dialogs for Mac OS X (left) and unix-like systems (right).

```
> basename("d:/ont/mention/the/war.txt")
[1] "war.txt"
> dirname("d:/ont/mention/the/war.txt")
[1] "d:/ont/mention/the"
```

See also: Problem 5.20 gives examples of the `file.choose` command. The `Sys.chmod` function changes file permissions.

5.20 LOCATE AND CHOOSE FILES INTERACTIVELY

Problem: You want to interactively locate and choose a file.

Solution: Several R functions — for example, for importing data from external sources or for exporting data or graphics — require an argument that specifies the name of a file.

The `file.choose` can be used in lieu of giving the file name in the function call. Instead the user is asked to interactively provide the name of the file. On unix-like systems, this will be a simple text input prompt, where the user types in the file name. However, under Windows and Mac OS X the `file.choose` function opens a graphical dialog box so the user can choose the file interactively (see Figure 5.1).

The `file.choose` function returns a vector of selected file name(s) or an error if none are chosen.

```
> mydata <- read.table(file.choose(), header=TRUE)
```

There exist packages to provide the same graphical interactivity under unix-like systems, but they are not a part of base R. The `tcltk` package uses the Tcl language and the tk toolkit to create platform-independent widgets and it provides the `tk_choose.files` function which works just like the Windows version of `file.choose`. It requires Tcl to be installed on the computer, but that is often the case for most unix-like machines.

```
> library(tcltk)
> mydata <- read.table(tk_choose.files(), header=TRUE)
```

See also: The `gfile` function from the `gWidgets` package also provides a file selection dialog, but requires installation of some additional graphic toolkit implementation.

5.21 INTERACT WITH THE OPERATING SYSTEM

Problem: You wish to interact with the operating system, for example to run system commands.

Solution: The `system` function runs system commands directly under the operating system. The main argument to `system` is a character string with the system command to invoke. For example, under linux we can use the `df` command to report disk space usage.

```
> system("df -h")              # File system disk space usage (unix)
Filesystem              Size  Used Avail Use% Mounted on
/dev/sda1                47G  7,6G   37G  18% /
none                    5,9G  332K  5,9G   1% /dev
none                    5,9G  208K  5,9G   1% /dev/shm
none                    5,9G  264K  5,9G   1% /var/run
none                    5,9G     0  5,9G   0% /var/lock
none                    5,9G     0  5,9G   0% /lib/init/rw
/dev/sda3               826G  4,0G  780G   1% /priv
joy:/priv_2/users       690G  534G  121G  82% /home/user_4
/dev/sr0                 44M   44M     0 100% /media/cdrom0
```

Under Windows, a somewhat similar result is given by the `fsutil` command (provided you have administrator privileges).

```
> system("fsutil volume diskfree c:")  # Disk usage (Windows)
Total # of free bytes        : 535560192
Total # of bytes             : 10725732352
Total # of avail free bytes  : 535560192
```

The output from `system` function can be stored in R as a character vector (one for each output line produced by the system command) if the `intern=TRUE` option is set.

```
> when <- system("date", intern=TRUE)
> when
[1] "Fri Jan  6 21:48:25 CET 2017"
```

5.22 GET SESSION INFORMATION

Problem: You want to get information on your session including the operating system and installed packages and version numbers in order to reproduce your results at a later date.

Solution: For all scientific projects it is important to obtain reproducible results, i.e., given the same dataset we should be able to recreate the exact same results. Since R packages may change over time — default values for arguments may change, algorithms are tweaked, and errors are corrected — it is important to store information about the actual packages and their version number as well as operating system in order to ensure that it is known exactly which packages have been used.

R has a built-in function, `sessionInfo`, that returns information about the operating system, the version of R, as well as attached (loaded) packages and their version numbers.

```
> sessionInfo()
R version 3.3.1 (2016-06-21)
Platform: x86_64-apple-darwin13.4.0 (64-bit)
Running under: OS X 10.9.5 (Mavericks)

locale:
[1] C/UTF-8/C/C/C/C

attached base packages:
[1] grid      tcltk     methods   stats     graphics  grDevices
[7] utils     datasets  base

other attached packages:
 [1] lme4_1.1-12            mongolite_0.8.1
 [3] microbenchmark_1.4-2.1 data.table_1.9.6
 [5] haven_0.2.1.9000       XML_3.98-1.4
 [7] rvest_0.3.2            xml2_1.0.0
 [9] MESS_0.4-8             geepack_1.2-0.2
   .
   .
```

```
  .
[73] timeDate_3012.100      zoo_1.7-13
[75] ggplot2_2.1.0          knitr_1.14

loaded via a namespace (and not attached):
 [1] plyr_1.8.4         selectr_0.2-3      sp_1.2-3
 [4] splines_3.3.1      digest_0.6.10      htmltools_0.3.5
 [7] gdata_2.17.0       magrittr_1.5       readr_1.0.0
  .
  .
  .
[61] fracdiff_1.4-2     parallel_3.3.1     quadprog_1.5-5
[64] coda_0.18-1        class_7.3-14       minqa_1.2.4
[67] shiny_0.13.2
```

The devtools package provides the session_info function which returns slightly different information from the sessionInfo function. In particular, session_info gives information about the location from where a package was installed and its build date.

```
> library(devtools)
> session_info()

Session info ---------------------------------------------------

 setting  value
 version  R version 3.3.1 (2016-06-21)
 system   x86_64, darwin13.4.0
 ui       X11
 language (EN)
 collate  C
 tz       Europe/Copenhagen
 date     2016-09-01

Packages -------------------------------------------------------

 package       * version    date       source
 DBI           * 0.5        2016-08-11 CRAN (R 3.3.0)
 DataCombine   * 0.2.21     2016-04-13 CRAN (R 3.3.0)
 Formula         1.2-1      2015-04-07 CRAN (R 3.3.0)
 KernSmooth      2.23-15    2015-06-29 CRAN (R 3.3.1)
 LearnBayes      2.15       2014-05-29 CRAN (R 3.3.0)
 MASS          * 7.3-45     2016-04-21 CRAN (R 3.3.1)
 MESS          * 0.4-8      2016-08-28 local (ekstroem/MESS)
 Matrix        * 1.2-7      2016-08-28 CRAN (R 3.3.1)
 MatrixModels    0.4-1      2015-08-22 CRAN (R 3.3.0)
 haven         * 0.2.1.9000 2016-08-31 Github (hadley/haven@58b7cda)
 .
```

```
 .
 .
 waffle        * 0.5.0      2015-12-15  CRAN (R 3.3.0)
 withr           1.0.2      2016-06-20  CRAN (R 3.3.0)
 xlsx          * 0.5.7      2014-08-02  CRAN (R 3.3.0)
 xlsxjars      * 0.6.1      2014-08-22  CRAN (R 3.3.0)
 xml2          * 1.0.0      2016-06-24  CRAN (R 3.3.0)
 xtable          1.8-2      2016-02-05  CRAN (R 3.3.0)
 zoo           * 1.7-13     2016-05-03  CRAN (R 3.3.0)
```

If a package name is given as an argument, then information on any dependencies are listed. For example, to get information about the MESS package we input the following where we also see which packages are actually loaded (the starred packages in both outputs).

```
> session_info("MESS")

Session info ---------------------------------------------------

 setting  value
 version  R version 3.3.2 (2016-10-31)
 system   x86_64, darwin13.4.0
 ui       X11
 language (EN)
 collate  C
 tz       Europe/Copenhagen
 date     2017-01-06

Packages -------------------------------------------------------

 package       * version     date
 MASS          * 7.3-45      2016-04-21
 MESS          * 0.4-12      2016-12-12
 Matrix        * 1.2-7.1     2016-09-01
 Rcpp            0.12.8      2016-11-17
 RcppArmadillo   0.7.500.0.0 2016-10-22
 codetools       0.2-15      2016-10-05
 foreach       * 1.4.3       2015-10-13
 geeM          * 0.10.0      2016-07-28
 geepack       * 1.2-1       2016-09-24
 glmnet        * 2.0-5       2016-03-17
 iterators       1.0.8       2015-10-13
 kinship2        1.6.4       2015-08-03
 lattice       * 0.20-34     2016-09-06
 mvtnorm       * 1.0-5       2016-02-02
 quadprog        1.5-5       2013-04-17
 source
 CRAN (R 3.3.2)
 local (ekstroem/MESS@NA)
```

```
CRAN (R 3.3.2)
cran (@0.12.8)
cran (@0.7.500)
CRAN (R 3.3.2)
CRAN (R 3.3.0)
local
cran (@1.2-1)
CRAN (R 3.3.0)
CRAN (R 3.3.0)
CRAN (R 3.3.0)
CRAN (R 3.3.2)
CRAN (R 3.3.0)
CRAN (R 3.3.0)
```

Bibliography

Adler, D. (2005). *vioplot: Violin plot.* R package version 0.2.

Adler, D. and Murdoch, D. (2016). *rgl: 3D Visualization Using OpenGL.* R package version 0.96.0.

Aragon, T. J. (2012). *epitools: Epidemiology Tools.* R package version 0.5-7.

Bates, D., Maechler, M., Bolker, B., and Walker, S. (2016). *lme4: Linear Mixed-Effects Models using 'Eigen' and S4.* R package version 1.1-12.

Borenstein, M., Hedges, L., Higgins, J., and Rothstein, H. (2011). *Introduction to Meta-Analysis.* Wiley.

Canty, A. and Ripley, B. (2016). *boot: Bootstrap Functions (Originally by Angelo Canty for S).* R package version 1.3-18.

Carstensen, B., Gurrin, L., Ekstrøm, C., and Figurski, M. (2015). *MethComp: Functions for Analysis of Agreement in Method Comparison Studies.* R package version 1.22.2.

Carstensen, B. and Plummer, M. (2016). *Epi: A Package for Statistical Analysis in Epidemiology.* R package version 2.0.

Champely, S. (2016). *pwr: Basic Functions for Power Analysis.* R package version 1.2-0.

Choirat, C., Honaker, J., Imai, K., King, G., and Lau, O. (2016). *Zelig: Everyone's Statistical Software.* R package version 5.0-12.

Christensen, R. H. B. (2015). *ordinal: Regression Models for Ordinal Data.* R package version 2015.6-28.

Conway, J., Eddelbuettel, D., Nishiyama, T., Prayaga, S. K., and Tiffin, N. (2016). *RPostgreSQL: R interface to the PostgreSQL database system.* R package version 0.4-1.

Cortez, P., Cerdeira, A., Almeida, F., Matos, T., and Reis, J. (1998). Modeling wine preferences by data mining from physicochemical properties. *Decision Support Systems*, 47(4):547–553.

Couture-Beil, A. (2014). *rjson: JSON for R.* R package version 0.2.15.

Crawley, M. (2007). *The R Book.* John Wiley & Sons, New York.

Croissant, Y. (2013). *mlogit: multinomial logit model.* R package version 0.2-4.

Dalgaard, P. (2008). *Introductory Statistics with R (2nd ed.).* Springer, New York.

Demidenko, E. (2013). *Mixed Models: Theory and Applications with R.* Wiley Series in Probability and Statistics. Wiley.

Dowle, M., Srinivasan, A., Short, T., with contributions from R Saporta, S. L., and Antonyan, E. (2015). *data.table: Extension of Data.frame.* R package version 1.9.6.

Dragulescu, A. A. (2014). *xlsx: Read, write, format Excel 2007 and Excel 97/2000/XP/2003 files.* R package version 0.5.7.

Ekstrøm, C. T. (2016). *MESS: Miscellaneous Esoteric Statistical Scripts.* R package version 0.4-8.

Ekstrøm, C. T. and Sørensen, H. (2010). *Introduction to Statistical Data Analysis for the Life Sciences.* Chapman & Hall/CRC, Boca Raton, FL.

Ekstrøm, C. T. and Sørensen, H. (2016). *isdals: Provides datasets for Introduction to Statistical Data Analysis for the Life Sciences.* R package version 2.0-4.

Elston, D. A., Moss, R., Boulinier, T., Arrowsmith, C., and Lambin, X. (2001). Analysis of aggregation, a worked example: numbers of ticks on red grouse chicks. *Parasitology*, 122(5):563–569.

Everitt, B. S. and Hothorn, T. (2010). *A Handbook of Statistical Analyses Using R.* Chapman & Hall/CRC, Boca Raton, FL.

Fox, J. and Weisberg, S. (2016). *car: Companion to Applied Regression.* R package version 2.1-3.

Fraley, C., Raftery, A. E., and Scrucca, L. (2016). *mclust: Gaussian Mixture Modelling for Model-Based Clustering, Classification, and Density Estimation.* R package version 5.2.

Friedman, J., Hastie, T., Simon, N., and Tibshirani, R. (2016). *glmnet: Lasso and Elastic-Net Regularized Generalized Linear Models.* R package version 2.0-5.

Gamer, M., Lemon, J., and Singh, I. F. P. (2012). *irr: Various Coefficients of Interrater Reliability and Agreement.* R package version 0.84.

Gerds, T. (2009). *Publish: Turn R Output into Publishable Format.* R package version 1.1.9.

Giraudoux, P. (2016). *pgirmess: Data Analysis in Ecology.* R package version 1.6.4.

Grolemund, G., Spinu, V., and Wickham, H. (2016). *lubridate: Make Dealing with Dates a Little Easier.* R package version 1.6.0.

Grosjean, P. and Ibanez, F. (2014). *pastecs: Package for Analysis of Space-Time Ecological Series.* R package version 1.3-18.

Gross, J. and Ligges, U. (2015). *nortest: Tests for Normality.* R package version 1.0-4.

Harrell, Jr., F. E. (2016). *rms: Regression Modeling Strategies.* R package version 4.5-0.

Højsgaard, S., Halekoh, U., and Yan, J. (2016). *geepack: Generalized Estimating Equation Package.* R package version 1.2-0.2.

Holst, K. K. (2014). *gof: Model-diagnostics based on cumulative residuals.* R package version 0.9.1.

Hothorn, T., Bretz, F., and Westfall, P. (2016). *multcomp: Simultaneous Inference in General Parametric Models.* R package version 1.4-6.

Hothorn, T., Hornik, K., van de Wiel, M. A., Winell, H., and Zeileis, A. (2015). *coin: Conditional Inference Procedures in a Permutation Test Framework.* R package version 1.1-2.

Hyndman, R. (2016). *forecast: Forecasting Functions for Time Series and Linear Models.* R package version 7.1.

Hyndman, R. J. and Athanasopoulos, G. (2013). *Forecasting: principles and practice*. OTexts, Melbourne, Australia.

Ishwaran, H. and Kogalur, U. B. (2016). *randomForestSRC: Random Forests for Survival, Regression and Classification (RF-SRC)*. R package version 2.2.0.

Jackman, S., Tahk, A., Zeileis, A., Maimone, C., and Fearon, J. (2015). *pscl: Political Science Computational Laboratory, Stanford University*. R package version 1.4.9.

Kleiber, C. and Zeileis, A. (2015). *AER: Applied Econometrics with R*. R package version 1.2-4.

Koenker, R. (2016). *quantreg: Quantile Regression*. R package version 5.26.

Kuhn, M., Wing, J., Weston, S., Williams, A., Keefer, C., Engelhardt, A., Cooper, T., Mayer, Z., Kenkel, B., the R Core Team, Benesty, M., Lescarbeau, R., Ziem, A., Scrucca, L., Tang, Y., and Candan, C. (2016). *caret: Classification and Regression Training*. R package version 6.0-71.

Lemon, J., Bolker, B., Oom, S., Klein, E., Rowlingson, B., Wickham, H., Tyagi, A., Eterradossi, O., Grothendieck, G., Toews, M., Kane, J., Turner, R., Witthoft, C., Stander, J., Petzoldt, T., Duursma, R., Biancotto, E., Levy, O., Dutang, C., Solymos, P., Engelmann, R., Hecker, M., Steinbeck, F., Borchers, H., Singmann, H., Toal, T., and Ogle, D. (2016). *plotrix: Various Plotting Functions*. R package version 3.6-3.

Ligges, U., Maechler, M., and Schnackenberg, S. (2016). *scatterplot3d: 3D Scatter Plot*. R package version 0.3-37.

Lin, D. Y., Wei, L. J., and Ying, Z. (2002). Model-checking techniques based on cumulative residuals. *Biometrics*, 58:1–12.

Lumley, T. (2013). *biglm: Bounded Memory Linear and Generalized Linear Models*. R package version 0.9-1.

Maechler, M., Rousseeuw, P., Struyf, A., and Hubert, M. (2016). *cluster: "Finding Groups in Data": Cluster Analysis Extended*. R package version 2.0.4.

McDaniel, L. and Henderson, N. (2016). *geeM: Solve Generalized Estimating Equations.* R package version 0.10.0.

Mersmann, O. (2015). *microbenchmark: Accurate Timing Functions.* R package version 1.4-2.1.

Mevik, B.-H., Wehrens, R., and Liland, K. H. (2015). *pls: Partial Least Squares and Principal Component Regression.* R package version 2.5-0.

Meyer, D., Zeileis, A., and Hornik, K. (2015). *vcd: Visualizing Categorical Data.* R package version 1.4-1.

Murdoch, D. and Chow, E. D. (2013). *ellipse: Functions for Drawing Ellipses and Ellipse-Like Confidence Regions.* R package version 0.3-8.

Murrell, P. (2011). *R Graphics.* CRC Press, Boca Raton, FL, 2nd edition.

Neuwirth, E. (2014). *RColorBrewer: ColorBrewer Palettes.* R package version 1.1-2.

Ooms, J., James, D., DebRoy, S., Wickham, H., and Horner, J. (2016a). *RMySQL: Database Interface and 'MySQL' Driver for R.* R package version 0.10.9.

Ooms, J., Temple Lang, D., and Hilaiel, L. (2016b). *jsonlite: A Robust, High Performance JSON Parser and Generator for R.* R package version 1.0.

Parsons, N. (2016). *repolr: Repeated Measures Proportional Odds Logistic Regression.* R package version 3.4.

Pinheiro, J., Bates, D., and R-core (2016). *nlme: Linear and Nonlinear Mixed Effects Models.* R package version 3.1-128.

Pinheiro, J. C. and Bates, D. M. (2000). *Mixed Effects Models in S and S-Plus.* Springer, New York.

R Core Team (2015). *foreign: Read Data Stored by Minitab, S, SAS, SPSS, Stata, Systat, Weka, dBase, ...* R package version 0.8-66.

R Development Core Team (2010). *R: A Language and Environment for Statistical Computing.* R Foundation for Statistical Computing, Vienna, Austria. ISBN 3-900051-07-0.

Ripley, B. (2016a). *MASS: Support Functions and Datasets for Venables and Ripley's MASS.* R package version 7.3-45.

Ripley, B. (2016b). *nnet: Feed-Forward Neural Networks and Multinomial Log-Linear Models.* R package version 7.3-12.

Ripley, B. and Lapsley, M. (2016). *RODBC: ODBC Database Access.* R package version 1.3-13.

Robin, X., Turck, N., Hainard, A., Tiberti, N., Lisacek, F., Sanchez, J.-C., and Müller., M. (2015). *pROC: Display and Analyze ROC Curves.* R package version 1.8.

Rubba, C. (2016). *htmltab: Assemble Data Frames from HTML Tables.* R package version 0.7.0.

Rudis, B. (2015). *waffle: Create Waffle Chart Visualizations in R.* R package version 0.5.0.

Ryan, J. A. and Ulrich, J. M. (2014). *xts: eXtensible Time Series.* R package version 0.9-7.

Sarkar, D. (2008). *Lattice: Multivariate Data Visualization with R.* Springer, New York.

Sarkar, D. (2015). *lattice: Trellis Graphics for R.* R package version 0.20-33.

Sarkar, D. and Andrews, F. (2016). *latticeExtra: Extra Graphical Utilities Based on Lattice.* R package version 0.6-28.

Scheike, T., Martinussen, T., Silver, J., and Holst, K. (2015). *timereg: Flexible Regression Models for Survival Data.* R package version 1.8.9.

Sharpsteen, C. and Bracken, C. (2016). *tikzDevice: R Graphics Output in LaTeX Format.* R package version 0.10-1.

Shumway, R. H. and Stoffer, D. S. (2010). *Time Series Analysis and Its Applications: With R Examples (Springer Texts in Statistics).* Springer, New York.

Temple Lang, D. (2014). *RJSONIO: Serialize R objects to JSON, JavaScript Object Notation.* R package version 1.3-0.

Temple Lang, D. and the CRAN Team (2016). *XML: Tools for Parsing and Generating XML Within R and S-Plus.* R package version 3.98-1.4.

Therneau, T. M. (2016). *survival: Survival Analysis.* R package version 2.39-5.

Tibshirani, R. and Leisch., F. (2015). *bootstrap: Functions for the Book "An Introduction to the Bootstrap".* R package version 2015.2.

Tufte, E. R. (2009). *The Visual Display of Quantitative Information.* Graphics Press, Cheshire, CT.

Venables, W. N. and Ripley, B. D. (2002). *Modern Applied Statistics with S (4th ed.).* Springer, New York.

Verzani, J. (2005). *Using R for Introductory Statistics.* Chapman & Hall/CRC, London.

Verzani, J. (2014). *gWidgets: gWidgets API for building toolkit-independent, interactive GUIs.* R package version 0.0-54.

Viechtbauer, W. (2015). *metafor: Meta-Analysis Package for R.* R package version 1.9-8.

Warnes, G. R., Bolker, B., Bonebakker, L., Gentleman, R., Liaw, W. H. A., Lumley, T., Maechler, M., Magnusson, A., Moeller, S., Schwartz, M., and Venables, B. (2016). *gplots: Various R Programming Tools for Plotting Data.* R package version 3.0.1.

Wickham, H. (2009). *ggplot2: Elegant Graphics for Data Analysis.* Springer, New York.

Wickham, H. (2014). *reshape2: Flexibly Reshape Data: A Reboot of the Reshape Package.* R package version 1.4.1.

Wickham, H. (2016a). *readxl: Read Excel Files.* R package version 0.1.1.9000.

Wickham, H. (2016b). *rvest: Easily Harvest (Scrape) Web Pages.* R package version 0.3.2.

Wickham, H. (2016c). *stringr: Simple, Consistent Wrappers for Common String Operations.* R package version 1.1.0.

Wickham, H. and Chang, W. (2016a). *devtools: Tools to Make Developing R Packages Easier.* R package version 1.12.0.

Wickham, H. and Chang, W. (2016b). *ggplot2: An Implementation of the Grammar of Graphics.* http://ggplot2.org, https://github.com/hadley/ggplot2.

Wickham, H. and Francois, R. (2016). *dplyr: A Grammar of Data Manipulation.* R package version 0.5.0.

Wickham, H. and Hester, J. (2016). *xml2: Parse XML.* R package version 1.0.0.

Wickham, H., Hester, J., and Francois, R. (2016). *readr: Read Tabular Data.* R package version 1.0.0.

Wickham, H. and Miller, E. (2016). *haven: Import and Export 'SPSS', 'Stata' and 'SAS' Files.* https://github.com/hadley/haven, https://github.com/WizardMac/ReadStat.

Wood, S. (2016). *mgcv: Mixed GAM Computation Vehicle with GCV/AIC/REML Smoothness Estimation.* R package version 1.8-14.

Wood, S. N. (2006). *Generalized Additive Models.* Chapman & Hall/CRC, Boca Raton, FL.

Wright, M. N. (2016). *ranger: A Fast Implementation of Random Forests.* R package version 0.5.0.

Yee, T. W. (2016). *VGAM: Vector Generalized Linear and Additive Models.* R package version 1.0-2.

Zamar, D., Graham, J., and McNeney, B. (2013). *elrm: Exact Logistic Regression via MCMC.* R package version 1.2.2.

Zeileis, A., Grothendieck, G., and Ryan, J. A. (2016). *zoo: S3 Infrastructure for Regular and Irregular Time Series (Z's Ordered Observations).* R package version 1.7-13.

Index

vignette, 369